PRINCIPLES OF ORGANIZATION IN ORGANISMS

Principles of Organization in Organisms

PROCEEDINGS OF THE WORKSHOP ON
PRINCIPLES OF ORGANIZATION
IN ORGANISMS,
HELD JUNE, 1990
IN SANTA FE, NEW MEXICO

Editors

Jay E. Mittenthal
University of Illinois

Arthur B. Baskin
University of Illinois

Proceedings Volume XIII

Santa Fe Institute
Studies in the Sciences of Complexity

 Routledge
Taylor & Francis Group

NEW YORK AND LONDON

Director of Publications, Santa Fe Institute: *Ronda K. Butler-Villa*
Publications Assistant, Santa Fe Institute: *Della L. Ulibarri*

First published 1992 by Westview Press

Published 2018 by Routledge
605 Third Avenue, New York, NY 10017
4 Park Square, Milton Park, Abingdon, Oxon OX14 4RN

Routledge is an imprint of the Taylor & Francis Group, an informa business

Library of Congress Cataloging-in-Publication Data

Workshop on Principles of Organization in Organisms (1990 : Santa Fe,
N. M.)
    Principles of organization in organisms : Proceeding of the
Workshop on Principles of Organization in Organisms, held June 1990
in Santa Fe, New Mexico / editors, Jay E. Mittenthal, Arthur B.
Baskin.
        p.   cm. — (Proceedings volume, Santa Fe Institute studies in
the sciences of complexity ; 13)
    Includes bibliographical references and index.
    ISBN 0-201-52765-0 (hb) — ISBN 0-201-58789-0 (pbk.)
    1. Adaptation (Biology)—Congresses.   2. Adaptation (Physiology)—
Congresses.   3. Evolution (Biology)—Congresses.   4. Evolution—
congresses.   I. Mittenthal, Jay E.   II. Baskin, Arthur B.
III. Title.   IV. Series: Proceedings volume in the Santa Fe
Institute studies in the science of complexity ; v. 13.
    [DNLM: 1. Adaption, Biological—congresses.   2. Adaptation,
Physiological—congresses.   3. Physiology, Comparative—congresses.
QT 4 W9267p 1990]
    QH546.W67   1990
    575—dc20
DNLM/DLC
for Library of Congress   92-17985
                                                              CIP

This volume was typeset using T_EXtures on a Macintosh II computer.

ISBN 13: 978-0-201-58789-0 (pbk)
ISBN 13: 978-0-201-52765-0 (hbk)

About the Santa Fe Institute

The *Santa Fe Institute* (SFI) is a multidisciplinary graduate research and teaching institution formed to nurture research on complex systems and their simpler elements. A private, independent institution, SFI was founded in 1984. Its primary concern is to focus the tools of traditional scientific disciplines and emerging new computer resources on the problems and opportunities that are involved in the multidisciplinary study of complex systems—those fundamental processes that shape almost every aspect of human life. Understanding complex systems is critical to realizing the full potential of science, and may be expected to yield enormous intellectual and practical benefits.

All titles from the *Santa Fe Institute Studies in the Sciences of Complexity* series will carry this imprint which is based on a Mimbres pottery design (circa A.D. 950–1150), drawn by Betsy Jones.

Santa Fe Institute Studies in the Sciences of Complexity

PROCEEDINGS VOLUMES

Volume	Editor	Title
I	David Pines	Emerging Syntheses in Science, 1987
II	Alan S. Perelson	Theoretical Immunology, Part One, 1988
III	Alan S. Perelson	Theoretical Immunology, Part Two, 1988
IV	Gary D. Doolen et al.	Lattice Gas Methods for Partial Differential Equations, 1989
V	Philip W. Anderson, Kenneth Arrow, & David Pines	The Economy as an Evolving Complex System, 1988
VI	Christopher G. Langton	Artificial Life: Proceedings of an Interdisciplinary Workshop on the Synthesis and Simulation of Living Systems, 1988
VII	George I. Bell & Thomas G. Marr	Computers and DNA, 1989
VIII	Wojciech H. Zurek	Complexity, Entropy, and the Physics of Information, 1990
IX	Alan S. Perelson & Stuart A. Kauffman	Molecular Evolution on Rugged Landscapes: Proteins, RNA and the Immune System, 1990
X	Christopher Langton et al.	Artificial Life II, 1991
XI	John A. Hawkins & Murray Gell-Mann	Evolution of Human Languages, 1992
XII	Martin Casdagli & Stephen Eubank	Nonlinear Modeling and Forecasting, 1992

LECTURES VOLUMES

Volume	Editor	Title
I	Daniel L. Stein	Lectures in the Sciences of Complexity, 1989
II	Erica Jen	1989 Lectures in Complex Systems, 1990
III	Lynn Nadel & Daniel L. Stein	1990 Lectures in Complex Systems, 1991

LECTURE NOTES VOLUMES

Volume	Author	Title
I	John Hertz, Anders Krogh, & Richard Palmer	Introduction to the Theory of Neural Computation, 1990
II	Gérard Weisbuch	Complex Systems Dynamics, 1990

REFERENCE VOLUMES

Volume	Author	Title
I	Andrew Wuensche & Mike Lesser	The Global Dynamics of Cellular Automata: An Atlas of Basin of Attraction Fields of One-Dimensional Cellular Automata, 1992

Dedication

This volume is dedicated to our mentors.

Contributors to This Volume

Arthur B. Baskin, University of Illinois

Neil W. Blackstone, Yale University

Jessica A. Bolker, University of California, Berkeley

Bertrand Clarke, Purdue University

C. G. DeGuzman, Florida Atlantic University

M. Ding, Florida Atlantic University

Brian Goodwin, The Open University, United Kingdom

Timothy L. Karr, University of Illinois

Stuart A. Kauffman, Santa Fe Institute

J. A. S. Kelso, Florida Atlantic University

Arnold J. Mandell, Florida Atlantic University

Jay E. Mittenthal, University of Illinois

Eric Mjolsness, Yale University

Thomas R. Nelson, University of California, San Diego

Stuart A. Newman, New York Medical College

Klause Obermayer, University of Illinois

John Reinitz, Columbia University

Robert E. Reinke, University of Illinois

H. Ritter, Universität Bielefeld, Germany

Karen A. Selz, Florida Atlantic University

G. Schöner, Florida Atlantic University

K. Schulten, University of Illinois

David H. Sharp, Los Alamos National Laboratory

David G. Stork, Ricoh California Research Center

Michael P. Stryker, University of California, San Francisco

David B. Wake, University of California, Berkeley

Contents

General Introduction   1

Physiology   7

Introduction   9

The Flexible Dynamics of Biological Coordination: Living
in the Niche Between Order and Disorder
C. G. DeGuzman and J. A. S. Kelso   11

Biological Organization and Adaptation: Fractal Structure
and Scaling Similarities
T. R. Nelson   35

Commentary   63

Physiology and Development   69

Introduction   71

Constancy and Variation in Developmental Mechanisms:
An Example from Comparative Embryology
J. A. Bolker   73

Heterochrony in Hydractiniid Hydroids: A Hypothesis
N. W. Blackstone   87

Adaptive Mechanisms that Accelerate Embryonic Development in *Drosophila*
T. L. Karr and J. E. Mittenthal   95

Principles of Organization in Organisms,
SFI Studies in the Sciences of Complexity, Proc. Vol. XIII,
Eds. J. Mittenthal & A. Baskin, Addison-Wesley, 1992   xIII

A Connectionist Model of the *Drosophila Blastoderm*
J. Reinitz, E. Mjolsness, and D. H. Sharp   **109**

Activity-Dependent Reorganization of Afferents in the Developing Mammalian Visual System
M. P. Stryker   **119**

A Model for the Development of the Spatial Structure of Retinotopic Maps and Orientation Columns
K. Obermayer, H. Ritter, and K. Schulten   **141**

Commentary   **167**

Physiology, Development, Evolution, and Their Evolution **171**

Patterns of Constancy and Change   **173**

Introduction   **173**

Homoplasy: The Result of Natural Selection, or Evidence of Design Limitations?
D. B. Wake   **175**

Preadaptation and Principles of Organization in Organisms
D. G. Stork   **205**

Critical Allosteric Brain Enzyme Kinetics as Phenotypic Microevolutionary Process
A. J. Mandell and K. A. Selz   **225**

Generic Physical Mechanisms of Morphogenesis and Pattern Formation as Determinants in the Evolution of Multicellular Organization
S. A. Newman   **241**

Deletions and Mirror-Symmetries in *Drosophila* Segmentation Mutants Reveal Generic Properties of Epigenetic Mappings
B. Goodwin and S. Kauffman   **269**

Commentary   **297**

General Principles of Organization   301

Introduction   301

The Sciences of Complexity and "Origins of Order"   303
    S. A. Kauffman

Patterns of Structure and Their Evolution in the Organization of Organisms: Modules, Matching, and Compaction
    J. E. Mittenthal, A. B. Baskin, and R. E. Reinke   321

Reliability of Networks of Genes   333
    B. Clarke and J. E. Mittenthal

Exploring the Role of Finiteness in the Emergence of Structure
    A. B. Baskin, R. E. Reinke, and J. E. Mittenthal   337

Commentary   379

General Conclusions   383

Appendix   395

Dynamic Pattern Formation: A Primer   397
    J. A. S. Kelso, M. Ding, and G. Schoner

Index   441

General Introduction

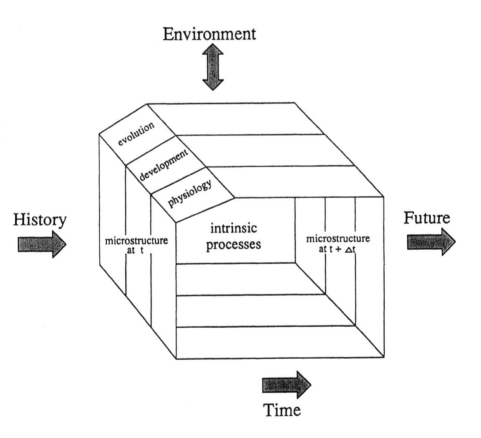

A unifying framework that incorporates and transcends views of the organization of organisms from molecular genetics, neo-Darwinism, and dynamical systems theory. Processes intrinsic to an organism, constrained by history and interacting with the environment, generate the changes in microstructure during the time interval $(t, t + \Delta t)$. Physiological, developmental, and evolutionary processes define the time scales Δt on which change occurs.

General Introduction

This book is an outgrowth of a workshop on Principles of Organization in Organisms, organized by Arthur Baskin, Stuart Kauffman, Jay Mittenthal, and David Wake and held at the Santa Fe Institute, Santa Fe, New Mexico, June 12–16, 1990. The essays presented here were largely written after the workshop. In many cases they embody a revision of concepts and a reinterpretation of preexisting data that reflect debates and conclusions of the workshop.

GOALS AND ORGANIZATION OF THE WORKSHOP

The workshop brought together researchers with widely varying perspectives who are pursuing a variety of research paradigms. The participants were united in seeking overarching principles of organization for biological systems. Some researchers brought detailed studies of developmental or physiological processes to the workshop; some brought summaries of evolutionary trends; and some brought mathematical models which, to varying degrees, could be related to specific biological processes. The interplay of these approaches led to spirited discussions and revealed several complementary points of view.

At present several views of biological organization dominate biology. In the molecular genetic view, stereotyped programs of gene activity underlie the phenomena of physiology and development. In the neo-Darwinian view of evolution,

Principles of Organization in Organisms,
SFI Studies in the Sciences of Complexity, Proc. Vol. XIII,
Eds. J. Mittenthal & A. Baskin, Addison-Wesley, 1992   **3**

gradual selection occurs among variant organisms. From the viewpoint of dynamical systems theory, the dynamics of processes within organisms channels physiological, developmental, and evolving structure into a limited set of pathways. Each of these viewpoints emerged at the workshop and influenced the chapters. These viewpoints are partially overlapping, partially inconsistent, and incomplete. At the workshop and in the preparation of this volume, we have searched for a unifying framework that would incorporate and transcend these viewpoints. This framework is summarized in Figure 1, and is discussed at length in the General Conclusions.

PRECEDENTS IN THE SEARCH FOR A FRAMEWORK

Studies of organisms have generated a wealth of detailed information, a wealth so vast that it seems unsurveyable. Detailed information about biological processes continues to accumulate rapidly. However, we lack a coherent framework for integrating this information, despite diverse insights and theories about the organization of organisms (especially Miller,[1] Riedl,[2] Whyte,[3] and Yates[4]).

Situations in which a wealth of data lacks a unifying framework are not uncommon in the history of science. For example, by the sixteenth century, astronomical observations had produced a vast body of data about the orbits of planets. The Ptolemaic theory of epicycles organized these data by describing each orbit as a set of circles rotating around one another, and the planets revolved around the earth. More precise data could be accomodated by adjusting the periods and phases of motion on the circles and by adding more circles. Kepler showed that a different viewpoint gave a much simpler description: Assuming that the planets revolved around the sun in elliptical orbits, predictions could be made that were as accurate as those made with the Ptolemaic theory. Kepler summarized these predictions in three laws relating the rate of traversal of an orbit to its geometry. These laws provided a compact set of principles that Newton inferred from his theory of mechanical dynamics. Newtonian mechanics could generate orbits (some not elliptical) for diverse initial conditions, and it explained a great variety of other phenomena.

This case, and other analogies from physics—in particular, analogies to field theories, statistical mechanics, and thermodynamics—suggest the desirability of seeking principles and theory for the organization of organisms. The principles would be analogs of Kepler's laws—compact summaries of patterns that emerge when constraints limit degrees of freedom. The theory would be a dynamical theory that generates these patterns and that makes further predictions. Ideally the principles and theory would unify the commonality in current approaches, resolve the inconsistencies, and fill in the gaps. Although this volume does not present a set of principles that emerge from a coherent theory, it does represent definite progress in that direction.

ORGANIZATION OF THE BOOK

Our overall goal is to promote the discovery of a comprehensive theory of organization for complex adaptive systems, by gathering and integrating such components as are presently available. We have focussed attention on the concept of structure, regarded both as material structure and as a dynamical structure of processes that the material structure supports. The central questions at issue are: What are the characteristic structures in organisms? What processes generate these structures? Figure 1 (the diagram on the cover page for this section) conveys this focus by representing a transition in the structure of an organism during a time interval. This transition depends on the history of the organism, on processes intrinsic to it during the interval, and on extrinsic processes in its environment. The theory must show how the interplay of these components shapes structure.

Biological structures are generated through processes on physiological, developmental, and evolutionary time scales. Physiological processes are most rapid; they occur, often repeatedly, during relatively brief intervals within the life cycle of an organism. Developmental processes mediate the overall traversal of the life cycle. Through evolutionary processes spanning many life cycles, structures at physiological and developmental time scales become modified.

These three time scales provided a natural way to organize the book, as a nested sequence of increasing time spans. The chapters in the first section deal with physiology; they interpret specific physiological processes in terms of relatively abstract models. In the second section the viewpoint encompasses physiological and developmental processes over a life cycle. The chapters of this section are the most specific; they provide a detailed foundation for the more abstract and general discussions that follow. Evolutionary concerns are most prominent in the third section, which span all three time scales. In the first half of this section, the chapters address generic biological problems in relatively specific contexts, whereas the chapters of the second half consider principles of organization that may characterize complex adaptive systems in general. Throughout the book the underlying viewpoint is that of dynamical systems theory; the Appendix provides a tutorial on this subject.

Interspersed among these chapters are essays that introduce and comment on the sections of the book. In these essays we have tried to orient the reader to the topics that emerged at the workshop, and to spotlight key points that relate to the emergence of principles and a theory of organization.

ACKNOWLEDGMENTS

We appreciate the money, time, and effort that many people contributed to make the workshop and this book happen. Funds for the workshop were provided by the Santa Fe Institute, including core funding from the John D. and Catherine T. MacArthur Foundation, the National Science Foundation (PHY-8714918), and the U.S. Department of Energy (ER-FG05-88ER25054); by the Program in Mathematics and Molecular Biology at the University of California, Berkeley (funded by the National Science Foundation, and administered by Sylvia Spengler); and by the Automation Support Center of the Department of Veterinary Biosciences at the University of Illinois (administered by Arthur Baskin). George Cowan and Mike Simmons offered the hospitality and facilities of the Santa Fe Institute for the workshop. We especially appreciate the invaluable aid that Ginger Richardson and Andi Sutherland of the Institute provided in planning and implementing the workshop.

At the Institute Ronda Butler-Villa and Della Ulibarri did many tasks necessary to make the book a reality, from fundraising to accepting graciously the vagaries of the editors, to whom most of these efforts were mercifully invisible. We are most grateful to them. Many secretaries worked hard to bring manuscripts from the authors' hands to the publisher. Of those we know, Buzz Swett deserves special thanks for persistently hounding the authors (including the editors) to get their contributions in; Sharol Hanson, Kathie Alblinger, and Judy Whittington helped us to do this.

We thank our colleagues, the participants in the workshop, and the contributors to the book, who made these supra-organismal phenomena possible.

REFERENCES

1. Miller, J. G. *Living Systems.* New York: McGraw-Hill, 1978.
2. Riedl, R. *Order in Living Organisms,* translated by R. P. S. Jefferies. New York: John Wiley & Sons, 1978.
3. Whyte, L. L. *Internal Factors in Evolution.* New York: George Braziller, 1965.
4. Yates, F. E. "Systems Analysis of Hormone Action." In *Biological Regulation and Development,* edited by R. F. Goldberger and K. Yamamoto. New York: Plenum, 1982.

Physiology

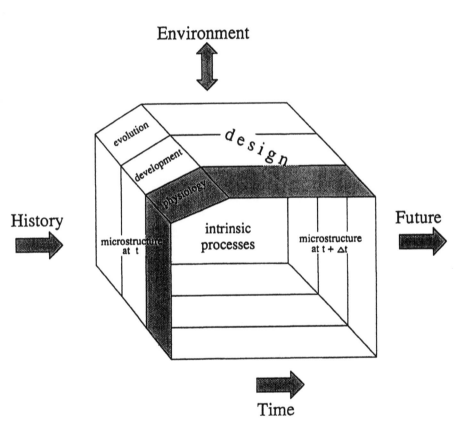

We begin with physiology because its processes occur on the most circumscribed time scale. The two chapters in this section look at two large and diverse classes of physiological systems, oscillators and fractal structures.

Physiology: Introduction

DeGuzman and Kelso discuss the problem of coordination among parts of an organism. Specifically, they consider interactions among oscillators—systems which, when unperturbed, generate a periodically varying output. Physiological oscillations include cell cycles, locomotor rhythms, and circadian rhythms. Of interest is the interaction of two or more oscillators, either with unidirectional coupling (as when an environmental light-dark cycle entrains a circadian rhythm) or with reciprocal coupling (as in the coordination of limbs in locomotion).

Nelson looks at organs and organ systems with fractal morphology or activity. Such organs include branching networks of epithelial tubes (such as vertebrate mammary and prostate glands, lungs, kidneys, and liver), and surfaces that are folded at multiple spatial scales (including vertebrate gut, cerebrum, and cerebellum). Nelson specifically considers systems with tree-like morphology that transport a resource between the trunk and the tips of the branches—the cerebellum, the bronchial airway, and the His-Purkinje network that activates contraction of a vertebrate heart. He also discusses the sinoatrial node of the heart, which generates a fractal pattern of variation in the heartbeat rate. In these systems Nelson identifies fractal patterns and considers processes that may generate them.

These chapters introduce major themes of the book. They show how models and data can contribute synergistically to understanding the organization of organisms. Many of the models are dynamic; they are formulated and analyzed

Principles of Organization in Organisms,
SFI Studies in the Sciences of Complexity, Proc. Vol. XIII,
Eds. J. Mittenthal & A. Baskin, Addison-Wesley, 1992

within the context of dynamical systems theory. Of special interest is the relation of macrodynamics—the description of a system at one level of organization, using a dynamical system with a few state variables—to the microdynamics of lower-level processes with many state variables. As the microdynamics vary gradually, the macrodynamics can exhibit distinct phases of behavior. Biological systems may often operate on boundaries between such phases. The Appendix, a primer on dynamics, provides background for understanding these issues, which recur at physiological, developmental, and evolutionary time scales.

G. C. DeGuzman and J. A. S. Kelso
Program in Complex Systems and Brain Sciences, Center for Complex Systems, Florida Atlantic University, Boca Raton, FL 33431

The Flexible Dynamics of Biological Coordination: Living in the Niche between Order and Disorder

1. INTRODUCTION

An essential feature of all living things is the high degree of coordination among their many parts. It is always tempting to assign the basis for this coordination to a causal agent residing inside the system that performs the duty of coordinating the parts. Yet there are many examples in which generic chemical or physical processes produce coordination among the elements in the sense of forming spatial and temporal patterns without any internal homunculus-like agent whatsoever. Notwithstanding the contributions of heredity, (self-organized) pattern formation in biology may emerge at levels far removed, say, from the level of single genes. On the other hand, it is possible, in principle, that cell types themselves may arise as a kind of spontaneous self-organization among the genes themselves.[11] Apparently, it is not until one reaches the level of the gene itself with its molecular "programs" and "switches" that the language of dynamics and the concepts of self-organization and

Principles of Organization in Organisms,
SFI Studies in the Sciences of Complexity, Proc. Vol. XIII,
Eds. J. Mittenthal & A. Baskin, Addison-Wesley, 1992

pattern formation get replaced by the language of the digital computer.[1] The "two sides" of Turing, as it were, show up. Because our interests lie in understanding the emergence of coordination, as an *a posteriori* consequence of certain dynamical laws rather than as an *a priori* program of instructions, we are inspired more by Turing's seminal work on morphogenesis[45] than by the metaphors provided by the Turing computing machine. Indeed, our research program, part of which is described here, may be appropriately described as one targeted upon the morphogenesis of behavior, i.e., the emergence of dynamic forms of coordination under specified boundary conditions.

In our view, the laws of coordination are (self-) organizational laws in the sense that they are not structure specific or dependent on physico-chemical processes *per se*. In this picture, there is no ontological priority of one level of description over another. The dynamics we shall define are of spatiotemporal patterns: even though such patterns are always realized and instantiated by physical structures, the laws are abstract and mathematical. Yet the dynamics of coordination are not simply mathematical curiosities. They have been shown to express spatiotemporal relationships between: (a) components of an organism; (b) organisms themselves; and (c) organisms and their environment. Such patterns are necessary features of everyday functions and, on longer time scales, are likely important in the evolution and development of organisms. Moreover, the present approach has an operational side to it: equations of motion are formulated for observable variables and often result in predictions that can be (and have been) experimentally tested. As we have stressed elsewhere,[26] the laws we seek at a chosen level of description deal specifically with *events*, pattern-switching bifurcations (order-order transitions), multistability, intermittency, and so forth. Variability plays a crucial role (see sections 2 and 3 of accompanying Primer[28]), since we want our descriptions to encompass the properties of stability and flexibility, both essential to biological function.

In this paper, we first (section 2) briefly review our approach to coordination, which, as we have intimated, is inspired by nonequilibrium pattern formation theory, especially Haken's[7] synergetics. The latter offers a kind of strategic physics that we aim to extend and elaborate by experimental and theoretical study of coordination. In section 3, we describe the basic coordination dynamics which reflect, on the one hand, the ubiquitous tendency of biological systems to exhibit phase and frequency synchronization among their interacting components, and, even more importantly, to switch between these collective states. In section 4, we stress the fundamental importance of symmetry breaking of the coordination dynamics which occurs when a difference between the eigenfrequencies of interacting units is introduced. The sources of such differences are manifold in biological systems. When symmetry is broken, much more varied forms of coordination are shown to be possible even though the dynamics are still low dimensional. In particular, a partial or relative coordination is possible when the individual components are allowed

---

[1] Of course, higher level brain and cognitive functions are often described in terms of rule-governed plans and programs. We do not delve into the relative merits of this language here (but see e.g., Kelso, Tuller and Saltzman,[17] Kugler and Turvey[29]).

to express their inherent spatial and temporal variability. Von Holst[46] coined the term relative coordination to describe occasional phase slippage or drift between coordinating components balanced by an intrinsic attraction to certain preferred phase relations (the so-called "Magnet" effect). Similar phenomena permeate cellular, organic, and systemic function levels in very different systems (see Kelso,[24] for review). We show that this less rigid, more flexible relative coordination may be understood within the broad class of critical behavior called (*Type I*) *intermittency* (see also Kelso and DeGuzman[25]). Such intermittency can be demonstrated in continuous flows (section 4) but it is easiest to explore in discrete maps (see accompanying Primer[28]). Intermittency is an important feature of DeGuzman and Kelso's[25] phase attractive circle map model of multifrequency coordination.[26] In section 5 we draw attention to new predictions and scaling relationships that arise from further theoretical analysis of this map that are now open to experimental test. Relative coordination corresponds to the intermittent regime of the coordination dynamics where self-organizing biological systems tend to live, right on the boundary between regular and irregular behavior.

2. DYNAMIC PATTERN APPROACH

The central hypothesis of the dynamic pattern approach to coordination[18,43] is that biological systems, on a chosen level of description, are subject to dynamics, that is, equations of motion that govern their coordination activity. These are not typically the physical mechanics of moving masses but equations of motion that generate the time course of states of the system itself. These time courses can be reconstructed from experimental measures on different levels of description. The key step in this reconstruction is to identify adequate *collective variables* (or *order parameters*, cf. Haken[8]) which capture the system's coordination activity and are directly measurable, most easily around phase transitions or bifurcations where different coordination patterns are clearly distinguishable. The choice of such variables is based on empirical insights: collective variables define stable and reproducible relationships among the components, e.g., of a behavioral, physiological, or neural system and are, in general, function or task specific. Different biological systems may be describable by the same type of collective variables in the same or similar functions, while different collective variables may characterize the same system in different functions or tasks. Choosing a set of variables fixes a level of description. Although the same coordination phenomena may be described at different levels of description, the goal is to find dynamic laws *within* a given level. Examples of such collective variables are relative phases which capture patterns of synchronization and entrainment in the nervous system and behavior (see section 2.7 of Primer[28]). Given a set of collective variables the coordination dynamics can be modeled by equations of motion of these collective variables. Theoretically, observed stable patterns correspond to attractors of the collective variable dynamics with biological boundary conditions such as task demands, environmental context, and inherent physiological constraints acting as parameters on such dynamics. If such constraints do not specify a particular pattern, the coordination dynamics are called *intrinsic dynamics*. Intrinsic dynamics need not be hardwired (e.g., studies of learning show how learning one task may effect the intrinsic dynamics of another). Rather, the term serves as a contrast with those influences—captured by the concept of *information*—that *do* specify a particular coordination pattern. Thus, the great advantage of identifying these intrinsic dynamics is that they allow one to specify *what* changes, due to, e.g., environmental, learned, intentional influences, etc., in terms of the *same collective variables* that are used to characterize coordination patterns. Such information can be formally expressed as a contribution to, a "force" acting upon, or a perturbation of the intrinsic dynamics. Information is part of the coordination dynamics attracting the system to a required pattern (see Kelso[23] and Schöner[41] for recent reviews).

What does it mean for biological systems to be subject to coordination dynamics and what are the experimental implications of this view? A major consequence of the theoretical approach is that stability is an essential property of a coordination pattern, most dramatically evident when stability is lost. Loss of stability ultimately leads to pattern change, and predictions regarding loss of stability can be tested before such change occurs, especially using measures of pattern fluctuations and relaxation time (see accompanying Primer[28]). The degree to which intrinsic coordination tendencies and behavioral information agree or conflict determines the patterns observed and their stability. This interplay between cooperative and competitive processes shows up in a number of contexts as a general pattern *selection* mechanism.[20,43]

3. BASIC COORDINATION DYNAMICS

To understand the dynamic approach to coordination better, we examine the model system of Kelso[12,14] which played a significant role in the development of the theory. The experiment deals with bimanual coordination and the subject's task is to move two homologous limbs rhythmically at a common frequency. Two patterns of coordination are found to be performed stably at different frequencies: (1) inphase (homologous muscles contracting in the same direction); and (2) antiphase (homologous muscles contracting in alternating fashion). If the cycling frequency is increased, the inphase pattern remains stable at sustainable frequencies. However, when the subject starts in the antiphase pattern, a spontaneous switch to inphase occurs when the frequency reaches a critical value. To describe the coordination patterns, the relative phase ϕ was chosen as the collective variable. This can be estimated from the trajectories of the two limbs by computing the angular arguments of the equivalent rhythmic but amplitude-normalized motions.[19] A discrete version, called the point estimate of the relative phase, is also sometimes used in place of the continuous estimate described above. The point estimate, ϕ_n, is a local measure of the phase relationship between the limbs in the nth cycle of their movements. As we will see later on, the discrete measure is more appropriate in some situations where the coordination exhibits discrete properties.[3] The choice of the collective variable is not unique and other measures (e.g., the integral overlap between the underlying muscle activities[15]) may be used as long as they accurately describe the functional behavior of the patterns at a chosen level of description.

The empirically stable patterns at $\phi = 0$ and $\phi = \pi$, corresponding to the inphase and antiphase modes of coordination, respectively, are then mapped onto the attractors of a theoretical model. A model consistent with the pattern behavior is given by the phase equation[8]

$$\dot{\phi} = -a \sin \phi - 2b \sin 2\phi \tag{1}$$

where a and b are parameters which can be obtained from the experimental data. If the right-hand side of Eq. (1) is expressed as the gradient of the function $V = -a \cos \phi - b \cos 2\phi$, we may graphically illustrate the motion of the system

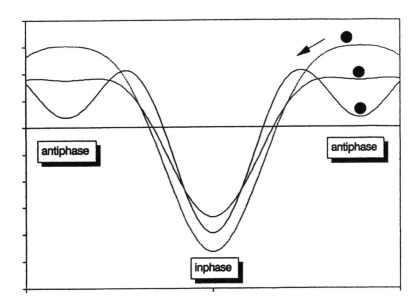

FIGURE 1 . Plot of the potential $V = -a \cos \phi - b \cos 2\phi$ for various values of the frequency parameter $\mid a/4b \mid$. An initial antiphase state of the system is represented by the ball on top of the corresponding minimum. (a) $\mid a/4b \mid = 0.3$ (solid curve). Both $\phi = 0$ and $\phi = \pi$ are local minima. (b) $\mid a/4b \mid = 0.8$ (dotted curve). Local minimum at $\phi = \pi$ starts to get unstable. (c) $\mid a/4b \mid = 1.1$ (dashed curve). Only the minimum at $\phi = 0$ remains. States initially located at the unstable critical point ($\phi = \pi$) can be induced to "switch" to $\phi = 0$ with the slightest addition of noise.

as that of a particle in the potential V (Figure 1). The ratio $|\,a/4b\,|$ corresponds to the frequency control parameter, ω in the experiment. For ratio values less than 1, both $\phi = 0$ and $\phi = \pi$ are stable fixed points of Eq. (1). This corresponds to the lower-frequency bistable coordination dynamics of this experimental model system. A phase transition, or pitchfork bifurcation from the bistable to monostable (in-phase) mode, occurs at the critical ratio $|\,a/4b\,| = 1$. Equation (1) is not just a good description of so-called absolute coordination[46] but contains additional properties that have been systematically explored. Typically, these properties address the behavior of systems around nonequilibrium transition points and the effects of noise. The phenomenon of hysteresis is one such property which has been verified experimentally. As the frequency of coordinated motion is increased gradually, there is a spontaneous switch from antiphase to inphase at some critical frequency ω_c. However, if the frequency is then decreased in the same run, no switching back to anti-phase occurs and the in-phase pattern remains stable[15] even below ω_c. Other experimentally confirmed behaviors near phase transitions include enhanced fluctuations[16] and increased relaxation times[39] (see Primer[28] for more detailed discussion of the terms). In addition, derivative predictions regarding the switching times between patterns have been obtained by using a stochastic form of Eq. (1).[42] The measured distribution times reflect the relative depths of the potential minima corresponding to each of the patterns, or, equivalently, the relative difficulty of climbing up the potential barrier from one minima to another.

4. SYMMETRY-BREAKING COORDINATION DYNAMICS: RELATIVE COORDINATION

The model system described by Eq. (1) exhibits phase attraction or a preference for a limited set of phase relations $(0, \pi)$ between the hands. The highly symmetric states are those of absolute coordination in which the oscillations are of the same frequency and at preferred phase relations. Let us formally consider for the moment the source of this symmetry. The functional form of V is constrained by the symmetry (similar limbs, same frequency) and explains the position of the natural phases 0 and π. More generally, if the dynamics depend only on the relative phase, one can write down the phase equation

$$\dot{\phi} = f(\phi) \tag{2}$$

where $f(-\phi) = -f(\phi)$, so that f can be expanded in series consisting only of $\sin n\phi$ terms. Equation (1) corresponds to taking the two lowest order terms, based on a minimality strategy (i.e., only observed states are modeled). When the symmetry of the dynamics is broken, however, it is possible to predict and observe other, perhaps more important, effects as we shall see. Symmetry breaking has been demonstrated to arise from several factors e.g., adjustment to environment,[23] learning,[47] intentional switching,[22] and multicomponent coordination.[27] It is likely also to play a role in other circumstances, e.g., coordination between people.[38] In this paper, we consider a case when the symmetry breaking can be accounted for by a simple correction to the basic coordination dynamics. Any influence that causes the eigenfrequencies of the components to be different acts as a source of symmetry breaking. In biological coordination, although coordinating components are able to adjust their intrinsic frequencies over some range, there are regions when entrainment is difficult if not impossible to maintain. In such cases the symmetry of the relative phase dynamics under the operation ϕ to $-\phi$ can no longer be assumed. For small enough bias and at low frequencies, this sometimes results in shift of the phase-locked positions (as seen, for instance, in the data of Rosenblum and Turvey[37]). For example, if one of the hands in our model system is loaded with a weight, that hand cannot sustain as high a frequency as the other. A similar condition occurs when human subjects must coordinate hand motion with an external periodic signal.[23] However, we note that when uncoupled, the components behave according to their respective eigenfrequencies. If $\delta\omega$ is the eigenfrequency difference, the rate of phase change is just this difference. We now assume that when coupled, the phase attractive dynamics of Eq. (1) still asserts itself so that the modified phase dynamics is given by

$$\dot{\phi} = \delta\omega - a\sin\phi - 2b\sin 2\phi. \tag{3}$$

It can be easily verified that for small values of $\delta\omega$, one can obtain phase locking, as before, but the fixed points are slightly shifted away from the pure inphase and antiphase patterns of absolute coordination described above. Equation (3) corresponds to motion in a potential

$$V = \phi\delta\omega - a\cos\phi - b\cos 2\phi \tag{4}$$

plotted in Figure (2). The effect of the linear term is to tilt the whole curve along the line $V = \phi\delta\omega$. For a given movement frequency, then, as $\delta\omega$ is increased, there is a point at which the curve loses its critical points, the system is no longer phase locked, and running solutions predominate. The exact point at which the detuning parameter $\delta\omega$ causes the running solution to appear and whether it appears at all depends on the other parameters a and b. This simply reflects the fact that both the frequency difference as well as the basic frequency of coordination affect the onset of the running solution. Figure 3 shows the solution $\phi(t)$ for fixed $\delta\omega$ and increasing frequency. The behavior exhibited in Figure 3 is of particular significance for it shows that although there is no strict mode locking, there is an "intermediate" behavior characterized by coordination occasionally interrupted by phase wandering. This form of partial coordination is termed "relative coordination" and was first explored experimentally in depth by von Holst[46] in his studies of fin movements in fish. Qualitatively similar results have been observed when one tries to synchronize with an external periodic signal.[23] If a finger cycles on the beat (or off the beat) with a metronome, stable inphase and antiphase patterns are possible at low frequencies (e.g., near 1 Hz). As the cycle frequency is increased, occasional phase wandering first occurs, until for high enough frequencies total desynchronization appears. The interpretation is that as the frequency is increased, the information processing limits of sustaining a given (syncopated) perception-action pattern are reached: as a result the pattern gradually moves out of a sustainable frequency range. Within this frequency range, however, although strict mode locking does not happen, the hand-metronome phase does not drift continuously but maintains an almost coordinated behavior.

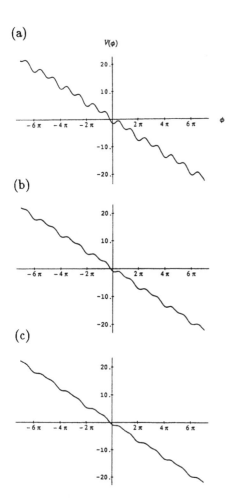

FIGURE 2 Potential with the corrective term $\phi\delta\omega$ representing the difference in the intrinsic frequencies of the coordinating components. The effect of the linear term is to tilt the curve of Figure 1 along the line $V = \phi\delta\omega$. For fixed $\delta\omega$, as the movement frequency is increased (top to bottom), the curve loses its critical points and the system is no longer phase locked.

(a)

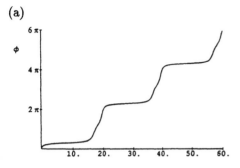

(b)

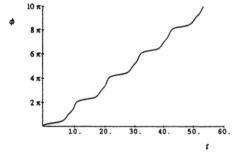

FIGURE 3 The solution $\phi(t)$ of Eq. (3) corresponding to the parameter values of Figure 2(b) (phase locked) and Figure 2(c) (unlocked), respectively.

Our discussion of relative coordination can easily be cast in the language of dynamical systems. Consider the plot of $\dot{\phi}$ vs. ϕ (Figure (4)) as the movement frequency is increased and let us examine the region near $\phi = 0$. Here the curve crosses the horizontal axis at two points, one stable and the other unstable. As the frequency change lifts the curve up, these two points coalesce into a saddle node and one ends up with a fixed point which is both an attractor and a repeller (see Primer,[28] section 2.8). As the curve is lifted up some more, a phantom fixed point appears. The motion hovers around this point most of the time but occasionally the system escapes out of the region along the repelling direction. This is termed intermittency and represents one of the generic processes found in low-dimensional systems near tangent bifurcations. Intermittency has been implicated as one of the possible routes to chaos[2] as well as transitions between periodic and quasi-periodic (type I intermittency). The identification of this more common form of relative coordination with intermittent behavior of periodic flows,[25] is consistent with the emerging view that biological systems tend to live near boundaries between regular and irregular behavior (e.g., Kaufmann[11] and Mandell[32]). A key concept in this view is that the strategic position near the boundaries allows for flexible switchings among metastable "states." Relative coordination may also be viewed as a manifestation of flexibility in biological systems for it allows the system to be partially in tune with the environment despite opposing tendencies.

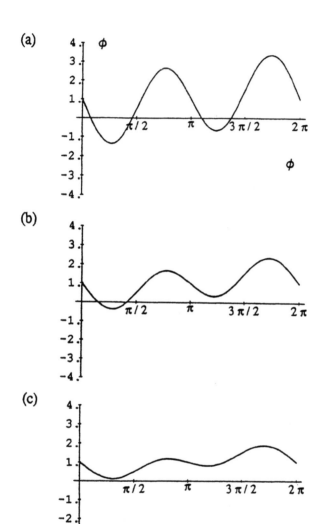

FIGURE 4
Plot of $\dot\phi$ vs. ϕ corresponding to the parameter values of Figures 2(a), (b), and (c). The running solution (c) occurs when a pair of stable-unstable fixed points coalesce.

In this section, we have emphasized the difference in the eigenmotion as an important source of symmetry breaking in biological coordination. The other sources briefly outlined above are equally important and have been explored in considerable detail (e.g., Kelso[23] for review and references therein).

5. DISCRETE COORDINATION DYNAMICS: THE PHASE ATTRACTIVE CIRCLE MAP

The motivation for using discrete dynamical systems rather than flows when studying biological coordination comes from both theoretical and experimental considerations. For our model system of bimanual coordination, the trajectories can be approximated by harmonic motion. The phase dynamics of Eq. (1) follow from the asymptotic properties of the limit cycle (harmonic) solutions of coupled oscillators.[8] Trajectory studies (e.g., Kay, Kelso, Saltzman, and Schöner;[19] and Kelso, Holt, Rubin, and Kugler[13]) show that finger movements can be mapped onto limit cycle dynamics. In this range of solutions, the mode-locking properties can in turn be described by circle maps (see, e.g., Glazier and Libchaber[6]). Discrete features of coordination are also found in experiments. This essentially means that for a given coordination pattern, the actual trajectories of the components are not as crucial as the points of synchronization between the components. Thus, the bursts of EMG signals from a muscle may be considered phase locked with the finger movements even though both may not exhibit pure harmonic oscillations.[15] Coordination with the external environment is often of a discrete nature. The phenomenon of "gating" suggests that only intervals around special points in a given action (e.g., the locomotory cycle) are important (so-called phase-dependent effects). From experimental studies of rhythmic multi-limb coordination in humans,[27] it has been shown that the four-limb relative phase patterns for a "pace" (homologous limbs cycling antiphase, ipsilateral limbs cycling in phase) and a "trot" (ipsilateral limbs cycling antiphase) converge only at certain points in their trajectories suggesting that the essential information for coordination is localized to a discrete region in the relative phase space.[1]

For two component systems, one can see this localization in another way by manipulating a frequency ratio parameter. In a recent study[25] this ratio was varied by training the right hand to cycle at a fixed frequency and amplitude while monotonically increasing the frequency of the mechanically driven left hand. The time series for both hands when the frequency ratio is near 2:1 is plotted in Figure 5. At certain points, both amplitude and frequency are modified and the result is phase attraction at the intrinsic points ($\phi \sim 0$; $\phi \sim \pi$).

The dynamic pattern approach discussed earlier can also be applied to the discrete case. As before, we choose the relative phase for the collective variable. Instead of the continuous measure $\phi(t)$, we use the point estimate ϕ_n. The discrete measure can be obtained by strobing the time series at characteristic points, e.g., time t_n of peak positions or peak velocities, and computing the local relative phase, ϕ_n (see Primer[28] for the complete procedure). The resulting point estimate is a Poincaré transformation of the trajectories projected onto a 2-Torus. The dynamics of the relative phase can then be inferred from the return map (plot of ϕ_{n+1} vs.

ϕ_n). In the return map, for 1:1 coordination, phase-locked modes appear as fixed-points, while for non-1:1 they are stable orbits of discrete points. [2] In the following, the fixed-point properties of bimanual coordination are the main requirement for the map. These are accommodated in the modified sine (or phase attractive) circle map[3]:

$$\phi_{n+1} = \Omega - \frac{K}{2\pi}(1 + A\cos 2\pi\phi_n)\sin 2\pi\phi_n \tag{5}$$

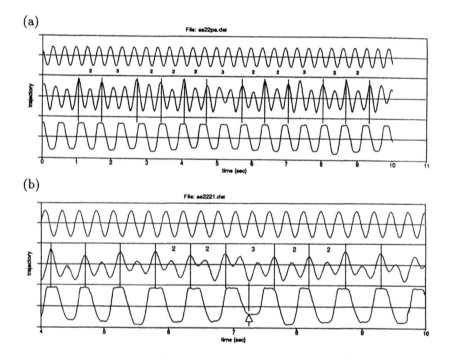

FIGURE 5 (a) Trajectories of the input signal used to drive the torque motor (top), the actual motion produced by the driven left hand (middle) and the "free" hand (bottom) when the required frequency condition is near 2:1. Enhancement of the peaks near inphase regions points to the discrete nature of the coordination, i.e., intermediate portions of the left-hand trajectories are adjusted to sustain the natural tendency to be in-phase with the right-hand (indicated by the vertical lines). Occasional slips occur when the position of the enhanced peak extends beyond the broadened peak of the right hand (three steps instead of two). (b) Similar plot near 2:1 from another time interval showing the slippage effect on the right hand.

[2] The first iterate, ϕ_{n+1} vs. ϕ_n of a 2:1 ratio is a stable orbit. The second iterate, ϕ_{n+2} vs. ϕ_n of the same ratio is a stable fixed point.

(a)

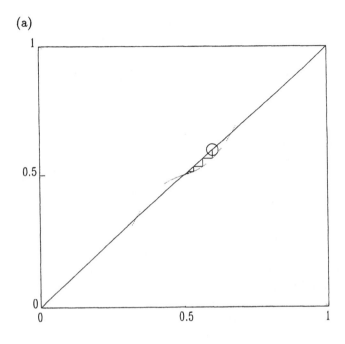

(b)

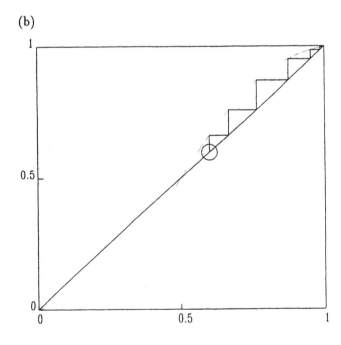

FIGURE 6 Phase transition in 1:1 case illustrated using the circle map (Eq. (5)). (a) The system is bistable (inphase and antiphase are coexisting locally stable fixed points) for $A > 1$. An initial phase ϕ_0 close to antiphase ($\phi = 1/2$) relaxes to it. Similarly, for ϕ_0 close to inphase ($\phi = 1$), (b) the system is monostable (in-phase stable and antiphase unstable). As A is decreased to below 1, all ϕ_0 not exactly at antiphase relaxes to inphase. It is clear that if, after settling down to inphase, A is then gradually increased back to its starting value in (a), no further switch occurs, thereby demonstrating hysteresis.

where the phase, ϕ, is normalized to the interval $[0,1)$. The meaning of the parameters are inferred partly from general properties of circle maps as well as the corresponding parameters for the flow (1). Thus Ω is the frequency ratio parameter, K is the strength of the coupling between the components, and, by analogy with Eq. (1), A determines the relative stability of the fixed points. The parameter A is correlated experimentally with the inverse of the basic frequency of motion. Thus $A < 1$ corresponds to cycling frequencies greater than the critical frequency (when bistability to monostability transition occurs in 1:1), $A > 1$ to lower frequencies, and $A = 1$ to the critical frequency. Figure 6 shows the pattern switching and hysteresis effects as A is varied. Let $A > 1$ and take an initial phase, ϕ_0 close to $1/2$. After the transient, the phase settles down to antiphase (Figure 6). If A is decreased to below 1, antiphase becomes unstable and the same initial condition, ϕ_0, relaxes to inphase (Figure 6(b)). It is clear that if, after settling down to inphase (Figure 6(b)), A is then gradually increased back to its starting value, no further switch occurs, demonstrating hysteresis. The change in relaxation times near the phase transition[3] can be demonstrated by computing the Lyapunov exponents at different parameter values (see Primer,[28] section 2.8). Increase in the relaxation times means a decrease in the Lyapunov exponent.

In the above discussion, we reproduced some of the qualitative features of the flow dynamics discussed in section 3 as they pertained to 1:1 coordination. The presence of the other parameters Ω and K allow for a wealth of other phenomena in addition to the bistability to monostability phase transition. Unlike the flow model, to which we have mapped stable 1:1 coordination patterns and small deviations from them, the circle map may be used in non-1:1 or multifrequency situations. These additional features may provide us with a different insight into some of the coordination patterns already explored experimentally in physiology, neurobiology, and psychology (see, e.g., the volume by Haken and Köpchen, 1991). For example, let us consider the ubiquity of rational frequency ratios. Numerical studies on coupled oscillators or circle maps often generate data on the extent of mode-locked regions, that is, regions in the parameter space corresponding to various frequency ratios. For circle maps the asymptotic value of the frequency ratio is termed a winding number (ρ) defined as

$$\rho = \lim_{n \to \infty} \frac{|\phi_n - \phi_0|}{n} \tag{6}$$

and is a well-defined limit when the map is invertible, i.e., contains no local inflection points (see e.g., Bergé, Pomeau, and Vidal[2] for some of the interesting properties of the winding number). The parameter boundary where non-invertibility sets in is called the *critical surface* of the map. In Figure 7 are examples of these regions, called Arnol'd tongues, computed using Eq. (6) for the non-1:1 coordination. For physical systems, the universality of rational frequency ratios when two systems

---

[3]Note that it is only a coincidence in the present context that the word *phase* appears as a collective variable (*relative phase*) and as a change in state (*phase transition*).

synchronize comes from the structural stability property of such states.[2] Structural stability means that slight modifications to the system do not alter the ratio states in the present case, and it is to be differentiated from dynamical stability which is tested by perturbing the motion, without altering the essential component makeup of the system. Coordinating biological systems seem to display similar tendencies (e.g., respiration vs. heartbeat, respiration vs. locomotion, stimulated nerve membranes, cells in different regions of visual cortex, etc.) although in such cases, one does not have the same resolution to differentiate frequency ratios as in well-controlled physical systems. On a behavioral level, the source of the difficulty in performing varied frequency ratio coordination can also be addressed using Figure 7. Studies by ourselves[20] and others[4,34,36] have shown that low-order ratios (i.e., 1:1, 2:1, etc.) are easier to perform than higher order ones (4:3, 5:3,...). This means that under a similar set of experimental controls, the variations of the control parameters that allow a given ratio to be performed reasonably well are greater for lower-order ratios than for higher-order ratios. Interestingly, there is evidence[30,31] that much greater neural activity (e.g., in frontal regions of cerebral cortex) is required to perform coordination patterns that are not 1:1. Our viewpoint is that the relative difficulty of tasks that require coordination and temporal pattern discrimination are intimately related to the widths of the mode-locked regions (the Arnol'd tongues) in the system parameter space (Figure 7). Thus, the wide-region modes, 2:1 and 3:1, are relatively easier to perform than the narrower ones (4:3, 5:2,...).

Transitions in coordination between two different frequency ratios are just a matter of crossing from one tongue to another. It is clear that a system buried deep inside one of the tongues is relatively stable as far as parametric or noise effects are concerned. This fact is significant for rationalizing the dominance of low-order ratios. The problem of flexibility may also be addressed using Figure 7. Here, the greatest flexibility is afforded the system when it is near the tongue boundaries where transitions to other modes are easily effected. Below the critical surface, transitions between strict mode lockings involve crossing thin regions containing quasi-periodic orbits. For motion on a torus, transition pathways are via intermittency, which here and elsewhere we have implicated as the essential process underlying relative coordination. Following Pomeau and Manneville,[35] we illustrate intermittent behavior using maps. Assuming a near 1:1 coordination, the local motion of the relative phase near a saddle node ϕ_c can be expressed as

$$\varphi_{n+1} = \epsilon + \varphi_n + \varphi_n^2 \tag{7}$$

where $\varphi_n = \phi_n - \phi_c$, and ϵ is the effective distance parameter that measures how far the system is parametrically from the given saddle node condition. The motion

(a)

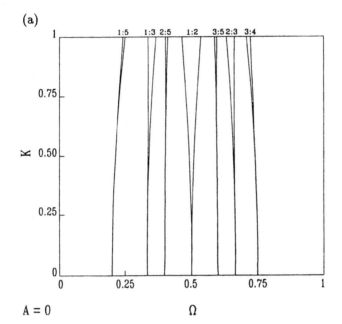

A = 0

(b)

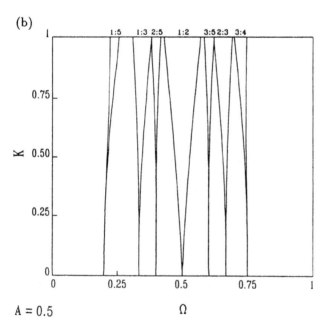

A = 0.5

FIGURE 7 Arnol'd tongues: Mode-locked regions computed for the circle map (Eq. (5)) for some of the lower order non-1:1 ratios. (a) $A = 0$ yields the familiar sine circle map tongues. (b) $A = .5$—we note the widening effect of a positive A parameter. In both cases, the relative widths of the tongues partly provide the basis for the differential stability and difficulty of performing the various frequency ratio coordination.

(a)

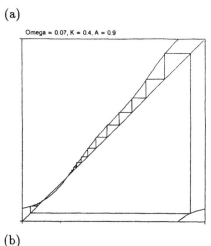

Omega = 0.07, K = 0.4, A = 0.9

(b)

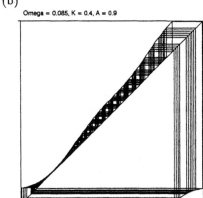

Omega = 0.085, K = 0.4, A = 0.9

FIGURE 8 Plot of ϕ_{n+1} (y-axis) vs. ϕ_n (x-axis) corresponding to the circle map (Eq. (5)) near a saddle node. (a) Mode-locked behavior at 1:1 showing a pair of fixed points, stable and unstable, close to the origin. (b) By increasing Ω, the curve may be lifted above the diagonal, producing intermittency.

consists of two qualitatively different behaviors. Near the point of tangency, a channel between the map and the diagonal line is formed (Figure 8), and the iterates concentrate within the channel for some characteristic time, τ. This is sometimes called the laminar region and corresponds to the regular, near mode-locking behavior of relative coordination. Because of the unstable direction, the system eventually escapes to explore other regions of the phase space. What happens after the escape depends on the attractor layout around the saddle node. If the motion wanders back into the channel, regular motion again ensues. One then observes a sequence of regular and irregular behavior. In the frequency scanning experiment described briefly above, near the 2:1 ratio we saw occasional phase slippage and the injection of additional cycles, strongly suggestive of intermittent dynamics. Additional predictions are discussed below.

5.1 DWELL TIME PREDICTION IN RELATIVE COORDINATION

For maps whose local motion near ϕ_c can be expressed as in Eq. (7), the characteristic time spent inside the narrow channel scales as (see, e.g., Bergé, Pomeau, and Vidal[2])

$$\tau \sim \epsilon^{-\frac{1}{2}} \tag{8}$$

so that the nearer the system is to the saddle node, the longer it stays around it. The characteristic exponent $(1/2)$ in Eq. (1) derives from the quadratic nature of the local expansion (Eq. (7)). In general, this varies with the order of the critical point. The distance parameter ϵ also depends on the other parameters of the map. For the saddle node bifurcation, one varies a single parameter. This means that for general maps, a one-dimensional path in the parameter space should be specified, and ϵ is a distance measure along that path. For example, in Eq. (5), any one of the control parameters Ω, K, A or some combination of them may be varied. When Ω is close to 1:1, we have seen that the monostable and the bistable dynamics of Kelso's model system are reproduced (e.g., Figure 6). By setting the appropriate conditions on the slope of the map at a fixed point, it can be shown that the surface

$$\Omega = \frac{K}{16A^2}\left(-1 + 10A^2 + (1 + 2A^2)\sqrt{1 + 8A^2}\right) \tag{9}$$

can generate saddle node solutions. A parameter change along a path intersecting this surface will induce intermittency in Eq. (5). Operationally, in our coordination experiments τ is proportional to the number of iterations of the measured relative phase before phase wrapping occurs; it is the time (in iteration units) the system spends near the mode-locked state before slipping away. In order to verify the scaling law (Eq. (8)), the threshold condition ($\epsilon = 0$) must be located, a step that requires very fine control of parameters. It is possible that a related quantity $P(\tau, \epsilon)d\tau$, defined as the probability that the experimentally measured relative phase has a dwell time between τ and $\tau + d\tau$ is more accessible experimentally. The scaling law (Eq. (8)) then becomes

$$\int_0^\infty \tau P(\tau, \epsilon)d\tau \sim \epsilon^{-\frac{1}{2}}. \tag{10}$$

A closed form expression for $P(\tau, \epsilon)$ can be found in Hirsch, Huberman, and Scalapino.[9] In the phase attractive map model of coordination, a parameter A is proportional to the driving frequency. A is the ratio of the coefficients for the $\sin\phi$ and $\sin 2\phi$ in the original HKB model (Eq. (1)). From another viewpoint it expresses the relative importance of the intrinsic phase-attractive states at 0 and π (we remind the reader that for $\mid a/4b \mid < 1$, there are two stable states; as $\mid a/4b \mid \to 1$, the anti-phase stable state loses stability, and for $\mid a/4b \mid > 1$, only the in-phase pattern is stable). Referring back to the surface Eq. (9), it can be shown that if the parameters A and K are fixed and only Ω is changed slightly, the scaling law can be expressed as

$$\tau \sim \left(\frac{\delta\Omega}{1 + \frac{3}{\sqrt{1 + 8A^2}}}\right)^{-\frac{1}{2}} \tag{11}$$

Equation (11) shows the directional effect of the parameter A (or equivalently of the cycling frequency) on how long a system can keep nearly mode locked in relative coordination. That is, the dwell time (in iteration units) is predicted to be greater for higher than for lower frequencies.

5.2 VARIABILITY AND NOISE EFFECTS

The role of noise in the stability of behavioral patterns and the ability to change the patterns have been examined in relation to the existence of different time scales in the system (see Primer,[28] section 2). A key question is how to reconcile observed multimodal patterns in terms of attractors of the collective variable dynamics. If the patterns are assumed to be asymptotic limits of these dynamics, how, then, are multiple patterns generated? As discussed in the Primer,[28] instabilities occur when certain time-scale relations are violated. However, not all changes are phase transitions; sometimes transitions are not stochastically induced but are really part of the deterministic dynamics. It has been recognized that variability in some biological systems can also be of chaotic origin.[5,33] Variability near fixed-point behavior (as well as for other types of behavior) can thus be from two sources: stochastic and (deterministically) chaotic. The effects of the two kinds of fluctuations are quite different. When noise is added to a state corresponding to a stable or unstable fixed point, the likelihood of the system jumping to a nearby state increases. Examples in phase transitions and annealing research are plentiful. In our coordination experiments,[40] we have demonstrated that small perturbations applied to the system as it approaches an instability induce transitions to a more stable state. In contrast, when the system is in an intermittent state, dwell times actually *increase* when noise is added to the system.[2,9] In the context of mode locking, this means that the average time the system spends near regular mode-locked state is enhanced in the presence of noise. This positive effect on the performance is in contrast to the traditional view that noise tends to destabilize states. If there is a stabilizing effect of noise, this is often because noise can induce transitions to other (deeper) minima so that subsequent stochastic forces do not affect the system's state as much.

6. CONCLUSION

Coordinated behavior is a necessary feature in a functioning biological system and manifests itself whenever there is an interaction between (a) components of an organism; (b) one organism with another; and (c) an organism with the environment. Because laws of coordination are (self-) organized laws, they are not necessarily rigidly bound to underlying physical and chemical processes. Like other cooperative phenomena in nature, coordination laws appear to be quite independent of the particular molecular machinery or material substrate that instantiates them. The fact that many physical structures realize the same pattern and that different patterns may be produced by the same material structure hints strongly that an understanding of coordination should be sought in terms of organizational principles rather than special mechanisms, *per se*. Simply put, the aim of the present approach to coordination is to understand "how things are put together."

The dynamic pattern approach we have used to unearth some of the general features of coordination include: (1) identifying collective variables that capture coordinated states, and (2) mapping of the observed stable patterns onto attractor states of a model, including their stability properties. These steps were demonstrated using Kelso's model system in bimanual coordination, which highlighted phase attraction or synchronization only at certain readily identifiable points under a wide range of conditions. Parallels exist with other patterns of coordination among very different systems. Our results and theoretical analysis as well as others obtained over the last decade or so suggest that one, perhaps the chief, way for complex biological systems to coordinate themselves is through phase and frequency synchronization. This *tendency* is ubiquitous in natural and artificial systems and represents a primitive form of self-organization. It may be that phase and frequency synchronization is one of nature's ways to connect "things," regardless of the level of description at which these "things" appear (genes, neurons, local fields, muscles, limbs, parts of an object, sensations about the same objects, ideas behind a theme,...). One cannot, it seems, understand life without confronting the problem of coordination.

Pure phase and frequency synchronization, however, is a kind of ideal form, too rigid for most biological functioning. Our early identification of in-phase and anti-phase collective states is due to the symmetry in the basic coordination dynamics (sections 2 and 3). Symmetries play a crucial role in classifying basic coordinative forms but, as we have stressed here, more flexible and varied forms of coordination arise when the symmetry of the basic coordination dynamics is broken. The demonstrated ability of components to work together despite intrinsic differences is strongly indicative of this flexibility. One form of flexible coordinative behavior is relative coordination. With the aid of a "detuning" parameter $\delta\omega$ added to the symmetric case (Eq. (1)), it has been shown that relative coordination can be identified with an intermittent (Type I) process. As a Type I intermittency, relative coordination represents pathways between ordered (mode-locked) and disordered behaviors and, if the motion is confined to a 2-Torus, between mode-locked and quasi-periodic behaviors.[2] A key point is that not only is there some degree of coordinated behavior, but transitions to other coordinated behaviors are easily effected. Biological systems in this flexible, intermittent regime clearly exhibit phase attraction but strictly speaking there are no longer any fixed points, at least in the sense defined here, i.e., asymptotic convergence of the collective variable trajectory to a mode-locked state. By living *near*, but not *in* mode-locked regions, biological systems are assured of a vital mix of flexibility and stability. Rather than an army marching perfectly in step, biological systems are more like the father and child who, in order to keep pace with each other slow down (the father) and skip steps (the child). They are poised near the "ghost" of fixed points surviving best, as it were, in the margins of instability.

ACKNOWLEDGMENTS

Research supported by NIMH (Neurosciences Research Branch) Grant MH 42900, BRS Grant RR07258, and the U.S. Office of Naval Research Contract N00014-88-J-119. Our thanks to Bill McLean for help with the figures and graphics, and Pamela Case for assistance with manuscript preparation.

REFERENCES

1. Beek, P. J. "Timing and Phase-Locking in Cascade Juggling." *Ecol. Psych.* **1** (1989): 55–96.
2. Bergé, P., Y. Pomeau, and C. Vidal. *Order Within Chaos: Towards a Deterministic Approach to Turbulence.* Paris: Hermann, 1984.
3. DeGuzman, G. C., and J. A. S. Kelso. "Multifrequency Behavioral Patterns and the Phase Attractive Circle Map." *Biol. Cyber.* **64/6** (1991): 485–495.
4. Deutsch, D. "The Generation of Two Isochronous Sequences in Parallel." *Percep. Psychophys.* **34** (1983): 331–337.
5. Glass, L., M. R. Guevara, A. Shrier, and R. Perez. "Bifurcation and Chaos in a Periodically Stimulated Cardiac Oscillator." *Physica* **7D** (1983): 89–101.
6. Glazier, J. A., and A. Libchaber. "Quasi-Periodicity and Dynamical Systems." *IEEE Transactions on Circuits and Systems* **35** (1988): 790–809.
7. Haken, H. *Synergetics, an Introduction: Non-Equilibrium Phase Transitions and Self-Organization in Physics, Chemistry, and Biology.* 2nd edition. Berlin: Springer, 1983.
8. Haken, H., J. A. S. Kelso, and H. Bunz. "A Theoretical Model of Phase Transitions in Human Hand Movements." *Biol. Cyber.* **51** (1985): 139–156.
9. Hirsch, J. E., B. A. Huberman, and D. J. Calapino. "Theory of Intermittency." *Phys. Rev. A* **25** (1982): 364–377.
10. Kay, B. A., J. A. S. Kelso, E. L. Saltzman, and G. Schöner. "The Space-Time Behavior of Single and Bimanual Movements: Data and Limit Cycle Model." *J. Exper. Psychol.: Human Perception and Performance* **13** (1987): 178–192.
11. Kauffman, S. A. "Antichaos and Adaptation." *Sci. Amer.* **265** (1991): 78–84.
12. Kelso, J. A. S. "On the Oscillatory Basis of Movement." *Bulletin of the Psychonomic Society* **18** (1981): 63.
13. Kelso, J. A. S., K. G. Holt, P. Rubin, and P. N. Kugler. "Patterns of Interlimb Coordination Emerge From the Properties of Nonlinear Limit Cycle Oscillatory Processes: Theory and Data." *J. Motor Behav.* **13** (1981): 226–261.
14. Kelso, J. A. S. "Phase Transitions and Critical Behavior in Human Bimanual Coordination." *Amer. J. Physiol.: Regulatory, Integrative and Comparative Physiology* **15** (1984): R1000–R10004.
15. Kelso, J.A.S., and J. P. Scholz. "Cooperative Phenomena in Biological Motion." In *Complex Systems: Operational Approaches in Neurobiology, Physical Systems and Computers*, edited by H. Haken, 124–149. Berlin: Springer-Verlag, 1985.
16. Kelso, J. A. S., J. P. Scholz, and G. Schöner. "Non-Equilibrium Phase Transitions in Coordinated Biological Motion: Critical Fluctuations." *Phys. Lett.* **A118** (1986): 279–284.
17. Kelso, J. A. S., E. L. Saltzman, and B. Tuller. "The Dynamical Perspective in Speech Production: Data and Theory." *Phonetics* **14** (1986): 29–60.

18. Kelso, J. A. S., and G. Schöner. "Toward a Physical (Synergetic) Theory of Biological Coordination." In *Lasers and Synergetics*, edited by R. Graham and A. Wunderlin, 224–237. Springer Proceedings in Physics, vol 19, 1987.

19. Kelso, J. A. S., G. Schöner, J. P. Scholz, and H. Haken. "Phase-Locked Modes, Phase Transitions and Component Oscillators in Biological Motion." *Physica Scripta* **35** (1987): 79–87.

20. Kelso, J. A. S., and G. C. DeGuzman. "Order in Time: How Cooperation Between the Hands Informs the Design of the Brain." In *Neural and Synergetic Computers*, edited by H. Haken. Berlin: Springer-Verlag, 1988.

21. Kelso, J. A. S., J. D. Delcolle, and G. Schöner. "Action-Perception as a Pattern Formation Process." In *Attention and Performance XIII*, edited by M. Jeannerod, 139–169. Hillsdale NJ: Erlbaum, 1988.

22. Kelso, J. A. S, J. P. Scholz, and G. Schöner. "Dynamics Governs Switching Among Patterns of Coordination in Biological Movement." *Phys. Lett. A* **134(1)** (1988): 8–12.

23. Kelso, J. A. S. "Phase Transitions: Foundation of Behavior." In *Synergetics of Cognition*, edited by H. Haken and M. Stadler, 249–268. Berlin: Springer, 1990.

24. Kelso, J. A. S. "Behavioral and Neural Pattern Generation: The Concept of Neurobehavioral Dynamical System (NBDS)." In *Cardiorespiratory and Motor Coordination*, edited by H. P. Köpchen, 224–238. Berlin: Springer, 1991.

25. Kelso, J. A. S., and G. C. DeGuzman. "An Intermittency Mechanism for Coherent and Flexible Brain and Behavioral Function." In *Tutorials in Motor Neuroscience*, edited by J. Requin and G. E. Stelmach. Dordrecht: Kluwer, 1991.

26. Kelso, J. A. S., G. C. DeGuzman, and T. Holroyd. "Synergetic Dynamics of Biological Coordination with Special Reference to Phase Attraction and Intermittency." In *Synergetics of Rhythms*, edited by H. Haken and H. P. Köpchen. Berlin: Springer, 1991.

27. Kelso, J. A. S., and J. Jeka. "Symmetry Breaking Dynamics in Human Multi-limb Coordination." *J. Exper. Psychol.: Human Perception and Performance*, in press.

28. Kelso, J. A. S., M. Ding, and G. Schöner. "Dynamic Pattern Formation: A Primer." This volume.

29. Kugler, P. N., and M. T. Turvey. *Information, Natural Law and the Self-Assembly of Rhythmic Movement*. New Jersey: Erlbaum, 1987.

30. Lang, W., M. Lang, I. Podreka, M. Steiner, F. Uhl, E. Suess, C. Muller, and L. Deecke. "*DC*-Potential Shifts in Human Visuomotor Learning." *Experimental Brain Research* **71** (1988): 353–364.

31. Lang, W., M. Lang, F. Uhl, C. Koska, A. Kornhuber, and L. Deecke. "Negative Cortical *DC* Shifts Preceding and Accompanying Simultaneous and Sequential Finger Movements." *Experimental Brain Research* **71** (1988): 579–587.

32. Mandell, A. J. "From Intermittency to Transitivity in Neuropsychobiological Flows." *Amer. J. Physiol.* **245** (1983): R484–R494.

33. Mpitsos, G. J., H. C. Creech, C. S. Cohan, and M. Mendelson. "Variability and Chaos: Neurointegrative Principles in Self-Organization of Motor Patterns." In *Dynamic Patterns in Complex Systems*, edited by J. A. S Kelso, A. J. Mandell, and M. F. Shlesinger, 162–190. Singapore: World Scientific, 1988.

34. Peper, C. E., P. J. Beek, and P. C. W. van Wieringen. "Bifurcations in Polyrhythmic Tapping." In *Tutorials in Motor Neuroscience*, edited by J. Requin and G. E. Stelmach. Dordrecht: Kluwer, 1991.

35. Pomeau, Y., and P. Mannevile. "Intermittent Transitions to Turbulence in Dissipative Dynamical Systems." *Commun. in Math. Phys.* **74** (1980): 189–197.

36. Povel, D. J. "Internal Representation of Simple Temporal Patterns." *J. Exper. Psychol: Human Percep. Perf.* **7** (1981): 3–18.

37. Rosenblum, L. D., and M. T. Turvey. "Maintenance Tendency in Coordinated Rhythmic Movements: Relative Fluctuations and Phase." *Neuroscience* **27** (1988): 289–300.

38. Schmidt, R. C., C. Carello, and M. T. Turvey. "Phase Transitions and Critical Fluctuations in the Visual Coordination of Rhythmic Movements Between People." *J. Exper. Psychol.: Human Perception and Performance* **16** (1990): 227–247.

39. Scholz, J. P., J. A. S. Kelso, and G. Schöner. "Non-Equilibrium Phase Transitions in Coordinated Biological Motion: Critical Slowing Down and Switching Time." *Phys. Lett. A* **123** (1987): 390–394.

40. Scholz, J. P., and J. A. S. Kelso. "A Quantitative Approach to Understanding the Formation and Change of Coordinated Movement Patterns." *J. Motor Behav.* **21** (1989): 122–144.

41. Schöner, G. "A Dynamic Theory of Biological Coordination: Phenomenological Synergetics and Behavioral Information." In *Biological Complexity and Information*, edited by H. Shimizu, 42–70. Singapore: World Scientific, 1990.

42. Schöner, G., H. Haken, and J. A. S. Kelso. "A Stochastic Theory of Phase Transitions in Human Hand Movement." *Biol. Cyber.* **53** (1986): 442–452.

43. Schöner, G., and J. A. S. Kelso. "Dynamic Pattern Generation in Behavioral and Neural Systems." *Science* **239** (1988): 1513–1520.

44. Schöner, G. S., P. G. Zanone, and J. A. S. Kelso. "Learning as a Change of Coordination Dynamics: Theory and Experiment." *J. Motor Behavior*, in press.

45. Turing, A. M. "The Chemical Basis of Morphogenesis." *Phil. Trans. Roy. Soc. London* **B237** (1952): 37–72.

46. von Holst, E. "Relative Coordination as a Phenomenon and as a Method of Analysis of Central Nervous Function." 1939. Reprinted in: *The Collected Papers of Erich von Holst*. Coral Gables, Fl: University of Miami Press, 1973.

47. Zanone, P. G., and J. A. S. Kelso. "The Evolution of Behavioral Attractors with Learning: Nonequilibrium Phase Transitions." *J. Exper. Psychol.: Human Perception and Performance*, in press.

Thomas R. Nelson, Ph.D.
Division of Physics and Engineering, Department of Radiology, University of California, San Diego, La Jolla, CA 92093-0610

Biological Organization and Adaptation: Fractal Structure and Scaling Similarities

Biological organisms exhibit complex organization and scaling relationships that are a natural and essential characteristic of their evolutionary and adaptational success. Such organizational characteristics have both a geometric and functional component with adaptation to the surrounding environment being an essential ingredient for both development and long-term viability of the organism in its ecological niche. Within this context, we can consider how fractals can be used to characterize biological organization and adaptation.

Development of our topic proceeds as follows: First, a basic overview of important factors related to organization and function is given. Second, the characteristic features of fractals in the context of biological organization are discussed. Third, the concept of a fractal dimension is introduced and related to more familiar ideas regarding dimension. Fourth, design criteria are considered as related to fractal structures, diffusion, and turbulence. Fifth, anatomic and physiologic examples showing how fractal analysis may yield insight are given. Sixth, an overview of how fractal models may be used to simulate complex anatomic or physiologic processes is presented. Finally, a brief summary of the key ideas is given.

Principles of Organization in Organisms,
SFI Studies in the Sciences of Complexity, Proc. Vol. XIII,
Eds. J. Mittenthal & A. Baskin, Addison-Wesley, 1992   **35**

---

1. OVERVIEW

Natural fractals often exhibit a structural hierarchy with detail present over a range of scales.[14] Their organization also conforms to local boundary constraints which dictate the large-scale features. An example is shown in Figure 1 for an electrical discharge in acrylic. While more classical descriptions of organization may necessarily focus on individual levels, more recent work employing concepts related to fractals has highlighted the importance of considering the full range of organizational hierarchy contributing to both functional and structural complexity.[5,14]

The hierarchical perspective implicit in fractal analysis is essential for developing a complete understanding of organization and function. An example of the problem of scaling is presented in Figure 2(a), which shows the metabolic rate as a function of body weight for several different mammalian species.[24] It is clearly appreciated that a power-law relationship exists between the mass of the animal and the metabolic rate of the individual. Figure 2(b), by contrast, shows the oxygen consumption rate per cell to be essentially constant over a range of mammalian species. The organizational challenge of developing a delivery system capable of supplying every cell in the body with the necessary amount of nutrient (in this case oxygen) while providing flexibility and uniformity over a range of body weights is formidable.

The relationship between physiologic complexity and organization may be addressed by considering the following questions: First, what is the underlying physiologic process? Second, how does that process relate to the structure; particularly the hierarchial organization of the support structure? Third, is there a relationship between local and global function? Fourth, is the physiologic process common to the entire population?

We know that some degree of organization is necessary to distribute a resource, such as a nutrient, through a system and provide for removal of toxic waste products. It has been suggested, from the perspective of complex systems analysis, that optimal organization and function occurs when the organism is living on the boundary between chaos, with many degrees of freedom, and extremely rigid organization, with limited options.[12] The resources committed to maintain the structure should be minimized to maximize the competitive advantage. Finally, and most importantly, the organism should have sufficient flexibility to adapt to a range of demands from the environment.

Disease often reflects a progressive loss of adaptability in the organism.[10] Such loss may be in an organ system context, such as atherosclerotic disease, or failure to provide a broad range of heart-rate response, or more globally, as in an organism's inability to adapt to changes in the ecosystem in which it lives.

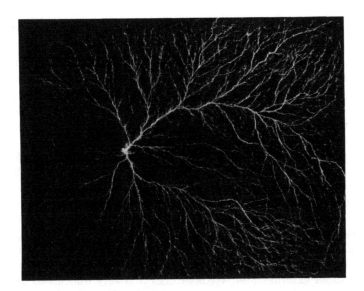

FIGURE 1   An electrical discharge pattern in acrylic. Note the progressive branching and the lack of cues regarding the size scale of the object while changing shape to correspond to the acrylic block boundary.

(a)

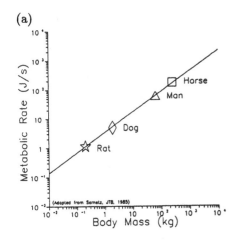

(b)

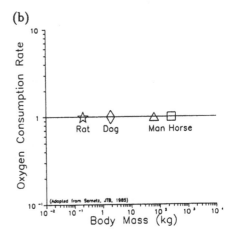

FIGURE 2   (a) Plot of the metabolic rate versus body mass for several mammalian species. A power-law relationship exists as the metabolic rate scales with body mass. (b) Plot of the oxygen consumption per cell versus body mass for leukocytes. Metabolic rate is essentially constant over a range of species.

2. FRACTAL CHARACTERISTICS

Fractal analysis, including concepts drawn from nonlinear dynamics and complex systems, offers an approach suitable for analyzing structural and functional organization with promise for developing a deeper understanding of natural phenomena. The principal features of fractal objects are: (1) a large degree of heterogeneity, (2) scaling similarity over many scales of observation, and (3) the lack of a well-defined (or characteristic) scale.[5,14]

The scaling properties of fractals are important because magnification does not reduce the apparent complexity of most objects. Traditional methods of classification presume a highest level of magnification above which no further detail or information is available. Within limits, such a premise is not representative of the natural world, thereby providing the motivation to obtain progressively higher magnifications. Object heterogeneity (natural objects are seldom regular or geometric in a "line" and "plane" sense) implies that organizational variability, along with scaling properties, leads to great difficulty in identifying the exact observation size without external cues. Without cues it is very difficult to determine the precise location or the degree of magnification. Furthermore, features observed at one magnification often may be similar to those observable at another scale with a rescaling or renormalization transformation relating the two. Extraction of these scaling relationships by analysis of the operational or structural organization, may offer greater insight into behavior and fundamental controlling mechanisms.

All organisms adapt to a range of stimuli by modifying their structure and/or function. The range of adaptation is limited by boundary constraints by which we mean any parameter affecting growth, development or function of the organism, such as geometric borders, diffusion distances, metabolic rates, etc. Since fractal objects have an inherent structural richness, with variability present at all scales, architectural or functional responses to various stimuli generally do not produce significant structural or functional deficits. Organisms possessing such inherent variability, thus, are better able to adapt to changing stimuli, incorporating a type of natural error tolerance.[9] The presence of many size, or time, scales in an object, as reflected in its organizational or functional scaling characteristics means that there is not a dominant, or optimum, mode that is preferred. That is, a more generalized system that is organized, or can function, equally well over a range is better able to accommodate the demands of a changing environment. The multiplicity of temporal or spatial scales is one manifestation of such design organization. In contrast, systems lacking flexibility are unable to adapt which may result in an important competitive disadvantage. The hypothesis has been offered that disease results from the progressive loss of adaptability in a system or organ.[10]

3. FRACTAL DIMENSION

The fractal dimension (D_F) is an important parameter which provides a measure of the organizational complexity, the capacity to occupy space, or the number of degrees of freedom in the behavior or function, or reflects the amount of information storage. Specific details for calculation of (D_F) have been considered in a number of references.[4,5,25] (D_F) relates the scaling of an object's bulk with its size adding to our intuitive notion of topological dimension (D_T); (D_F) may have fractional values.

$$\text{Bulk} = \text{Size}^{D_F}.$$

As (D_F) increases, the structure occupies more space, as shown in Figure 3 for several simple objects. Even though there are the same number of segments, the area covered by the object is not constant as would be the case were we only measuring (D_T). A consequence of this fact is that if we have a measuring unit (ϵ), then as ϵ becomes smaller, the number of units $(N(\epsilon))$ required to characterize the object increases as:

$$N(\epsilon) = \left(\frac{1}{\epsilon}\right)^{D_F}$$

where (D_F) is the fractal dimension. For a fractal, as ϵ becomes smaller, $N(\epsilon)$ becomes extremely large. While (D_F) is often used in the context of structural organization, such as for a branching network, such approaches are equally applicable to study of time-varying phenomena related to fluid dynamics and turbulence.

Measurements of D_F can be useful in understanding the space and volume filling characteristics of organ structures. We know that organs are three-dimensional (3-D) objects; however, the constituent sub-units (arteries, veins, nerves, alveoli, etc.) are often one- or two-dimensional and do not completely fill the organ. If we define the embedding dimension (D_E) to represent the number of independent parameters necessary to describe the space occupied by the object with (D_T) the dimension of the constituent units of the object (e.g., the branches in a tree), then D_F can be thought of as a measure of the actual amount of space occupied by the object.

Partial volume filling as measured by D_F is of significant importance to understanding organ function and can be appreciated by a simple analogy. Consider the volume of a 3-D object such as a cube with $D_E = 3$, the volume of which is the product of the X, Y, and Z lengths of the perpendicular sides (three independent parameters). If the cube were filled completely with water, D_E, D_F, and D_T would all equal three. On the other hand, if the sides of the cube were the same but the cube were a porous matrix consisting of a patchwork of small surfaces $(D_T = 2)$ producing voids throughout the cube, resulting in a large surface area but small volume, then the actual volume occupied by the solid material in the cube would be less than the volume measured by the product of X, Y, and Z but greater than the surface area patches being somewhere between two and three, such that:

$$D_E \geq D_F \geq D_T.$$

A physiological example of this situation occurs in the lung where the surface area of all the alveoli is hundreds of square meters contained within a volume of less than ten liters. Also, arterial, venous, neural, and bronchial structures all will have a fractal dimension of less than three based on the architecture of the object in relation to the overall organ geometry. Measurements of the bronchial airway have shown that the fractal dimension ($D_F \approx 2.7$) is remarkably consistent between normal individuals.[18]

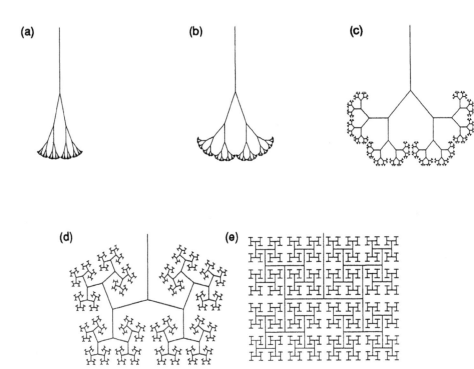

FIGURE 3 Self-similar branching networks. There are nine orders of branching with 2^9 branches in the terminal order. The scaling factor (α) relating the segment length to the preceding order varies in each network. α is constant for all branches within a given structure. (a) $\alpha = 0.51, D_F = 1.03$, (b) $\alpha = 0.53, D_F = 1.09$, (c) $\alpha = 0.59, D_F = 1.31$, (d) $\alpha = 0.66, D_F = 1.67$, (e) $\alpha = 0.71, D_F = 2.00$. The branching angle ($\theta$) is chosen to produce the most compact object without overlap of the terminal segments: $(\theta°) = 482.6 \ln(\alpha) + 349.79$.

4. ORGANISM DESIGN CRITERIA

In a global context a fractal design describes the dissemination pathway for the resource throughout the organ with the entire network, or structure, providing the support base for the distribution system. Organs such as the lung, kidney, or heart must provide a pathway to transport the essential quantity (blood, air, etc.) close enough to each cell or subregion to ensure that the final diffusion step occurs freely to maintain cell viability. Often the functional part of an organ may be only a thin surface (or interface) between the constituent networks, possessing a large surface area in a small volume (e.g., the area of the alveolar bed of the lung). The entire structure serves as a platform to distribute the resource throughout the system with the objective of facilitating diffusion down to the smallest levels, such as the cell.

Diffusion is the final step in the distribution of a resource throughout the organ although it occurs primarily on the smallest scales (both time and distance). The mixing of two liquids (e.g., tea and a small amount of milk) is an example of diffusion. A drop of milk will slowly diffuse throughout the tea becoming uniformly distributed after a long time. The rate of mixing may be accelerated by stirring the tea with a circular motion, thereby setting up laminar flow currents that increase the rate of diffusion. Movement of the stirrer in a back and forth motion will set up turbulent vortices and rapidly distribute the milk through the entire volume.

Turbulence exhibits structure over a range of scales with the turbulent vortices becoming progressively smaller as energy is dissipated to the surroundings. The distribution of scales associated with turbulent vortices is the same type of scaling hierarchy, or cascade, that is observed in structural fractals. Fully developed turbulence also exhibits temporal and frequency scaling characteristics over a broad range, with inverse power-law features. Structural fractals, which are static objects, and turbulence, which is a dynamic process, are therefore closely related. Indeed, from the standpoint of a dissipative system, we can view structural fractals as objects in which the fractal structure possesses the features of turbulence with an important difference: the turbulence is a consequence of the structural organization rather than a dynamic process. Organ systems optimize their use of resources and energy by employing fractal structural organization to essentially produce large-scale turbulence (offering improved energy efficiency via laminar flow in vessels) while using local diffusion processes to complete the transfer of resources to the cellular levels.

The fractal dimensions of many biologic structures, such as the bronchial tree and the arterial bed, are in the range of 2.5–2.7 which corresponds to values reported for fully developed hydrodynamic turbulence. Thus the relationship between structural turbulence and hydrodynamic turbulence dimensionality, while possibly coincidental, accentuates the striking similarities between both phenomena.

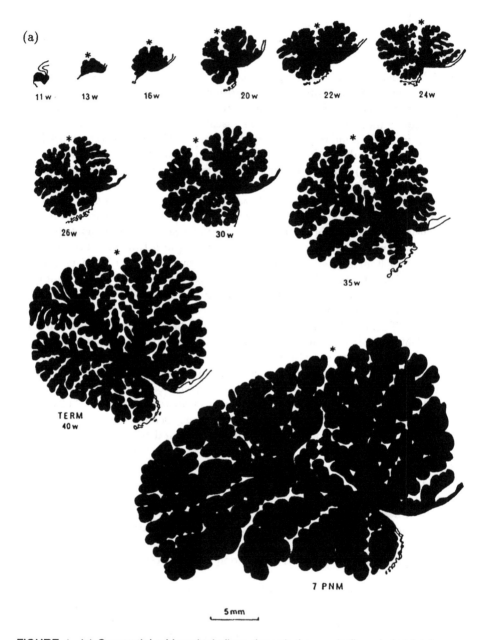

FIGURE 4   (a) Sequential mid-sagittal slices through the cerebellum during fetal development and early neonatal period (Adapted from Rakic and Sidman[20]).

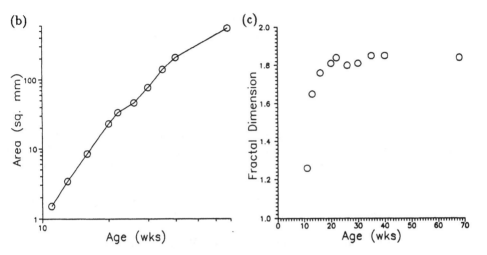

FIGURE 4 (cont'd.) (b) Change in area of mid-sagittal slices versus age showing log-log scaling relationships. (c) Change in D_F with age showing rapid change early in development with structural complexity stabilizing after approximately 20 weeks even though detail refinement continues to occur.

5. FRACTAL ORGAN ANATOMY AND PHYSIOLOGY

The evolution of organ systems from the initial zygote through early embryogenesis and organogenesis with progressive differentiation and increasing structural complexity is of particular interest to developmental biologists. Complete understanding of the evolutionary sequence leading to a fully developed embryo is crucial to obtaining an accurate description of each stage of development particularly in the context of developmental anomalies. We will present several examples illustrating how concepts discussed in this paper may be applied to study such phenomena.

5.1 CEREBELLUM

Figure 4(a) shows a series of sagittal images of the cerebellum throughout fetal and neonatal development, showing a progressive increase in the arbor structure and complexity of the cerebellar surface.[23] The accompanying graph (Figure 4(b)) shows the scaling relationship of cerebellar area as a function of age. Note the change in fractal dimension (Figure 4(c)), as the cerebellum becomes increasingly complex. A hallmark feature of this type of organization, shown in fine detail in an adult cerebellum (Figure 5), is the tree-like structure, with white matter penetration minimizing the distance to the cortical surface in an efficient manner while preserving a significant perimeter-to-length relationship.

5.2 BRONCHIAL AIRWAY

The morphology of the bronchial airway has features that lend themselves to fractal analysis. The fetal bronchial tree develops *in utero* prior to any requirement for oxygen transport. In contrast to other developing organs, the architectural development of the bronchial tree is dependent only on external constraints provided by the chest cavity and diaphragm for its final geometry; suggesting that airway morphology is the result of an evolutionary process with structural characteristics transmitted via the gene pool (Figure 6). The geometry of the lung is remarkably consistent between individuals; however, specific conditions existing *in utero*, during development, render each individual unique. Prior to birth the fetal lung serves many important functions, including the production of hormones to regulate fetal growth and development.

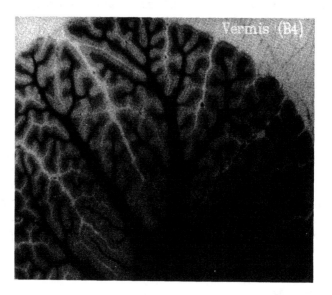

FIGURE 5 A slice through the cerebellum showing the progressive branching structure as the white matter is distributed throughout the cerebellar volume. The geometric complexity of these structures provides for rapid dissemination of information (or resource) via a large surface area in a compact space. This feature is a hallmark of fractal structures which maximize the surface area within a finite volume.

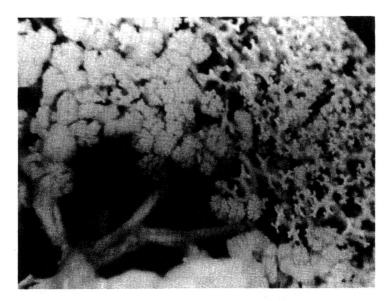

FIGURE 6   An erosion cast of a human fetal bronchial airway. The progressive
dichotomous branching structure is clearly apparent through a series of scales covering
several orders of magnitude in size.

Airway architecture is tree-like, beginning with an initial limb bud, subsequently developing an increasingly complex branching structure up to the level
of the respiratory bronchioles and alveolar ducts, and terminating in the alveolar
sacs. The architecture of this system provides for distribution of oxygen throughout
the entire lung. Design optimization is dependent on several factors with theoretical
models considering minimization of the total luminal surface area, the total luminal volume, the power to move fluid through the system or sheer force extended
through the walls. While all of these factors are interrelated, generally, energy costs
are minimized with branching angles of approximately 40–50. Morphometric data
show that the bronchial airway branching angles approximate these values although
a wide range of branching angles are found.

Early work studying the morphometry of the human lung by Weibel[27] showed
that relationships between branch generation, diameter and length could be expressed by models employing several exponentials, depending on the particular
region of the lung. While this approach provided a good model for understanding
airway organization, difficulties in relating the more distal branch segments to the
more proximal segments has limited its application. More recently, fractal analysis
using renormalization group methods has shown a relationship between branch order, length, and diameter that may be expressed in terms of an inverse power-law

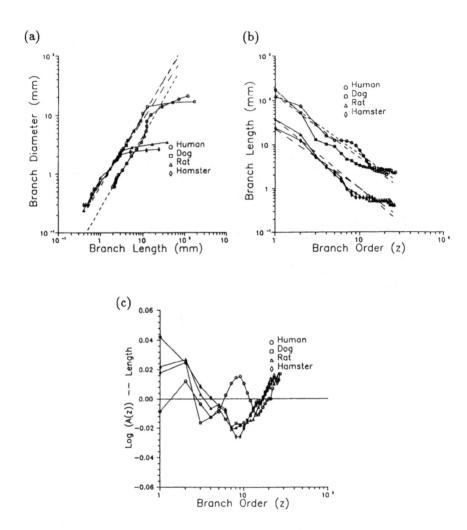

FIGURE 7 (a) Plot of the mean branch diameter versus the mean branch length for the four species. Each point represents one bronchial order (z) where the trachea $(z = 0)$ is the rightmost point. After the initial few orders all species demonstrate similar inverse power-law scaling relationships between mean branch diameter and length. (b) Plot of mean branch length versus branch order for four species. All four data sets show a harmonic modulation around the inverse power-law regression (broken line) with a fit of the form: Length = Intercept* Order^{Slope}. Note: a pure inverse power-law behavior will be represented as a straight-line plot on a log-log graph. The harmonic modulation is accounted for by the renormalization model. (c) Detailed plot of harmonic variation in mean branch length from (b) with the inverse power-law behavior removed. There is a marked difference in phase between human airway data and that of the other three species. Note the approximately 180° discrepancy in mean phase between the data points from the human lung and the other three species.

relationship modulated by a harmonic function.[19,24,28] Results of fractal analysis also show a remarkably consistent relationship between mean branch length and diameter for several mammalian species (Figure 7(a)) with D_F in humans being approximately 2.76. Renormalization analysis shows an inverse power-law scaling with a harmonic modulation between branch order, length, and diameter (Figures 7(b) and (c)).

As can be seen from these data, there is a significant difference in the phase between the harmonic terms of the human and the other three mammalian species which persists in both diameter and length measurements and was previously unrecognized. Possible explanations include evolutionary adaptation of the organism to differences in postural orientation between the human bipedal motion and the quadrupedal motion of the dog, rat, and hamster.

5.3 CARDIAC HEART RATE AND ACTIVATION

A number of cardiac structures have a self-similar or fractal-like appearance: the coronary arterial and venous trees, the chordae tendineae, certain muscle bundles, and the His-Purkinje network.

One consequence of cardiovascular disease is that a progressive loss of adaptability is often observed in patients with significant cardiac disease.[2,6,8,9,13,16,26] Figure 8(a) shows the beat-to-beat variability in a normal patient, along with the frequency spectrum, indicating a broad inverse power-law distribution. Figure 8(b) shows a patient in heart failure with a slow variability in the heart rate and much smaller variation in the beat-to-beat interval. Frequency analysis shows a very strong line corresponding the the slow variability rate consistent with a narrower range of functional response.

Cardiac activation also is an essential part of efficient heart pumping and muscle contraction with the His-Purkinje system being an important component of myocardial activation. The His-Purkinje conduction system is organized as a progressive branching network that extends from the AV node and bundle of His to the right and left bundle branches and the subsequent ramified network of the Purkinje fibers as they ennervate the endocardial surface of the ventricular myocardium. The Purkinje fibers distribute the excitation pulse to the ventricular myocardium with a broadened arrival time distribution resulting in the onset of contraction varying regionally according to the temporal distance from the atrial ventricular node and the bundle of His (Figure 9). A broad distribution of myocardium excitation leads to efficient ventricular depolarization and pumping. Ultimately, muscle activation based on the His-Purkinje distribution leads to the externally recorded electrocardiogram (ECG) signal of which the QRS portion reflects myocardial depolarization.

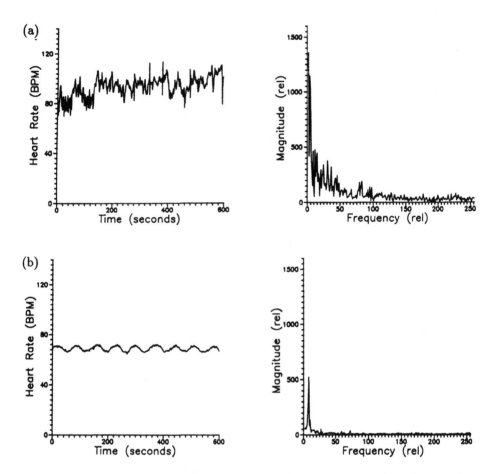

FIGURE 8 Heart-rate time series plots for two patients (A and B). The left curve
shows the temporal behavior. Trends are often visible only over 10s of minutes.
The right curve shows the frequency spectrum for each time series. The frequency
spectrum reflects the dominant heart-rate beat periodicity. Normal individuals (a) exhibit
a broadband inverse power-law scaling behavior discussed previously with no specific
resonance or preferred heart rate. The presence of cardiac pathology, however, often
produces an increased periodicity at specific frequencies (b) which is indicative of a
system less able to meet a variety of workload requirements (data courtesy Dr. A. L.
Goldberger).

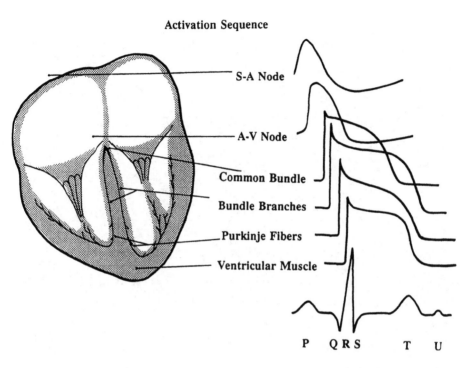

Activation Sequence

- S-A Node
- A-V Node
- Common Bundle
- Bundle Branches
- Purkinje Fibers
- Ventricular Muscle

P   Q R S   T   U

FIGURE 9 Stylized diagram of His-Purkinje conduction system. Activation proceeds from the sino-atrial (SA) node across the atria to the atrio-ventricular (AV) node progressing via the common bundle to the right and left bundle branches and through the Purkinje cells to the myocardium. Transmembrane action potentials are shown for each of the primary regions as activation progresses in relation to the QRS portion of the normal external electrocardiogram. (Adapted from F. H. Netter, *The Heart*, CIBA, 1987.)

The precise structure of the His-Purkinje system, while not a pure theoretical fractal, shares may characteristics with natural fractals. Figure 10 shows a more detailed example of the anatomy of the His-Purkinje conduction system in relation to the myocardial muscle. Interconnection between pathways of the His-Purkinje system, does not invalidate the applicability of the fractal model. The heterogeneity present in fractal networks incorporates a multiplicity of pathways which cross connections do not fundamentally alter. Indeed, it is heterogeneity of the fractal distribution that gives the system the robustness and redundancy crucial to the error tolerance inherent in a fractal system.

FIGURE 10   Canine myocardium with selective iodine stain of Purkinje cells
demonstrating progressive branching. (Courtesy Dr. Ian LeGrice, University of Auckland,
NZ.)

Changes in the myocardial activation distribution are reflected in the ECG,
which is an important source of diagnostic information regarding depolarization,
although the precise relationship between conduction system disruptions and the
QRS is often not clear. A theoretical argument has been made that a broadband
power-law spectrum is consistent with a model of depolarization of the myocardium
resulting from an irregular, self-similar branching network.[9] Thus, according to
fractal theory, the frequency content of the QRS is related to the macroscopic
structure of the His-Purkinje system in addition to the microscopic nature of the
Purkinje-myocardial cell interactions and local wavefront propagation.

Spectral analysis of normal QRS complexes reveals a broadband frequency spectrum. While most of the frequency content of the QRS is comprised of frequencies
below 20–30 Hz, there is a small but important contribution of higher frequencies
that go up to several hundred Hz.[2,11,30,22] Changes in the geometry or propagation
properties of the conduction network may alter the frequency content of the QRS
complexes, independent of any changes in myocardial conduction.[7] A loss of fractal properties by the system including a loss of 1/f-like behavior has been noted

in response to pathology. The transition from fractal to non-fractal behavior thus may represent a possible indicator for identifying early changes in the conduction system of potential diagnostic value associated with loss of scaling heterogeneity and complexity.

6. FRACTAL MODELS OF BIOLOGICAL ORGANIZATION AND FUNCTION

Models based on fractals offer a means of relating microscopic and macroscopic organization in a unifying framework whose characteristics may increase our understanding of the underlying physiology. Furthermore the significance of differences between a fractal model and the physiologic system may be more clearly understood by beginning with a simple model and refining it to facilitate convergence between theoretical and experimental data. Additionally, computer models that simulate biological organization and function may increase our understanding of how organization and function are related. This section will discuss some recent developments in the area of applying theoretical fractal models to the simulate morphogenic and functional properties of these systems.

Classical methods of analysis suggest that simple relationships lead to predictable and simple behavior. Developments in the area of complex systems and nonlinear dynamics, which share many common features with fractals, suggest alternative ways of viewing how simple relationships may lead to very complex behaviors. In fact, many heretofore complex behaviors that appear to be stochastic processes may, in fact, result from recursive relationships of nonlinear functions. A recursive process is one in which the current level of evolution or development is used as input to the next round of development. Nonlinear recursion is a type of feedback in which small variations early in development may produce substantial changes in the long-term evolution of the system or be entirely damped out. Clearly either of the two extremes of this situation, erratic and unpredictable or steady-state behavior, is sub-optimal. More optimal would be the case in which the delicate balance between erratic and moribund behavior is maintained by the organism living on the boundary between chaos, with many degrees of freedom, and extremely rigid organization with limited options.

An example will demonstrate how a simple nonlinear model may lead to complex and apparently unpredictable results with analogy to our focus on organization and function. Figure 11(a) shows an example of long-term heart-rate patterns in a normal individual resulting in a marked degree of heterogeneity and variability, appearing almost to be a random process. A similar type of behavior is produced by the Weierstrass function ($W(t)$) shown below (Figure 11(b)), where a and b are constants ($a, b > 1$).[5,14]

$$W(t) = \frac{(a-1)}{a} \sum_{n=0}^{\infty} \frac{1}{a^n} \cos(b^n t).$$

$W(t)$ consists of adding a series of cosine functions of increasing frequency scaled by a decreasing amplitude. In the infinite limit $W(t)$ is a function everywhere continuous but nowhere differentiable with

$$D_F = \frac{\ln(a)}{\ln(b)}.$$

$W(t)$ is fractal with the result that progressive magnification does not simplify the structural organization rendering it difficult to determine the scale of the object without additional information.

Branching networks from a recursive algorithm exhibit similar features to the Weierstrass function. As we saw in Figure 3, the principle features of single-scale fractal networks showed that D_F increased as the network filled greater amounts of space, with $1 < D_F < 2$ for two-dimensional networks. While an optimal design would fill the entire two-dimensional surface, it is readily appreciated that a simple

(a)   (b)

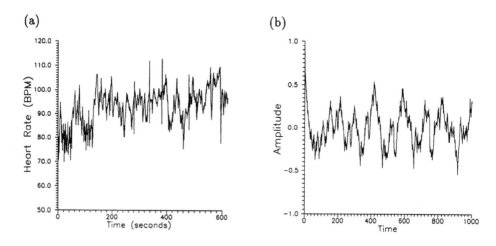

FIGURE 11   (A) Heart-rate time-series plot for normal patient. Note that even with a long time series recording the behavior is highly uncorrelated. (B) Plot of a Weierstrass function exhibiting similar characteristics to the normal patient. Even though the long-term behavior appears to be stochastic, it is based on a precisely described recursive generating function ($D_F \approx 1.5$).

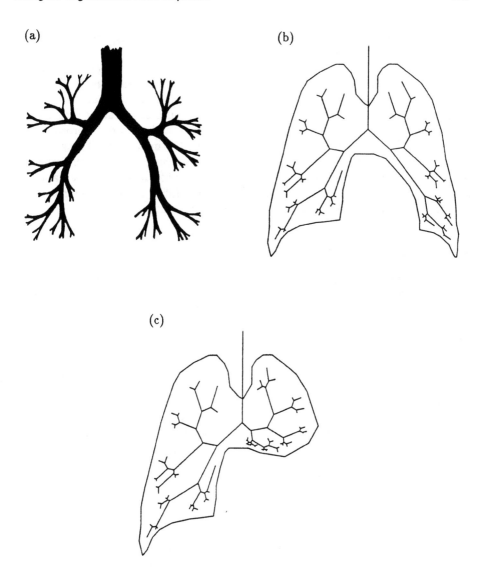

FIGURE 12 (a) Stylized diagram of the human bronchial airway derived from a chest radiograph. (b) Recursive fractal model of the airway based on boundary data for chest wall. There is good agreement between model and morphometric data. (c) The consequences of growth with a severely restricted left lung border comparable to the situation found in fetal diaphragmatic hernia.

fractal network with $D_F = 2$ does not accurately reflect the architecture of anatomic fractals. Further, if we study the propagation characteristics of such a network, we see that a pulse arrives at all terminal segments simultaneously, thus introducing no decorrelation into the distribution.

A modified recursive algorithm incorporating a response to boundary conditions has been shown to exhibit properties including structural complexity and a large degree of heterogeneity, with $D_F \approx 1.8$ comparable to the anatomic fractals already considered.[19,20] Additionally, an excitation pulse propagating through the network produces a broad decorrelation distribution, and, in turn, a broad power spectrum with an inverse power-law characteristic. Optimum space-filling characteristics, based on maximizing D_F, result with branching angles in the range of 40–60. Fractal model results are in good agreement with morphometric data and cost minimization calculations suggesting that space filling plays an important role in branching angle optimization and biological selection.

6.1 BRONCHIAL AIRWAY

Modeling of lung development with the possibility of simulating abnormal developmental conditions was explored using a fractal model. Figure 12(a) is a stylized diagram of the bronchial tree in the human. Figure 12(b) shows a fractal model of the bronchial tree resulting from an algorithm that used a boundary constraint derived from a contour based on the radiographic projection in a normal adult.[18] The branch length segments show a very high degree of positional correlation with those observed in the normal lung. Figure 12(c) presents the results with a restricted development space similar to what would be encountered in a case of diaphragmatic hernia or other growth restriction. All branch segments are present, although the geometry is significantly altered, in good agreement with morphometric data derived from radiographic studies of afflicted individuals. Model complexity may be increased by including additional branch orders to simulate the tracheobronchial tree more completely. Figure 13 shows a model with 16 orders of branching containing approximately 65,000 segments.

6.2 CARDIAC ACTIVATION

Fractal models offer powerful methods to simulate cardiac activation.[1] A 3-D recursive fractal algorithm was used to produce a model of the His-Purkinje conduction system (Figure 14(a)). The location and propagation characteristics are uniquely defined for each segment with $D_F \approx 1.7$. The model incorporates both network and cellular conduction propagation velocity constant to network termination (Figure 14(b)). Spherical wavefront progression from each terminating segment continues "cellular" activation at a reduced velocity compared to primary network conduction velocity. Segment propagation delays are computed independently. Propagation distributions are similar to clinically observed measurements with depolarization proceeding from the apex and endocardial surface toward basal and lateral regions (Figure 14(c)) in agreement with clinical data.[3] QRS complexes are simulated by projecting the instantaneous vector onto a reference lead system approximating the standard ECG leads with 2-D and 3-D simulations possible (Figure 14(d)).

Simulation of clinical conduction pathology is accomplished by selectively altering network segment properties with distal elements of the branching network affected by all perturbations in parent segments. Severing a conduction pathway in a second- or third-order pathway affects 50° (bundle branch) or 25° (segmental distribution) of the conduction system. Conduction blocks for any order or location in the network may cover areas from < 1% to 50° of the myocardial volume.

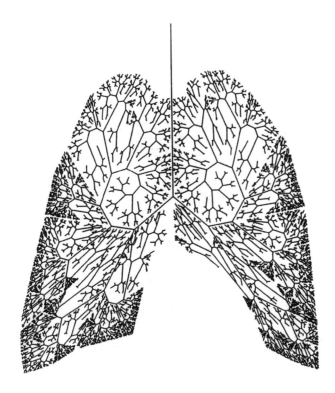

FIGURE 13 A more detailed model of the airway structure resulting from 16 levels of branching with the same algorithm presented previously. There are approximately 65,000 segments throughout the entire structure. Note that incorporation of more branching levels while filling in the lung field continues to produce a highly heterogenous structure. Models with this and even greater levels of complexity begin to approach the level of detail encountered necessary to simulate physiologic processes.

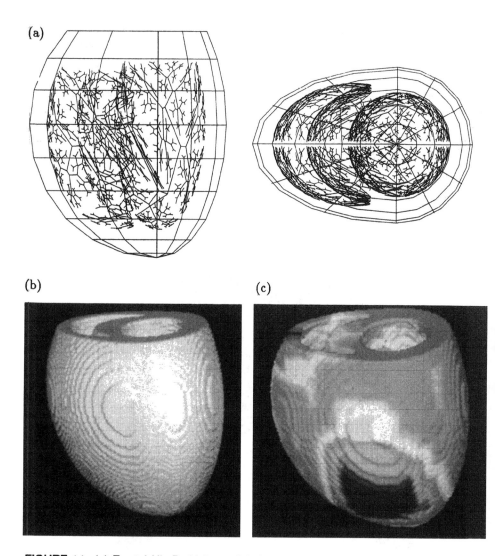

FIGURE 14 (a) Fractal His-Purkinje model showing 3D branching fractal network. (b) 3D myocardial model showing exterior surface of model. (c) Myocardial surface and activation sequence time delay based on network propagation is shown.

(d)

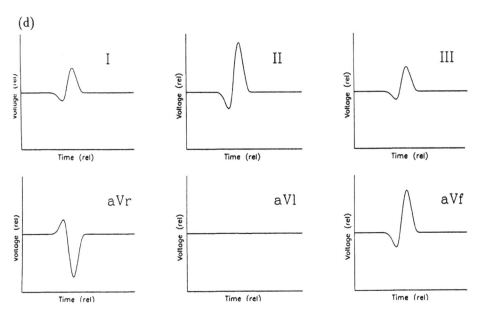

FIGURE 14 (cont'd.) (d) Simulated ECG recordings produced by projecting the instantaneous electrical activity in the myocardium onto a reference system approximating the clinical ECG lead placement.

Conduction by "cellular" diffusion continues from segment ends into blocked areas. Simulation of local conduction defects produces propagation changes apparent in QRS projections and spectral data. Perturbations of the same order have different consequences depending on their location. Perturbing a high-order (smallest) segment has a very localized effect and minimal impact on the QRS signal. Spectral analysis also represents an alternative method for analyzing disruptions over a range of length scales more completely. In regions with a defect, propagation is more coherent spatially, consistent with a diffusion model.

The conduction properties of the myocardium represent a complex system of which the His-Purkinje system is only one part. However, the distribution and depolarization pulse characteristics imparted by the His-Purkinje fibers are crucial to the shape of the resultant mechanical contraction pattern in the ventricle. Diseases often result from breakdown or disruption of this distribution system. Fractal models may lead to more precise methods of localizing perturbations.

7. SUMMARY

Fractals offer several advantages to the study of biological organization and adaptation. A fractal design has improved error tolerance due to its inherent variability with adaptability extending over a broad range of stimuli consistent with organization at the edge of chaos. A fractal design also represents a compact, efficient coding representation of inherently complex natural phenomena which, if we can understand the code, will offer greater insight regarding organization and function.

Fractal models offer a way to simulate complex phenomena in a controlled environment while using experimental data to refine model validity. Fractal models allow investigation of tradeoffs among reliability, flexibility, economy, and speed through variation of parameters controlling fractal growth. Generation of fractal structures is an economical means of communicating design codes through recursion mechanisms based on single dynamical modules that interact with the environment adaptively. Furthermore, the inherent simplicity of fractal modules imparts robustness to the resultant structure or process, thus ensuring error tolerance and adaptability necessary to long-term success. Perturbation of a fractal model presents a means whereby the full range of hierarchial organization and function may be studied to evaluate the effect of compromise on the functional performance. Finally, a fractal model provides an essential link between organizational and functional complexity and modular simplicity.

In summary, we have discussed the application of fractal concepts to biological organization and adaptation drawing on examples from analysis of experimental data to theoretical modeling. Dynamic and structural optimization play an important role in the evolution of organ systems with economic utilization of available resources resulting in superior adaptation for survival and superiority expressed via the gene pool. However, the adaptability to a changing environment is essential for long-term survival of the individual, and species, and appears to be best accomplished by individuals who maintain adaptability through complex hierarchical organization and function bordering on the edge of chaos.

ACKNOWLEDGMENTS

This work supported in part by American Heart Association (California Affiliate) Grant-in-Aid #90-119.

ADDITIONAL READING

The following books represent a source of general and detailed information regarding fractals, scaling, and the physical and mathematical foundations and applications to biological and physiological sciences.

1. Barnsley, M. F. *Fractals Everywhere*. New York: Academic Press, 1988.
2. Feder, J. *Fractals*. New York: Plenum Press, 1988.
3. Glass, L., and M. C. Mackey. *From Clocks to Chaos*. Princeton, NJ: Princeton University Press, 1988.
4. Koslow, S. H., A. J. Mandell, and M. F. Shlesinger, eds. "Perspectives in Biological Dynamics and Theoretical Medicine." *Ann. NY Acad. Sciences* **504** (1987): 1–313.
5. MacDonald, N. *Trees and Networks in Biological Models*. New York: Wiley, 1984.
6. Mandelbrot, B. B. *The Fractal Geometry of Nature*. New York: W. H. Freeman, 1983.
7. Moon F. C. *Chaotic Vibrations*. New York: Wiley, 1987.
8. Peitgen, H. O., and P. H. Richter. *The Beauty of Fractals*. Berlin: Springer-Verlag, 1986.
9. Reimann, H. A. *Periodic Diseases*. Philadelphia: F. A. Davis, 1963.
10. Thompson, D. A. *On Growth and Form*, 2nd edition. Cambridge: Cambridge University Press, 1963.

REFERENCES

1. Abboud, S., O. Berenfeld, and D. Saadeh. "Simulation of High-Resolution QRS Complex Using a Ventricular Model with a Fractal Conduction System." *Circ Res.* **6** (1991): 1751–1760.
2. Bhargava, V., and A. L. Goldberger. "Effect of Exercise in Healthy Men on QRS Power Spectrum." *Am. J. Physiol.* **243** (1982): H964–H969.
3. Durrer, D., R. T. van Dam, G. E. Freud, M. J. Janse, F. L. Meijler, and R. C. Arzbaecher. "Total Excitation of the Isolated Human Heart." *Circulation* **61** (1970): 899–912.
4. Farmer, J. D., E. Ott, and J. A. Yorke. "The Dimension of Chaotic Attractors." *Physica* **7D** (1983): 153–180.
5. Feder, J. *Fractals*. New York, NY: Plenum Press, 1988.
6. Glass, L., and M. C. Mackey. *From Clocks to Chaos*. Princeton, NJ: Princeton University Press, 1988.

7. Goldberger, A. L., A. L. V. Bhargava, V. Froelicher, and J. Covell. "Effect of Myocardial Infarction on High-Frequency QRS Potentials." *Circulation* **64** (1981): 34–42.

8. Goldberger, A. L., L. J. Findley, M. R. Blackburn, and A. J. Mandell. "Non-linear Dynamics in Heart Failure: Implications of Long-Wavelength Cardiopulmonary Oscillations." *Am. Heart J.* **107** (1984): 612–615.

9. Goldberger, A. L., V. Bhargava, B. J. West, and A. J. Mandell. "On a Mechanism of Cardiac Electrical Stability: The Fractal Hypothesis." *Biophys. J.* **48** (1985): 525–528.

10. Goldberger, A. L., B. J. West, and V. Bhargava. "Non-Linear Mechanisms in Physiology and Pathophysiology: Toward a Dynamical Theory of Health and Disease." In *Proceedings of the 11th International Association for Mathematics and Computers in Simulation. World Congress,* edited by B. R. Wahlstrom, R. Henriksen, and M. P. Sundby. Oslo, Norway, North-Holland, 1985.

11. Golden, D. P., R. A. Wolthuis, and G. W. Hoffler. "A Spectral Analysis of the Normal Resting Eectrocardiogram." *IEEE Trans. Biomed. Eng.* **20** (1973): 366–372.

12. Kauffman, S. A. Personal communication, Symposium on the Organization of Organisms, Santa Fe Institute, Santa Fe, NM, 1990.

13. Kobayashi, M., and T. Musha. "1/f Fluctuation of Heartbeat Period." *IEEE Trans. Biomed. Eng.* **29** (1982): 456–457.

14. Mandelbrot, B. B. *The Fractal Geometry of Nature.* New York: W. H. Freeman, 1983.

15. McAllister, R. E., D. Noble, and R. W. Tsien. "Reconstruction of the Electrical Activity of Cardiac Purkinje Fibres." *J. Physiol.* **251** (1975): 1–59.

16. Modanlou, H. D., and R. K. Freeman. "Sinusoidal Fetal Heart Rate Pattern: Its Definition and Clinical Significance." *Am. J. Obstet. Gynecol.* **142** (1982): 1033–1038.

17. Montroll, E. W., and M. R. Shlesinger. "Maximum Entropy Formalism, Fractals, Scaling Phenomena, and 1/f Noise: A Tale of Tales." *J. Stat. Physics* **32** (1983): 209–230.

18. Nelson, T. R., and D. K. Manchester. "Modeling of Lung Morphogenesis Using Fractal Geometries." *IEEE Trans. Med. Img.* **7** (1988): 321–327.

19. Nelson, T. R., B. J. West, and A. L. Goldberger. "The Fractal Lung: Universal and Species Related Scaling Patterns." *Experentia* **46** (1990): 251–254.

20. Nelson, T. R. "Pulse Propagation on a Fractal Network I: Structural and Temporal Scaling Characteristics." *Physica D* in press.

21. Nelson, T. R. "Pulse Propagation on a Fractal Network II: The Effect of Localized Perturbations." *Physica D,* in press.

22. Nygards, M. E., and J. Hulting. "Recognition of Ventricular Fibrillation Utilizing the Power Spectrum of the ECG." In *Computers in Cardiology.* Piscataway, NY: IEEE Computer Society, 1977

23. Rakic, P., and R. L. Sidman. "Histogenesis of Cortical Layers in Human Cerebellum, Particularly the Lamina Dissecans." *J. Comp. Neuro.* **139** (1972): 473–500.
24. Sernetz, M., B. Gelleri, and J. Hofman. "The Organism as Bioreactor. Interpretation of the Reduction Law of Metabolism in Terms of Heterogeneous Catalysis and Fractal Structure." *J. Theor. Biol.* **117** (1985): 209–230.
25. Theiler, J. "Estimating Fractal Dimension." *J. Opt. Soc. Am. A* **7** (1990): 1055–1073.
26. Waddington, J. L., M. J. MacCulloch, and J. E. Sambrooks. "Resting Heart Rate Variability in Men Declines With Age." *Experentia* **35** (1979): 1197–1198.
27. Weibel, E. R., and D. M. Gomez. "Architecture of the Human Lung." *Science* **137** (1962): 577–585.
28. West, B. J., V. Bhargava, and A. L. Goldberger. "Beyond the Principle of Similitude: Renormalization in the Bronchial Tree." *J. Appl. Physiol.* **60** (1986): 1089–1097.
29. West, B. J. "Fractals, Intermittency and Morphogenesis." In *Chaos in Biological Systems*. New York: Plenum, 1987.

Physiology: Commentary

The preceding chapters show how a combination of generative and design approaches can offer insight into the properties of organisms. Here we first characterize each of these approaches, and then argue that applying the two approaches jointly provides a broad avenue to principles of organization.

A generative approach proceeds from basic principles and building blocks, through the dynamics of processes, to generate structure. A generative approach to complex structures poses formidable problems. The processes that generate organisms occur in dynamical systems that are composites of organisms and their environments. These composite systems are large and diverse, to a degree that taxes or exceeds the capacity of present computation. The hierarchy of processes within an organism makes dynamical analysis still more difficult. These problems are more severe at longer time scales, on which the interactions among levels of organization, and among intrinsic and extrinsic processes, are more complex.

To some extent an approach through design can circumvent these difficulties. Some aspects of the environment do not change, or change very slowly. This is especially likely to be the case for physiological systems that deal directly with features of the environment associated with fundamental physical laws. For example, a visual system must deal with the occurrence of objects, which all have shape, position, movement, and reflectivity. Relatively invariant aspects of the environment map into criteria for designing structures that can deal with these aspects. In a design approach the designer (say, an engineer) begins with performance criteria and

Principles of Organization in Organisms,
SFI Studies in the Sciences of Complexity, Proc. Vol. XIII,
Eds. J. Mittenthal & A. Baskin, Addison-Wesley, 1992   **63**

a class of system that is likely to meet them. The designer selects a suitable design by estimating the performance of alternative systems, and seeing which systems meet the performance criteria well. The estimation is done through an approximation, or model, of the generative process. The process of design often involves an alternation between refining the characterization of a trial system and estimating its performance to evaluate it.

In practice the generative approach typically looks in parallel at the generation of many alternative systems. The design approach tends to look at one system at a time until a satisfactory design is found. For example, in the search for drugs with particular geometry and affinity, a generative approach through computer simulation now allows an extensive exploration of many candidate molecules. As practiced formerly, the design approach involved a trial-and-error search through a relatively small set of likely candidate molecules, based on associations from previous experience. These approaches are equivalent if unlimited resources and time are available for generating and evaluating alternatives. However, in reality there are limits on resources and time, so that neither approach can be explored fully. Often one or the other approach is chosen for convenience, recognizing that the two approaches, incompletely realized, may give different results.

Reasoning either from a generative or a design approach, one can begin to identify principles that favor particular families of structures. For example, often the dynamical behavior of a system approaches an attractor. If the same dynamic governs behavior for a long time, structures associated with attractors are likely to be generated. Reasoning from design, one can identify families of structures that preferentially match design constraints. Perhaps the most fundamental constraint in biology is that a lineage of living organisms has persisted since the origin of life, even though individual organisms cannot persist indefinitely. Structures that are self-reproducing are therefore needed.

Either approach may suggest a principle of organization, or generative and design approaches may converge on the same principle. The latter conjunction provides a powerful argument that a structure is likely to occur, if prerequisites for its generation are available and if it favors the persistence of organisms having it.

PRINCIPLES OF PHYSIOLOGICAL ORGANIZATION THAT EMERGE FROM A GENERATIVE APPROACH

As the preceding chapters show, a generative approach to physiological processes yields phenomena often encountered in nonlinear dynamical systems. There may be multiple attractors of various types—stable points, limit cycles, and chaotic attractors. Dynamical modes may cooperate or compete, producing patterns of transition between modes that include hysteresis and intermittency. Changes in parameters do not change the attractors of a structurally stable system, but can produce a pattern-switching bifurcation in other systems.

DeGuzman and Kelso clarify the relation between micro- and macrodynamics. At the micro-level many variables interact in a network of processes. In the oscillators considered, neurons interact in a neural network. At the macro-level a few variables, or *order parameters*, suffice to describe the macroscopic behavior; the phase of a neural oscillator is an order parameter. The dynamics of the order parameters, with accompanying noise, emerge from the dynamics of microscopic variables. Order parameters also emerge from the behavior of many cells in physiological systems with fractal behavior, such as the sinoatrial node of the heart.

Generative approaches to microdynamics point to the principle that physiological systems often operate on a boundary between distinct phases of behavior. As the frequency of a driving oscillator is varied, the response of a driven oscillator shows phase transitions. Transitions occur between orderly behavior, with mode locking, and disorderly behavior, in which the oscillator has an irregularly or chaotically fluctuating phase. At the boundary between the orderly and disorderly phases, the oscillator exhibits relative coordination. The phase transition emerges as a generic feature from the microdynamics.

Nelson suggests that morphology or activity that is fractal—self-similar over a large range of scales—also represents a boundary between order and disorder. The beat-to-beat variability of cardiac activation illustrates this idea. In a normal individual the variability has an inverse power-law distribution associated with a fractal. In pathological conditions the system's performance can deviate from this boundary in either direction; the spectrum of variations may be sharply narrowed (corresponding to rigid or orderly behavior) or abnormally broad (corresponding to disorder). Loss of a fractal branching pattern, which occurs when a branch of the His-Purkinje system is blocked, is often pathological.

PRINCIPLES OF PHYSIOLOGICAL ORGANIZATION THAT EMERGE FROM A DESIGN APPROACH

A design approach can use concepts and tools to reason from performance criteria to principles of organization. It is useful to recognize that many performance criteria fall into four classes—criteria of reliability, flexibility, economy, and timing. In general, criteria are not independent; there are tradeoffs among them, in that a change in one allows changes in others. Among the tools a designer uses, one of the most basic is associational linkages among features—linkages between properties of the system under design, or between a property and a performance criterion. Some parameters of a design are linked, either through generative processes or through *ad hoc* design heuristics. Such linkages allow one to implement tradeoffs through a tool called constraint propagation: Fixing a parameter to a constrained range of values limits the values of other parameters; these limitations in turn limit the ultimate design that can meet performance criteria.[1]

The preceding chapters suggest the progression from performance criteria to constraints on designs. Operation at a phase boundary between order and disorder is desirable, in that it provides reliability (through its proximity to ordered behavior) and flexibility (through its proximity to disorder). An oscillator may meet constraints on timing: A physiological system may be needed to track variations in an external periodic signal, and to maintain periodic behavior of the organism in the absence of the signal. (Note that a non-oscillatory process would suffice to meet the first of these constraints, but the second requires an entrainable endogenous oscillator.) A space-filling fractal structure may meet constraints on transport: A physiological system may be needed to transport a resource to a region. The system should effectively mix the resource throughout the region, and should work despite variations in the shape of the region. Nelson argues that transport along a fractal branching tree, with diffusion at its terminals, resembles the mixing that occurs in turbulence, a dynamic fractal process.

MATCHING THE OUTCOMES OF GENERATIVE AND DESIGN APPROACHES IN PHYSIOLOGY

The generative and design approaches converge in predicting that oscillators and fractal structures are likely to occur. Both classes of structures meet performance criteria and have dynamics that are available from universal laws. Relative coordination of oscillators, at a phase boundary between mode-locked entrainment and desynchronization, offers stable yet flexible behavior.

Fractal structures have available dynamics and may meet all four classes of performance criteria—reliability, flexibility, economy, and timing. The variability of a fractal increases its reliability and flexibility: A small change in a generative process at one scale, during its development or activity, can often be compensated by changes at other scales in ways that still leave the system fractal. For example, in the development of a lung, failure of an early branching process can be compensated by extra branching later. A fractal system is also flexible in that it operates at a boundary between erratic and rigid behavior.

A fractal can also meet criteria of economy and timing. A branching structure is a relatively economical way to transport a resource between the trunk and the region of termini. The economy can be measured in terms of the cost of transport, and the cost of developing and maintaining the branching structure. A fractal is also economical in that it can be generated recursively from an algorithm that can be concisely encoded. The distribution of branches can be adjusted to meet criteria of timing; for example, the His-Purkinje network distributes electrical activation to the wall of the ventricle with a spatiotemporal pattern such that each contraction of the wall normally ejects a large stroke volume. Here a fractal structure is performing a coordination.

PHYSIOLOGY: CONCLUSIONS

The preceding discussion shows that it is possible to identify likely structures (properties, or principles of organization) from approaches through generation or through design. Concurrence of the two approaches increases the likelihood that the structure will occur. Several properties and principles emerge in this section: Fractal structures have scale-invariant features, can be generated by recursion of a single process, and can provide reliable, flexible, economical, and temporally adequate solutions to problems of transporting resources. Reliable and flexible coordination of oscillators can emerge from the dynamics of their interaction. Biological systems may be likely to operate at a boundary between orderly and disorderly behavior.

These structures have several features that are desirable characteristics for principles of organization. They characterize levels of organization intermediate between microscopic and organismal. They are valid at various levels and time scales, and they are not tied to any particular physicochemical processes.

REFERENCES

1. Stefik, M. "Planning with Constraints (Molgen: Parts 1, 2)." *Artificial Intelligence* **16(2)** (1981): 111–170.

Physiology and Development

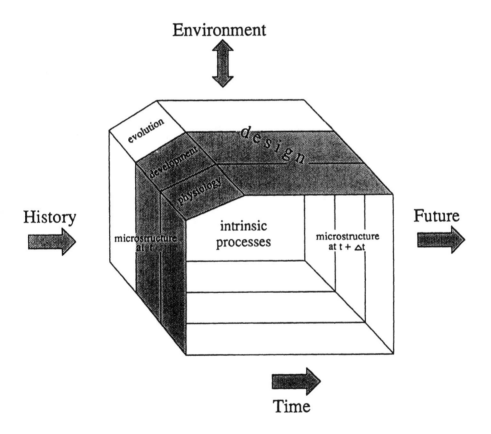

Here the contributors show how developmental processes make material structures that can meet performance criteria. The chapters treat interactions among intrinsic processes at different levels, relations between these processes and history (in the sense of prior events during the generation of gametes and the development of a new organism), and the modulation of intrinsic processes by environmental factors. These chapters support the following section, which broadens the viewpoint to evolution and to general principles of organization.

Physiology and Development: Introduction

As an organism develops from a fertilized egg to an adult, molecules interact in networks that mediate its material transformation. These networks operate at several levels. Each cell has interacting macromolecules in the nucleus (in networks of genes), in the cytoplasm, and in the extracellular space. As cells proliferate, networks of interactions among cells also become important. In early development the interacting cells may be similar and relatively unspecialized. For example, in an embryo of the fruit fly *Drosophila*, segments develop in a sheet of morphologically similar cells covering the surface of the embryo. A segmental pattern of gene activity emerges from interactions among genes, within and between cells. Reinitz et al. model these networks of interactions.

As differentiation proceeds, cells become more specialized and diverse. They deform themselves and each other and rearrange, changing the shape of the embryo. Bolker shows how differences in morphology of similar embryos (frog, sturgeon) are correlated with differences in the use of similar morphogenetic processes early in gastrulation.

Intracellular and intercellular networks, and material processes that shape an embryo, are intrinsic to it. Factors extrinsic to the embryo can modulate the intrinsic processes; these factors include chemicals, mechanical forces, and electromagnetic fields. Two chapters explore the influence of chemicals and light on the interacting neurons of a developing visual system. Stryker shows how vertebrate visual organization develops through a path that proceeds from generic cell activities to

Principles of Organization in Organisms,
SFI Studies in the Sciences of Complexity, Proc. Vol. XIII,
Eds. J. Mittenthal & A. Baskin, Addison-Wesley, 1992

processes that depend on electrical activity of neurons, prior to experience. Stryker and Obermayer et al. use models of neural networks with modifiable connections to account for developmental changes in synaptic connectivity in terms of the activity of neurons.

These chapters pose questions about history on a developmental time scale. What are the initial conditions for a period of development, and how important are they? What are the relative times at which various processes make their contributions to a particular outcome?

Reliability, flexibility, economy, and timing are performance criteria for development as well as for physiology. Each type of organism has a life history that makes particular combinations of these criteria relevant to it. These chapters explore several issues about the criteria: What micro-level organizations can generate macro-level performance that meets the criteria? What tradeoffs among criteria limit the combinations of criteria that can be attained? Karr and Mittenthal examine the tactics of molecular organization through which the fruit fly *Drosophila* develops with remarkable speed, and the tradeoffs accompanying this specialization for speed. The same issues of design arise in a colony, which develops through the proliferation and interaction of individual organisms. Blackstone examines tradeoffs of resource allocation in colonial hydroids.

Using models of networks with modifiable connections among units, one can infer connectivity patterns that provide favorable performance. Neural networks can be sought that perform a task adequately while using resources (such as the number of neurons or the extent of arborizations) economically. Analogous models are appropriate for analyzing the molecular networks underlying development. Reinitz et al. show how adjustable connections in a network of genes can be used to fit the performance of the network to data about spatiotemporal patterns of gene activity. Obermayer et al. show that optimization procedures can generate the spatial layout of neural circuits, including topographic mappings; economical mappings that enable a two-dimensional array of model neurons to signal the presence of a multi-dimensional feature can be inferred.

J. A. Bolker
Department of Molecular and Cell Biology and Museum of Vertebrate Zoology, University
of California, Berkeley, California 94720

Constancy and Variation in Developmental Mechanisms: An Example from Comparative Embryology

Patterns of morphological organization in mature organisms must reflect underlying principles in ontogeny. Such principles, in turn, arise from interactions of cells and tissues during morphogenesis. Organisms employ a variety of morphogenetic mechanisms, and they may use new combinations of shared processes to generate diverse outcomes, or to achieve similar results from different initial conditions. Comparative studies of morphogenetic mechanisms can help define fundamental processes, as well as determine the different ways these processes can be used. A comparison of gastrulation in the African clawed frog, *Xenopus laevis*, and the white sturgeon, *Acipenser transmontanus*, reveals similarities in blastopore lip formation and in the use of involution to move dorsal axial material into the archenteron roof. However, the lip forms at the equator in *Acipenser*, in contrast to its more vegetal position in *Xenopus*. As a result, the mechanical context of the early stages of involution differs radically in the two organisms: while dorsal involuting material in *Xenopus* can converge and extend simultaneously, in *Acipenser* a substantial amount of extension must precede convergence. This extension in *Acipenser* may result from radial intercalation of cells in the thick dorsal marginal zone, which thins as involution occurs. Comparison of gastrulation in these two organisms

Principles of Organization in Organisms,
SFI Studies in the Sciences of Complexity, Proc. Vol. XIII,
Eds. J. Mittenthal & A. Baskin, Addison-Wesley, 1992

illustrates both the conservation of basic morphogenetic mechanisms such as convergence and extension, and the reorganization of such processes to generate qualitatively different results.

INTRODUCTION

In this paper I compare gastrulation in the white sturgeon, *Acipenser transmontanus*, and a frog, *Xenopus laevis*, to show how comparative studies can contribute to understanding morphogenetic processes and their evolution. Many of the morphological similarities between early stages of sturgeon and amphibian embryos were noted in early descriptions.[33] What I have begun to investigate is how the two organisms compare morphogenetically: I compare the process of generating shapes, rather than just the final result.[4,5] Specifically, I am focusing on how gastrulation can be so similar in *Acipenser* and *Xenopus* despite their different starting points (Figure 1). My results suggest that these two organisms share some morphogenetic mechanisms, but that changes in the relative timing of specific processes have evolved to accomodate differences in the mechanical context in which the shared "tools" are used.

Gastrulation in *Xenopus* is well studied[24,30]; gastrulation in *Acipenser* is less thoroughly understood, in part because of the difficulty of obtaining research material. Development in the sturgeon has been studied by scientists with access either

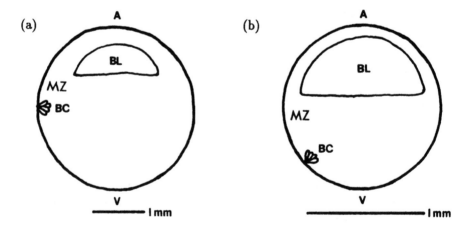

FIGURE 1 Schematic diagrams of mid-sagittal cross sections of the early gastrula of *Acipenser* (a) and *Xenopus* (b), showing relative positions of the animal (A) and vegetal (V) poles, marginal zone (MZ), blastocoel (BL), and bottle cells (BC). Dorsal is to the left; 1 mm bars indicate scale.

to the sturgeon hatcheries of the Soviet Union,[2,11,18,33,34] or to the once-thriving commercial sturgeon fisheries of the United States.[10] Recently, increasing commercial and scientific interest in establishing an aquaculture program for the white sturgeon on the west coast of the U.S.[3,9,12] has again made the sturgeon available as a research system for comparative embryology and developmental biology in general.

Meaningful comparative studies of morphogenesis require (1) a well-founded phylogenetic hypothesis about the relationship between the organisms being compared, and (2) developmental patterns that are sufficiently similar to have directly comparable characters (and be amenable to similar experimental techniques), but differ in significant ways. The chondrostean fishes (sturgeon and paddlefish) and the Amphibia meet these criteria. First, the relationships among the major groups of vertebrates are well established.[13,15,31,32] The Actinopterygii (ray-finned fishes, including both chondrosteans and teleosts) and Sarcopterygii (tetrapods and lung-fishes) diverged from a common ancestor. Most of their shared characters are presumed to be plesiomorphic, or retained from the common ancestor (although see Wake[39]), while their differences represent evolutionary modifications of the ancestral ontogeny. Second, the early development of chondrostean fishes (such as the sturgeon) closely resembles that of most amphibians, rather than that of teleosts. This allows direct comparison of many specific features in *Acipenser* and *Xenopus* (for example, the formation of the blastopore lip), as well as the application of many of the same experimental techniques.[5,7]

Comparative studies provide clues about where to look for generality in the organization of organisms, and where to look for variation: they are thus one route to discovering any organizational principles that may exist. Comparisons of morphogenesis in *Acipenser* and *Xenopus* reveal generality at the level of cell and tissue behaviors such as bottle cell formation, and radial and mediolateral intercalation (see below). However, the timing and mechanical context of these behaviors differ in the two organisms, so that the same morphogenetic mechanisms can produce different effects.[5]

COMPARISON OF PROCESSES
BLASTULA STRUCTURE

Development up to and during gastrulation in the white sturgeon, *Acipenser transmontanus*, and the African clawed frog, *Xenopus laevis*, is strikingly similar. In this respect, these groups resemble each other more than their respective sister taxa (most closely related groups: Teleostei for the sturgeon, and Amniota for amphibians). The similarities begin at the earliest stages of development: the dorsal-ventral axis of the embryo is established by cortical rotation and the formation of a grey crescent,[8,11,14,38] and cleavage is holoblastic, with furrows extending completely through the egg.[10,33] This pattern contrasts with the meroblastic or incomplete cleavage seen in non-mammalian amniotes and teleosts.

Just before gastrulation, the embryo of a sturgeon or of *Xenopus* consists of a hollow ball of cells, with a blastocoel occupying much of the animal hemisphere (Figure 1). There is a gradation of cell size from the micromeres of the animal cap to the large macromeres of the vegetal hemisphere, which contain most of the yolk. The marginal zone, a band of intermediate-sized cells located between the micromeres and macromeres, is the site of the first events of gastrulation. This region consists of cells which will come to form mesodermal and endodermal structures in both *Xenopus*[20,21] and the sturgeon.[2]

BOTTLE CELL FORMATION

Gastrulation begins in both organisms with the formation of a pigment line on the dorsal side of the marginal zone. This line consists of the contracted apices of bottle cells, named for their bottle- or flask-like shape, with a contracted apex and a long "neck," that form in the surface epithelial layer (Figure 2). In *Xenopus*, the bottle cells lead (but are not absolutely required for) the involution that generates the archenteron, and later respread as the anterior lining of the archenteron.[23] Bottle cells in *Acipenser* appear quite similar to those in *Xenopus* (Figure 2); however, this does not necessarily mean that their function is exactly the same in the two organisms. Bottle cell function, like other aspects of morphogenesis, is known to be context sensitive,[16,17,29] and one of the major differences between gastrulation in *Xenopus* and *Acipenser* is the location (and, hence, the mechanical context) in which the bottle cells first appear: at the equator in *Acipenser*, and about 50 degrees up from the vegetal pole in *Xenopus*.

INVOLUTION OF THE MARGINAL ZONE

The pigment line is the first indication of the formation of the dorsal lip of the blastopore, the site at which involution of the marginal zone during gastrulation begins. Most or all of the prospective axial mesoderm (notochord and somites) and the endoderm that will line the archenteron involutes: that is, it rolls around the lip of the blastopore as it moves to the interior of the embryo.[18,20,21,23] There are differences between *Acipenser* and *Xenopus* in the initial location of the material that will involute (for example, prospective axial mesoderm is in the deep layer in *Xenopus* before gastrulation,[21] and on the surface in *Acipenser*[2]). Nevertheless, in both organisms involution plays an important role in establishing the primary germ layers in the embryo and setting up the geometry of subsequent inductive interactions.

FIGURE 2   Scanning electron micrograph of bottle cells (white pointers) at the dorsal blastopore lip of a sturgeon embryo (fractured approximately at the dorsal midline). Dorsal is to the left of the figure, animal up. 403x magnification.

CONVERGENCE AND EXTENSION

As involution occurs, the dorsal side of the embryo lengthens dramatically. In *Xenopus*, this increase in length is driven by convergent extension of the dorsal, axial mesoderm.[24,26,27,28,29,41] The ability to converge and extend is intrinsic to the axial mesoderm: explants removed from the dorsal side of the *Xenopus* embryo at the start of gastrulation extend dramatically in culture.[25,26,27,28,29,41] Such explants have been extensively studied in *Xenopus*[35,36,37,41]; similar explants from sturgeon embryos also extend.[7] Sturgeon explants show two distinct zones of extension, one in the mesodermal region and another in the prospective neural ectoderm,[7] as do *Xenopus* explants.[25]

The principal cell behaviors that generate length increases in dorsal explants—and in embryos—are radial and mediolateral intercalation[24,41] (Figure 3). In radial intercalation, cells move between neighbors above and below them, so the tissue as a whole becomes thinner (flatter) and longer, while maintaining or increasing its width (Figure 3(a)). In mediolateral intercalation, cells move between their neighbors to the sides, so the tissue becomes longer, narrower, and sometimes thicker (Figure 3(b)). These processes are involved in several stages of gastrulation.[24,28,29]

In *Xenopus* and several other amphibians,[24] radial intercalation is responsible for the epiboly of the animal cap early in the blastula and gastrula stages. It also produces the extension and thinning of the dorsal marginal zone of *Xenopus* in the first half of gastrulation.[41] Radial intercalation appears to be more extensive in *Acipenser*: it produces relatively more extension, and lasts longer than in *Xenopus*.[7]

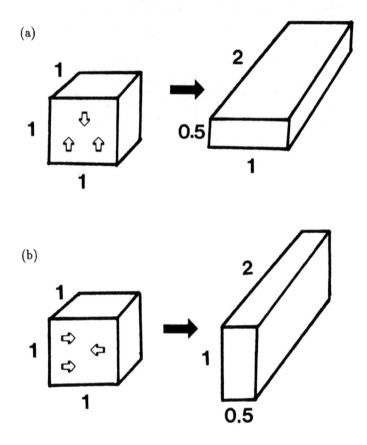

FIGURE 3   Schematic illustration of the effects of radial (a) and mediolateral (b) intercalation in deforming a fixed volume of tissue. In radial intercalation the tissue thins and lengthens, but maintains its original width. In mediolateral intercalation, the tissue becomes longer and narrower without flattening. Neither process requires an increase in cell number or volume.

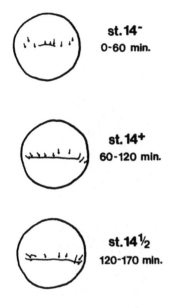

st.14⁻
0-60 min.

st.14⁺
60-120 min.

st.14½
120-170 min.

FIGURE 4 Tracings from time-lapse films of the dorsal side of a sturgeon embryo during the early stages of gastrulation (the animal pole is at the top of each diagram). Horizontal line indicates position of the dorsal lip of the blastopore at the start of each interval; arrows show the path and direction of movement of points on the surface of the embryo. Note that the movement of points near the dorsal midline remains parallel; medial convergence begins after the lip is well formed, and occurs only at its lateral ends.

During the first three to five hours of gastrulation in *Acipenser* (stages 14-14.5)[6,11] there is little or no convergence (Figure 4). Even as late as stage 15 (approximately 11-12 hours after the start of gastrulation), when the dorsal lip extends halfway around the embryo, most movement of surface points is directed vegetally rather than medially.[7] While mediolateral intercalation generates most of the narrowing and length increase along the dorsal axis in the last half of gastrulation in *Xenopus*, in the sturgeon a significant part of the dorsal extension appears to be generated by radial intercalation, and the onset of mediolateral intercalation is relatively later.[7]

RADIAL INTERCALATION DRIVES EXTENSION IN *ACIPENSER*

The initial phase of extension in *Acipenser* may result from radial intercalation of cells in the thick dorsal marginal zone, which thins as the blastopore lip moves vegetally during early stages of gastrulation. Support for this hypothesis comes from two sources: time-lapse films of the dorsal side of gastrulating embryos (made with an Arriflex news camera using 16mm Kodak Tri-X and Plus-X film, controlled by an intervalometer that exposed one frame per minute), and scanning electron microscopy (SEM) of the animal cap and marginal zone at different stages (before, during, and at the end of gastrulation). (SEM specimens were fixed in 2% glutaraldehyde diluted in 0.125M sodium cacodylate buffer, and prepared using standard methods before viewing on an ISI-DS130 scanning electron microscope. Micrographs were taken in Ilford FP4 film.)

(a)

(b)

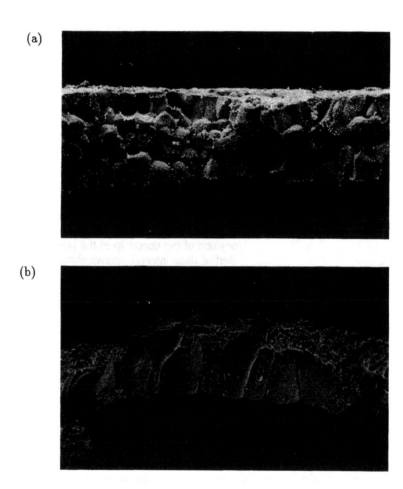

FIGURE 5 Scanning electron micrographs of cross sections of the blastocoel roof just before the start of gastrulation (a: stage 13),[6,11] and after most radial intercalation and dorsal extension have been completed (b: stage 15-16). The outer surface of the embryo is toward the top of the figure; the blastocoel is at the bottom. The number of cell layers in the roof decreases from approximately five to one or two during this period. Note difference in magnification: 5a, x181; 5b, x780.

Time-lapse films show overall morphogenetic changes in the embryo, and also allow one to follow the paths taken by individual points on the surface[22] (in *Acipenser*, these points are naturally occurring pigment concentrations in surface cells). Figure 4 illustrates the position of the blastopore lip at successive time points during gastrulation (horizontal lines), and the paths followed by several points on the dorsal side (arrows). The surface points move straight down toward the vegetal pole during the early stages of gastrulation: their paths remain parallel, and do not converge toward the midline even though the midline is lengthening. Convergence toward the midline does not begin until about stage 14.5, by which time the dorsal lip of the blastopore has moved well down below the equator into the vegetal hemisphere[7,11] (and is in fact at a latitude comparable to that at which the lip forms in *Xenopus*). Mediolateral convergence is clearly not the mechanism driving the initial stages of dorsal extension in *Acipenser*.

SEM studies strongly suggest that radial intercalation, rather than convergence, generates the first part of the dorsal extension during gastrulation in the sturgeon. The marginal zone is proportionally thicker in *Acipenser* than in *Xenopus*, and the blastocoel is smaller and higher (Figure 1). Moreover, the animal cap and marginal zone thin dramatically during gastrulation in the sturgeon. In the late blastula stage in the sturgeon (stage 12),[11] the blastocoel roof is approximately 8-10 cells thick at the animal pole. It starts thinning just before the onset of gastrulation (stage 13; Figure 5(a)); by stage 17-18, when the blastopore closes (nominally marking the end of gastrulation), the roof thickness is reduced to 1-2 cells (Figure 5(b)).

SIGNIFICANCE OF DIFFERENCES IN EXTENT AND TIMING OF CONVERGENCE AND EXTENSION IN THE STURGEON AND *XENOPUS*

A critical difference between *Acipenser* and *Xenopus* is the initial location of the dorsal lip of the blastopore: in *Acipenser* the bottle cells and pigment line form at the equator, while in *Xenopus* the lip forms much lower on the embryo, about 50 degrees up from the vegetal pole (Figure 1). Involution begins at the dorsal blastopore lip—that is, at the equator in *Acipenser*, and well down in the vegetal hemisphere in *Xenopus*. This difference means that the early stages of involution occur in radically different mechanical contexts in the two organisms. In *Xenopus*, dorsal involuting material could conceivably converge and extend simultaneously at the onset of gastrulation (although thinning and extension in fact occurs first). In contrast, in *Acipenser* a substantial amount of extension *must* precede convergence, in order to move the lip below (vegetal to) the equator before the dorsal tissue begins narrowing.

If convergence began at the equator in *Acipenser*, the embryo would constrict in the middle and exogastrulate, rather than involuting surface material from the animal hemisphere and enclosing the yolk as the blastopore closes. This scenario is supported by preliminary experiments in which the blastocoel roof was removed from sturgeon embryos at the start of gastrulation, and convergence occurred approximately at the equator.[7] The resulting embryos had an hourglass shape: convergence of the marginal zone occurred despite the absence of the blastocoel roof, and constricted the embryo in the middle.

CONCLUSIONS

The comparison between *Xenopus* and *Acipenser* presented here illustrates variation in the use of common (shared) morphogenetic mechanisms. Both organisms use radial and mediolateral intercalation in the course of gastrulation, but in subtly different ways that have important mechanical consequences. A difference in starting geometry is accomodated by a shift in the timing and degree of extension with thinning, driven by radial intercalation of cells, and extension and convergence, driven by mediolateral intercalation. The ability to make such accomodations and rearrangements—to use shared, ancestral mechanisms in new or different ways—is the basis for variations in morphogenesis, both within individuals and in lineages, and thus for evolutionary changes in ontogeny. Morphological evolution proceeds primarily by modifications of existing structures (although these modifications need not be gradual transformations), rather than by the wholesale invention of new ones. This pattern reflects the underlying processes that produce structures, and the way morphogenesis itself evolves: through adjustments, rearrangements, and "tinkering"[19] with existing mechanisms.

By the end of gastrulation the germ layers are established and arranged for subsequent development. However, the existence of variation in starting conditions and in the processes of gastrulation[1,40,42] suggests that events leading to this point may not be conserved. Thus, progress in understanding gastrulation as a process, and as an evolutionary phenomenon, must rest on detailed studies of the morphogenetic mechanisms involved rather than simply on descriptions or analyses of their results. Moreover, comparative studies of morphogenesis can show which aspects of ontogenies are stable and which are not. This is the first step towards a mechanistic understanding of why some elements of the process are more strongly conserved than others; such an understanding may in turn help make the connection between evolution and development, by bringing to light biases in the flexibility of morphogenetic systems. Evolutionary patterns in morphology may well reflect such biases (see Wake[39]).

ACKNOWLEDGMENTS

Sturgeon embryos used in this study were generously provided by the Sea Farm, Washington, Inc. California Sturgeon Project (Herald, CA). This research was supported by a Cellular and Molecular Biology training grant at the University of California, Berkeley, and NSF DCB89052 to R. E. Keller, in whose lab the work was done. The author thanks D. B. Wake and R. E. Keller for their comments on the manuscript.

REFERENCES

1. Ballard, W. W. "Morphogenetic Movements and Fate Maps of Vertebrates." *Amer. Zool.* **21** (1981): 391–399.
2. Ballard, W. W., and A. S. Ginsburg. "Morphogenetic Movements in Acipenserid Embryos." *J. Exp. Zool.* **213** (1980): 69–103.
3. Beer, K. E. "Embryonic and Larval Development of the White Sturgeon (*Acipenser transmontanus*)." Master's thesis, University of California, Davis, 1981.
4. Bolker, J. A. "Gastrulation in the White Sturgeon, *Acipenser transmontanus.*" (Abstract.) *Amer. Zool.* **29(4)** (1989): 86A.
5. Bolker, J. A. "Comparison of Gastrulation in the White Sturgeon (*Acipenser transmontanus*) and an Anuran (*Xenopus laevis*)." (Abstract.) *Amer. Zool.* **30(4)** (1990): 43A.
6. Bolker, J. A. "A Scanning Electron Microscopy Study of Gastrulation in the White Sturgeon, *Acipenser transmontanus.*" In preparation.
7. Bolker, J. A. "The Mechanisms of Gastrulation in the White Sturgeon." In preparation.
8. Clavert, J. "Symmetrization of the Egg of Vertebrates." *Advances in Morphology* **2** (1962): 27–60.
9. Conte, F. S., Doroshov, S. I., Lutes, P. B., and E. M. Strange. "Hatchery Manual for the White Sturgeon, Acipenser transmontanus Richardson, with Application to other North American Acipenseridae." Publication 3322, Cooperative Extension of the University of California, Division of Agriculture and Natural Resources (c. UC Regents), 1988.
10. Dean, B. "The Early Development of Gar-pike and Sturgeon." *J. Morph.* **11** (1895): 1–62.
11. Dettlaff, T. A., and A. S. Ginsburg. *Embryonic Development in Sturgeon (Russian sturgeon, sevruga, and beluga) in Connection with Problems of Artificial Propagation.* (In Russian.) Moscow: USSR Acad. Sci. Press, 1954.

12. Doroshov, S. I., W. H. Clark Jr., P. B. Lutes, R. L Swallow, K. E. Beer, A. B. McGuire, and M. D. Cochran. "Artificial Propagation of the White Sturgeon, *Acipenser transmontanus* Richardson." *Aquaculture* **32** (1983): 93–104.
13. Forey, P. "Blood Lines of the Coelacanth." *Nature* **351** (1991): 347–348.
14. Ginsburg, A. S. "The Origin of Bilateral Symmetry in the Eggs of Acipenserid Fishes." (In Russian.) *C.R. Acad. Sci. URSS* **90** (1953): 477–480.
15. Gorr, T., T. Kleinschmidt, and H. Fricke. "Close Tetrapod Relationships of the Coelacanth *Latimeria* Indicated by Haemoglobin Sequences." *Nature* **351** (1991): 394–397.
16. Hardin, J. "Context-Sensitive Cell Behaviors During Gastrulation." *Seminars in Developmental Biology* **1(5)** (1990): 335–346.
17. Hardin, J., and R. E. Keller. "The Behaviour and Function of Bottle Cells During Gastrulation of *Xenopus laevis*." *Development* **103** (1988): 211–230.
18. Ignatieva, G. M. "Relationship Between Epibole and Invagination in Sturgeon Embryos During the Period of Gastrulation." *Proc. Acad. Sci. USSR—Biol. Sect.* (transl.) **65** (1965): 712–715.
19. Jacob, F. *The Possible and the Actual.* New York: Pantheon Books, 1982.
20. Keller, R. E. "Vital Dye Mapping of the Gastrula and Neurula of *Xenopus laevis*. I. Prospective Areas and Morphogenetic Movements of the Superficial Layer." *Devel. Biol.* **42** (1975): 222–241.
21. Keller, R. E. "Vital Dye Mapping of the Gastrula and Neurula of *Xenopus laevis*. II. Prospective Areas and Morphogenetic Movements in the Deep Region." *Devel. Biol.* **51** (1976): 118–137.
22. Keller, R. E. "Time-Lapse Cinemicrographic Analysis of Superficial Cell Behavior During and Prior to Gastrulation in *Xenopus laevis*." *J. Morph.* **157** (1978): 223–248.
23. Keller, R. E. "An Experimental Analysis of the Role of Bottle Cells and the Deep Marginal Zone in Gastrulation of *Xenopus laevis*." *J. Exp. Zool.* **216** (1981): 81–101.
24. Keller, R. E. "The Cellular Basis of Amphibian Gastrulation." In *Developmental Biology: A Comprehensive Synthesis*, edited by L. Browder, Vol. 2, 241–327. New York: Plenum Press, 1986.
25. Keller, R. E., and M. Danilchik. "Regional Expression, Pattern and Timing of Convergence and Extension During Gastrulation of *Xenopus laevis*." *Development* **103** (1988): 193–209.
26. Keller, R. E., M. Danilchik, R. Gimlich, and J. Shih. "Convergent Extension by Cell Intercalation During Gastrulation of *Xenopus laevis*." In *Molecular Determinants of Animal Form*, UCLA Symp. Mol. Cell. Biol., edited by G. M. Edelman, **31** 111–141. New York: Alan R. Liss, 1985.
27. Keller, R. E., M. Danilchik, R. Gimlich, and J. Shih. "The Function and Mechanism of Convergent Extension During Gastrulation of *Xenopus laevis*." *J. Embryol. exp. Morph.* **89** (1985) (supplement): 185–209.
28. Keller, R., J. Shih, and P. Wilson. "Cell Motility, Control and Function of Convergence and Extension During Gastrulation in *Xenopus*." In *Gastrulation: Movements, Pattern, and Molecules*. New York: Plenum Press, 1991.

29. Keller, R., J. Shih, P. Wilson, and A. Sater. "Pattern and Function of Cell Motility and Cell Interactions During Convergence and Extension in *Xenopus*." 49th Symposium of The Society of Developmental Biologists, Cell-Cell Interactions in Development, 1991.
30. Keller, R. E., and R. Winklbauer. "The Cellular Basis of Amphibian Gastrulation." *Current Topics in Developmental Biology*, in press.
31. Meyer, A., and A. C. Wilson. "Origin of Tetrapods Inferred from Their Mitochondrial DNA Affiliation to Lungfish." *J. Mol. Evol.* **31** (1990): 359–364.
32. Rosen, D. E., P. L. Forey, B. G. Gardiner, and C. Patterson. "Lungfishes, Tetrapods, Paleontology, and Plesiomorphy." *Bull. Am. Mus. Nat. Hist.* **167** (1981): 159–276.
33. Salensky, A. "Recherches sur le Developpement du Sterlet (*Acipenser ruthenus*)." *Arch. Biol.* **2** (1881): 233–341.
34. Sawadsky, A. M. "Untersuchungen zur Entwicklungsgeschichte des Sterlets." *Zeit. fur Anat. und Entw.* **78** (1926): 26–65.
35. Shih, J., and R. E. Keller. "Induction of Dorsal Morphogenesis and Tissue Differentiation by the Dorsal Epithelium of the *Xenopus* Gastrula." In preparation.
36. Shih, J., and R. E. Keller. "The Cellular Mechanism Underlying Convergence and Extension in Explants of *Xenopus laevis* Dorsal Marginal Zone." In preparation.
37. Shih, J., and R. E. Keller. "The Spatial and Temporal Patterns During the Convergence and Extension of the Developing Neurectoderm (NIMZ)." In preparation.
38. Vincent, J. -P., and J. C. Gerhart. "Subcortical Rotation in *Xenopus* Eggs: An Early Step in Embryonic Axis Specification." *Devel. Biol.* **123** (1987): 526–539.
39. Wake, D. This volume.
40. Waddington, C. H. "Modes of Gastrulation in Vertebrates." *Quart. J. Microsc. Sci.* **93(2)** (1952): 221–229.
41. Wilson, P. A., and R. E. Keller. "Cell Rearrangement During Gastrulation of *Xenopus*: Direct Observation of Cultured Explants." *Development*, in press.
42. Wolpert, L. "The Evolution of Development." *Biol. J. Linn. Soc.* **39(2)** (1990): 109–124.

N. W. Blackstone
Department of Biology, Yale University, New Haven, Connecticut 06511

Heterochrony in Hydractiniid Hydroids: A Hypothesis

Two species of hydractiniid hydroids show a classic pattern of heterochrony. *Podocoryne carnea* colonies grow rapidly, reproduce and senesce early, disperse widely, and colonize readily as compared to *Hydractinia echinata* colonies. Complementing this progenetic life history, *Podocoryne* exhibits a juvenilized morphology, including a hydrodynamically inefficient gastrovascular architecture. Differences in energy allocation are suggested; until reproduction, *Podocoryne* channels energy into outward growth and a high rate of gastrovascular circulation, while throughout the ontogeny *Hydractinia* diverts energy into building an elaborate somatic morphology. Treatment of *Podocoryne* colonies with 2,4-dinitrophenol supports this interpretation: treated colonies show "loose-coupling" of oxidative phosphorylation, reduced gastrovascular circulation, and induction of the features of adult morphology. Treated *Podocoryne* colonies, thus, morphologically resemble *Hydractinia* colonies. Other clonal organisms show differences similar to those observed in these species, and the proposed hypothesis underlying these differences may have wide applicability as well.

Principles of Organization in Organisms,
SFI Studies in the Sciences of Complexity, Proc. Vol. XIII,
Eds. J. Mittenthal & A. Baskin, Addison-Wesley, 1992

Heterochrony, an evolutionary change in the timing of development, is considered a major mechanism for producing dramatic morphological transformations with a minimum of genetic change.[1,8,13,19,24,25,28] Historically, studies of heterochrony have focused on describing patterns of comparative morphology and embryology.[13] Despite increasingly mechanistic[25] and cladistic[11] approaches, this pattern-oriented tradition persists. In the words of Bonner,[8] "everyone agrees that *the* most effective way to elicit big phenotypic changes is by heterochrony. Nevertheless, we seem unable to find the key of how to attack the problem."

A major aspect of finding this "key" will involve elucidating the way in which an organism uses time.[12] For empirical measures of the timing of development, it is convenient to use chronological time, but clearly the organism itself must rely on intrinsic time scales.[3,4,12] In this context, it is useful to re-examine the extensive literature implicating "metabolic activity" in developmental processes.[10,14,21] It may be that the organism used metabolic activities as measures of time.[16,26] Alterations in aspects of metabolic activity may, thus, alter the timing of development. Gould,[13] in fact, explicitly suggests such a connection between heterochrony and metabolic rate: "... fecundity may be enhanced in tiny progenetic animals if a physiological juvenilization accompanies paedomorphic morphology. Young animals have higher metabolic rates than related (but not progenetic) adult organisms of the same size." Gould is suggesting that taxa exhibiting progenesis (i.e., precocious sexual reproduction) and the associated morphological paedomorphosis (i.e., "child-shape") may, in fact, exhibit an accelerated metabolic rate.

While Gould's hypothesis is likely an oversimplification, it nonetheless succinctly summarizes the possibilities inherent in considering links between metabolic activity and heterochrony. Surprisingly, these possibilities have received little attention, despite the potential that such links may have powerful explanatory value at several levels of the biological hierarchy. Here I will outline a case of heterochrony in hydractiniid hydroids[5] which can be related to energy allocation intuitively. I will then summarize a series of experiments[6] with 2,4-dinitrophenol (DNP) which support these intuitive notions.

Hydractiniid hydroids encrust a variety of types of hard substrata in the sea. *Hydractinia echinata* and related species are commonly found on the shells of hermit crabs and often exhibit a species-specific correlation with host hermit crabs. *Podocoryne carnea* also encrust hermit crab shells, but commonly inhabit other surfaces as well. Colony development in both taxa begins with the metamorphosis of the planula larvae into a primary polyp. Runner-like stolons extend from the primary polyp. Stolons encase fluid-filled, vascular canals which are continuous with the gastrovascular cavity of the polyp. *Podocoryne* continues to develop in this way, i.e., by lineal extension of the stolons, initiation of new stolonal tips, and iteration of feeding polyps on the stolons. Stolons in *Hydractinia*, however, quickly fuse to form a continuous stolonal mat, a closely knit complex of anastomosing stolons capped by a continuous layer of ectoderm, which shows sheet-like growth, and from which extend relatively few peripheral stolons. At sexual maturity, *Hydractinia* colonies completely cover the available substratum with a spiny, chitinous stolonal mat.

While *Podocoryne* forms a closely knit network of stolons at sexual maturity, the ectodermal cell layers do not fuse and chitinous spines are not formed.

Morphological and life history characteristics of these taxa (Table 1) show a pattern of heterochrony; in every aspect, *Podocoryne* has features which are paedomorphic, specifically progenetic, relative to *Hydractinia*. Compared to the latter, *Podocoryne* colonies grow rapidly, reproduce and senesce early, disperse widely, and colonize readily. Further, as described above, *Podocoryne* exhibits a paedomorphic morphology.

TABLE 1 Morphological and life history differences in hydractiniid hydroids.

	Podocoryne	Hydractinia
Morphology	Runner-like; older colonies will pack stolons closely but they do not fuse; polyps widely spaced	Sheet-like; forms fused stolonal mat with some peripheral stolons; polyps closely packed
	Does not form chitinous mat, forms few spines	Forms a spiny, chitinous basal mat
Growth	Rapid	Slow; colonies which form only mat (with no peripheral stolons) usually have the slowest growth
Maturation	Rapid; reproduces before forming stolonal mat or spiny basal mat	Slow; correlated with covering the surface
Reproduction	Energetically inexpensive asexual medusae	Energetically costly gametes
Dispersal	Wide; swimming medusae and planulae	Local; no medusae, slow-crawling planulae
Senescence	Rapid; lack of chitinous basal mat usually leads to disappearance from shell	Slower

Morphological differences depend on the flow of the gastrovascular fluid through the stolonal canals. It is primarily the muscular polyps that drive the gastrovascular flow by active contractions[27]; because of the elasticity of their tissue, polyps and stolon tips can also act as mechanical oscillators, storing energy when filling and releasing it when emptying. Unlike polyps, however, stolons are encased in a rigid periderm which limits their stretching and thus limits the contribution of mechanical forces to their contractions.

Viewing the colony from a hydrodynamic perspective, a clear relationship between morphology and gastrovascular energetics is suggested. Many polyps connected by short segments of stolons (such as those provided by a closely knit complex of anastomosing stolons) can circulate gastrovascular fluid with relatively little per-polyp exertion of pressure; fewer polyps separated by longer segments of stolons have to exert a correspondingly greater per-polyp pressure to maintain the same volumetric rate of circulation.

An important aspect of the volumetric rate of gastrovascular circulation can be measured by the emptying and filling of the distal tips of the peripheral stolons. Since these blind tips are essentially dead-end channels, gastrovascular flow must be reversed in each tip. Stolon tips fill as fluid enters; the velocity of the fluid then decreases to zero. Tips then empty, and the fluid velocity again decreases to zero. The difference between the width of the stolon lumen when it is full (and velocity is zero) and when it is empty (and velocity is again zero) can provide a measure of volumetric rate if it is measured over time. Higher rates or amplitudes of lumen contractions (or both) indicate greater displacement of the gastrovascular fluid and greater flow to the peripheral stolon tips. Microvideo image analysis of the filling and emptying of distal stolon tips show that the lumen of *Podocoryne* exhibits contractions of roughly 40% of the stolonal width, while *Hydractinia* exhibits contractions of only 20% of the stolonal width. Nevertheless, the frequency of stolonal contractions was indistinguishable between the two taxa. These patterns were consistent in nine different strains of each species.[6] Thus, *Podocoryne* pumps gastrovascular fluid at a greater volumetric rate than *Hydractinia*.

In *Podocoryne*, the greater volume of circulation provided to stolon tips is likely the basis for its ability to maintain a runner-like morphology and greater amounts of peripheral stolons. Further, *Podocoryne*'s greater physiological activity is likely supported by a greater proportion of the available metabolites.

To further investgate the relationship between morphology, gastrovascular circulation, and energetics, I began experimenting with DNP, an uncoupler of oxidative phosphorylation, at the suggestion of Stuart Newman at this workshop. In both *Hydractinia* and *Podocoryne*, dilute DNP solutions initially reduce the volume of gastrovascular fluid delivered to peripheral stolon tips. Treated colonies initiate new polyps and stolon tips and prematurely develop the fully adult somatic morphology (i.e., a closely knit network of stolons with closely packed polyps), although reproduction is not initiated. Subsequently, gastrovascular flow returns to normal and these colonies survive and appear perfectly healthy for months in DNP solutions; apparently the adult somatic morphology can be maintained indefinitely in this state of "loose coupling" of oxidative phosphorylation.[6]

Reproduction in these taxa likely demands most of the metabolites produced by respiration; hence, the adult somatic morphology may result as a consequence of the diversion of these metabolites into reproduction. *Hydractinia* diverts energy into building an elaborate somatic morphology (thick ectodermal cell layers, chitinous spines, large feeding polyps), so the adult morphology (i.e., the stolonal mat) appears early in the ontogeny. *Podocoryne* does not build an elaborate somatic morphology and channels energy into outward growth and gastrovascular circulation until reproduction; hence, the juvenile morphology is apparent until reproduction.

This interpretation is supported by the quantities of mitochondria seen in the two taxa. Using rhodamine 123, mitochondria can be visualized *in vivo*.[17] Compared to *Podocoryne*, *Hydractinia* colonies have more mitochondria in their polyps and stolons. Thus, Gould's hypothesis is somewhat of an oversimplification. It is not that *Podocoryne* has an absolutely greater metabolic rate; rather, the two taxa likely differ in the allocation of the aerobic metabolites. *Hydractinia* uses aerobic metabolites to differentiate and elaborate somatic structures, maintaining gastrovascular flow with a relatively small amount of energy. *Podocoryne* invests relatively little energy in building somatic structures, while gastrovascular circulation consumes a large proportion of its energy budget until reproduction.

The data suggest that the morphological and life history heterochronies seen in hydractiniid hydroids are related to differences in how the products of aerobic metabolism are allocated during the ontogeny. This hypothesis may apply to other clonal taxa as well. Runner- and sheet-like forms, with associated life history traits, are common in several phyla of clonal organisms.[9] Many of these taxa rely on circulatory systems which are superficially similar to those in hydractiniid hydroids, and similar hydrodynamic principles may apply. Experiments in progress[7] suggest that *Eirene viridula*, a thecate hydroid, responds to DNP in a manner similar to hydractiniids; additional experiments on other clonal taxa are planned.

The observed morphogenetic effects of DNP can be accommodated by models of gradients of diffusible regulatory substances.[18,20] Müller and coworkers[20] have suggested that pattern formation in colonial hydroids is regulated by activating and inhibiting factors which emanate from the stolon tip. It is unlikely that DNP mimics these "morphogens"; rather, DNP likely alters the transmission of the regulatory substances. In fact, since inhibitory factors are thought to be transmitted within the stolon,[20] the effects of DNP on gastrovascular circulation may produce the observed morphogenetic effects. The actions of the regulatory substances may be controlled by the hydrodynamics of the gastrovascular circulation, i.e., by the internal physical environment of the colony.[22]

Uncouplers may alter the timing of development in complex, aclonal taxa as well. For instance, tributyltin has been introduced into the marine environment in antifouling paints, and, in dilute quantities, it has been shown to affect the development of bivalves, gastropods, and fish.[2,23] There may be general opportunities to employ the diverse array of uncouplers[15] as experimental tools for assessing links between aerobic metabolism and recognized morphological heterochronies.

ACKNOWLEDGMENTS

Comments from A. Baskin, D. Bridge, L. Buss, M. Dick, J. Mittenthal, S. Newman, and B. Schierwater aided the development of these ideas. The National Science Foundation (BSR-88-05961) provided support.

REFERENCES

1. Alberch, P., S. J. Gould, G. F. Oster, and D. B. Wake. "Size and Shape in Ontogeny and Phylogeny." *Paleobiology* 5 (1979): 296–313.
2. Bailey, S. K., and I. M. Davies. "The Effects of Tributyltin on Dogwhelks (*Nucella lapillus*) from Scottish Coastal Waters." *J. Mar. Biol. Ass. U.K.* 69 (1989): 335–354.
3. Blackstone, N. W. "Size and Time." *Syst. Zool.* 36 (1987): 211–215.
4. Blackstone, N. W., and P. O. Yund. "Morphological Variation in a Colonial Marine Hydroid: A Comparison of Size-Based and Age-Based Heterochrony." *Paleobiology* 15 (1989): 1–10.
5. Blackstone, N. W., and L. W. Buss. "Shape Variation in Hydractiniid Hydroids." *Biol. Bull.* 180, in press.
6. Blackstone, N. W., and L. W. Buss. "Treatment with 2,4-Dinitrophenol Mimics Ontogenetic and Phylogenetic Changes in a Hydractiniid Hydroid." *Proc. Natl. Acad. Sci.*, submitted.
7. Blackstone, N. W., and L. W. Buss. "Redox Control of Heterochrony in Three Species of Colonial Marine Hydroid." In preparation.
8. Bonner, J. T., ed. *Evolution and Development.* Berlin: Springer-Verlag, 1982.
9. Buss, L. W., and N. W. Blackstone. "An Experimental Exploration of Waddington's Epigenetic Landscape." *Phil. Trans. R. Soc. Lond B* 332 (1991): 49–58.
10. Child, C. M., *Patterns and Problems in Development.* Chicago: University of Chicago Press, 1941.
11. Fink, W. L. "The Conceptual Relationship Between Ontogeny and Phylogeny." *Paleobiology* 8 (1982): 254–264.
12. Gerhart, J. C., S. Berking, J. Cooke, G. F. Freeman, A. Hildebrandt, H. Jokusc P. A. Lawrence, C. Nusslein-Volhard, G. F. Oster, K. Sander, H. W. Sauer, G. S. Stent, N. K. Wessells, and L. Wolpert. "The Cellular Basis of Morphogenetic Change." In *Evolution and Development*, edited by J. T. Bonner, 87–114. Berlin: Springer-Verlag, 1982.
13. Gould, S. J. *Ontogeny and Phylogeny.* Cambridge, MA: Harvard Press, 1977.
14. Gustafson, T., and P. Lenique. "Studies of Mitochondria in the Developing Sea Urchin Egg." *Exp. Cell. Res.* 3 (1952): 251–274.

15. Heytler, P. G. "Uncouplers of Oxidative Phosphorylation." In *Inhibitors of Mitochondrial Functions*, edited by M. Erecinska and D. F. Wilson, 199–210. New York: Pergamon Press, 1981.

16. Itow, T. "Inhibitors of DNA Synthesis Change the Differentiation of Body Segments and Increase the Segment Number in Horseshoe Crab Embryoes (Chelicerata, Arthropoda)." *Roux's Arch. Dev. Biol.* **195** (1986): 323–333.

17. Johnson, L. V., M. L. Walsh, and L. B. Chen. "Localization of Mitochondria in Living Cells with Rhodamine 123." *Proc. Natl. Acad. Sci.* **77** (1980): 990–994.

18. MacWilliams, H. K., F. C. Kafatos, and W. H. Bossert. "The Feedback Inhibition of Basal Disk Regeneration in *Hydra* has a Continuously Variable Intensity." *Dev. Biol.* **23** (1970): 380–398.

19. McKinney, M. L., ed. *Heterochrony in Evolution: An Interdisciplinary Approach.* New York: Plenum Press, 1988.

20. Müller, W. A., A. Hauch, and G. Plickert. "Morphogenetic Factors in Hydroids: I. Stolon Tip Activation and Inhibition." *J. Exp. Zool.* **243** (1987): 111–124.

21. Newman, S. A. "Reversible Abolition of Normal Morphology in *Hydra*." *Nature New Biology* **244** (1973): 126–128.

22. Newman, S. A., and W. D. Comper. "'Generic' Physical Mechanisms of Morphogenesis and Pattern Formation." *Development* **110** (1990): 1–18.

23. Pinkney, A. E., D. A. Wright, and G. M. Hughes. "A Morphometric Study of the Effects of Tributyltin Compounds on the Gills of the Mummichog Fundulus Heteroclitus." *J. Fish. Biol.* **34** (1989): 665–677.

24. Raff, R. A., and G. A. Wray. "Heterochrony: Developmental Mechanisms and Evolutonary Results." *J. Evol. Biol.* **2** (1989): 409–434.

25. Raff, R. A., ed. "Heterochronic Changes in Development." *Seminars in Dev. Biol.* **1** (1990): 233–297.

26. Reiss, J. O. "The Meaning of Developmental Time: A Metric for Comparative Embryology." *Amer. Natur.* **134** (1989): 170–189.

27. Schierwater, B., L. W. Buss, and B. Piekos. "Hydroid Stolonal Contractions Mediated by Contractile Vacuoles." *J. Exp. Biol.*, in press.

28. Wake, D. B., P. Mabee, J. Hanken, and G. Wagner. "Development and Evolution—The Emergence of a New Field." In preparation.

Timothy L. Karr† and Jay E. Mittenthal‡

†Department of Biochemistry, 1209 W. California St. (and Beckman Institute), University of Illinois, Urbana, Illinois 61801, U.S.A.; ‡Department of Cell and Structural Biology, 505 S. Goodwin St. (and Center for Complex Systems Research, Beckman Institute; and College of Medicine), University of Illinois, Urbana, Illinois 61801, U.S.A.

Adaptive Mechanisms that Accelerate Embryonic Development in *Drosophila*

INTRODUCTION

Why does an organism have particular structures? One answer is historical: Processes within its ancestors generated structures that helped their users to survive the gauntlet of natural selection. In general it is difficult to reconstruct the historical sequence of interactions between self-organization and selection convincingly. An alternative approach to understanding structures considers them as designs that meet criteria of performance.[25] Here we take an approach through design to understanding the molecular organization of development in the fruit fly *Drosophila melanogaster*.

To survive in an ecological niche, organisms must display an appropriate combination of flexibility, reliability, economy, and speed. Organisms meet these performance criteria by developing a suitable material and behavioral structure, through a suitable life history. It is desirable to understand how internal organization meets performance criteria. In general the criteria are not met independently: Because an

Principles of Organization in Organisms,
SFI Studies in the Sciences of Complexity, Proc. Vol. XIII,
Eds. J. Mittenthal & A. Baskin, Addison-Wesley, 1992   **95**

organism is finite, in attaining some performance capabilities, it is likely to sacrifice others. The analysis of design in organisms should help us to understand these tradeoffs.

Features that contribute to performance, and tradeoffs among them, are likely to be especially evident in organisms that show extremes of performance. *Drosophila* is an organism specialized for speed. It is among the organisms that live on ephemeral resources, reproducing rapidly and generating many progeny during the limited period that a resource is available.[9,27] Such organisms achieve a high rate of reproduction through rapid development and a short life cycle. Because they proliferate rapidly, these organisms are suitable for genetic analysis of their internal organization; such analysis has been done extensively in *Drosophila*. We have used the resulting information to see how the development of *Drosophila* is specialized to promote speed, and to look for possible sacrifices in performance or in other mechanisms of optimization accompanying this specialization.

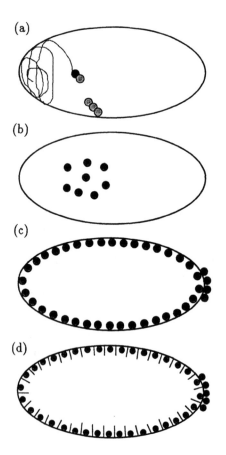

FIGURE 1 Diagram of early embryonic development in *Drosophila*. Anterior is to the left. Four stages (a-d) depict fertilization (a), early zygotic cleavages (b), formation of the syncytial blastoderm (c), and cellularization (d). Fertilization occurs *in utero* as a single sperm (coiled line), enters the egg and the pronuclei approach each other (filled and stippled circles) as shown in (a). Following karyogamy and formation of a diploid zygote nucleus, extremely rapid nuclear divisions occur (stippled circles in b). Successive rapid divisions and directed migration of nuclei to the egg periphery lead to the formation of pole cells (filled circles at the posterior end in c) and the formation of the syncytial blastoderm (stippled circles). Following three more nuclear divisions at the blastoderm surface, approximately 5000 nuclei synchronously cellularize (short straight lines depict invaginating membrane furrow during this process in d).

Development in *Drosophila* is remarkably rapid and reliable. As regards speed, the organism hatches from the egg as a larva 22 hours after fertilization. After three larval stages it forms a pupa, and emerges as an adult about ten days after fertilization. As for the reliability of development, more than 95% of mature eggs are fertilized, and an extraordinarily high percentage of fertilized eggs develop fully.[2] This is even more striking when it is realized that a single *Drosophila* female can lay up to a thousand eggs in her lifetime. To achieve such reliability the rates of errors in molecular processes must remain below threshold levels, beyond which errors are amplified until the organism cannot survive.[8]

To see how *Drosophila* develops with such speed and reliability, we first review the development of *Drosophila* briefly. We then look at tactics that implement the strategy of rapid development, first surveying general tactics and then looking more closely at two processes that are central to rapid development—DNA replication and mitosis. These specifics suggest generalizations about tactics for attaining speed, and implications of these tactics for economy, reliability, and flexibility. Finally we see what program of research follows from these considerations.

As we shall see, the speed of development depends both on the chemical kinetics of individual molecular interactions and on the supramolecular organization of these interactions. The latter is a network of coupled processes. During evolution the operation of the network can become faster through deletion of processes that occur sequentially, performance of processes in parallel, and addition of redundant processes. All of these tactics have accelerated development in *Drosophila*, perhaps to the limit of a threshold for reliability.

WHAT PROCESSES CONTRIBUTE TO THE SPEED OF DEVELOPMENT?

Insects develop in very diverse ways.[2,21] One classification of these ways is based on the temporal pattern of segmentation of the developing body of the insect, the germ band.[26] In one extreme pattern, short germ-band development, segments are generated in anteroposterior sequence from the caudal end of the germ band. Grasshoppers develop in this way. *Drosophila* shows the opposite extreme, long germ-band development; all segments are generated almost simultaneously. Thus segments are generated in parallel in long germ-band embryos, and this process is typically much more rapid than the serial generation of segments in short germ-band embryos.

In *Drosophila*, during fertilization the sperm enters the egg through a specific point, the micropyle. Remarkably, a single 1.8 mm long sperm fertilizes an egg only 0.5 mm in length. (The recent demonstration that the entire sperm tail enters the egg and then persists intact as a distinct structural element during embryogenesis suggests that the reliability of fertilization may be due in part to the presence of this structure.[11]) As shown in Figure 1, entry of the sperm activates the egg to undergo a second meiotic division, and one of the four meiotic products becomes the female pronucleus. The male and female pronuclei fuse near the anterior end of the embryo.[10] Nuclei undergo a series of rapid and virtually synchronous mitoses. The early mitoses occur deep within the egg and are not accompanied by cytokinesis; all nuclei initially share a common cytoplasm. By the time nuclei have divided nine times (yielding 512 nuclei), approximately six to eight nuclei arrive at the posterior end of the egg and cellularize to form the pole cells, which are destined to be germ-line progenitor cells. One nuclear division cycle later, cycle 10, the majority of remaining nuclei complete migration and populate the entire surface of the embryo, forming the syncytial blastoderm. After three more mitoses at the surface, approximately 5000 blastoderm nuclei cellularize synchronously, during an interval of only 30-45 minutes at about 3.5 hours after fertilization. The embryo is now ready to begin the all-important events of gastrulation in which the three germ layers, endoderm, mesoderm, and ectoderm, are formed. Elaboration of the body plan follows.

GENERAL TACTICS THAT ACCELERATE EMBRYOGENESIS

This brief review points to several features that increase the speed of development but that may affect other performance criteria.

1. The genome has been streamlined and is small by eukaryotic standards. There are only four chromosomes and the total genome size is about 2×10^8 base pairs.[18]
2. The early nuclear division cycles are short; they lack G1 and G2 phases.[6]
3. Generalized zygotic transcription does not occur until the blastoderm forms. The egg comes invested with all of the macromolecular precursors, either in the form of mRNA or proteins, to complete development up to the cellular blastoderm stage. Although zygotic transcription of specific genes does occur prior to cellularization,[13] vigorous transcription begins during cellularization that follows the 13th nuclear division.[7]
4. Cell cycles without cytokinesis more rapidly generate the nuclei needed to form a blastoderm. Apparently molecules associated with individual nuclei need not be confined within a cell membrane. The cytoplasm does not mix freely; each nucleus is surrounded by a coherent cytoplasm organized by microtubules and actin filaments.[12] Appreciable resources and time are saved by delaying cellularization. For example, the membrane turnover needed during each cell cycle would be expected to lengthen significantly the time needed for each cycle.
5. The germ layers are produced rapidly during gastrulation. The onset of gastrulation occurs with astonishing suddenness. The ventral furrow, the cephalic furrow, and the anterior and posterior midgut invaginations form in minutes. The entire process of gastrulation takes about one hour at room temperature.

Further morphogenetic events of segmentation, organogenesis, and neural development are remarkably rapid.

SPECIFIC TACTICS: DNA REPLICATION

A major factor in the speed with which *Drosophila* develops is the extraordinary rapidity of its DNA replication. During the early cleavage stages, the chromosomes replicate and undergo mitosis every 8-10 minutes. 2×10^8 base pairs of DNA are replicated in 3-4 minutes (at 5×10^7 base pairs/minute); the chromosomes are condensed, then organized on a mitotic spindle, and segregated in about 5 minutes; and a new interphase nucleus is constructed in about 2 minutes. This time interval may approach a theoretical limit for doubling time for an organism of this complexity. The remarkable speed of this process is evident when one compares *Drosophila* to *E. coli*, which has a genome two orders of magnitude smaller and a doubling time, under laboratory conditions, of 20 minutes—about twice the duration of a nuclear cycle in an early embryo of *Drosophila*.

DNA replication occurs at discrete points on the DNA molecule, at replication forks. A highly simplified diagram of a fork is shown in Figure 2. To appreciate the time-consuming events at a fork, note that for the simplest of forks (e.g., *E. coli*), no fewer than 20 proteins have been identified.[17] At a fork these proteins assemble on the DNA, forming a large replication complex, and then sequentially polymerize nucleotide triphosphates onto an existing 3'-OH of the growing strand. Many other essential steps are involved in the initiation of synthesis (using an RNA primer), the elongation steps (using gyrase molecules to relieve the torsional strain of the double helix), and the formation and subsequent ligation of the Okazaki fragments formed on the lagging strand.[1,3]

A fork moves along the DNA at about $1 - 5 \times 10^3$ base pairs/minute. It would take a single replication fork about 10^5 minutes (1667 hours or 69 days) to replicate a genome of 10^8 base pairs! Obviously, replication is faster if many forks copy the DNA in parallel. At least 10^4 forks (one fork for every 10^4 base pairs) are needed to complete replication as fast as it occurs in *Drosophila*. Electron microscopy of chromatin spreads of *Drosophila* nuclei confirm this calculation.[20] Similar chromatin spreads of other species with substantially greater doubling times, such as mammals, birds, and amphibians, show that replication forks are separated more widely, by about $1 - 3 \times 10^5$ base pairs. Thus, one adaptive mechanism in the *Drosophila* embryo that decreases synthesis time is an increase in the density of replication origins.

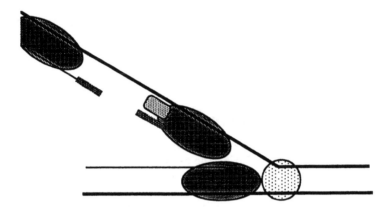

FIGURE 2 Schematic depiction of a replication fork. The two strands of DNA in the process of replication are represented by straight lines. The template strand is the heavier line; the lighter line represents newly synthesized DNA. Since DNA can only be synthesized in the 5′ to 3′ direction, two sites of synthesis are needed, one on the leading strand (horizontal lines) and one on the lagging strand (angled lines). Filled figures represent proteins at the fork. Ellipses represent DNA polymerases and associated proteins that directly carry out DNA synthesis. DNA gyrase, and other forms of helicases, are needed to relieve torsional strain as the DNA molecule unravels; these lie at the front of the advancing fork as represented by the filled circle. Lagging strand synthesis occurs discontinuously; it is synthesized by an additional protein system (filled rectangle) that uses an RNA primase to synthesize a short RNA primer (thick stippled line).

SPECIFIC TACTICS: MITOSIS

Available evidence suggests that in the early mitoses of a *Drosophila* embryo, the chromosomes from a nucleus segregate to form two daughter nuclei at about the maximum rate that the machinery for segregation can produce. Early in mitosis a mitotic spindle, a bipolar array of apposing microtubules, forms. Microtubules emanate inward from each spindle pole toward the kinetochores of the condensed metaphase chromosomes, which become arrayed on the mid-plane of the spindle, as shown schematically in Figure 3. Chromosomes actively move toward each spindle pole; their movement is directly correlated with progressive shortening of the spindle pole-kinetochore microtubules. Microtubules are polymers of tubulin; they shorten by disassembly into tubulin subunits, as diagrammed in Figure 4. Thus the kinetics of microtubule disassembly during chromosome segregation strongly influence the dynamics of mitosis.

As with DNA replication, most of our knowledge of microtubule dynamics has come from *in vitro* studies of purified mirotubules. To estimate the maximum rate of shortening of microtubules, we will consider only the reaction rate for disassembly, assuming that during chromosome separation no assembly of microtubules reduces the effective rate of disassembly. More importantly, we will assume that the measured rate *in vitro* determines the maximal rate possible *in vivo*. This assumption is probably valid because the rates *in vitro* have been measured in a purified system in the absence of other cellular factors. These factors are undoubtedly present *in vivo*, and they almost certainly regulate (and so retard) disassembly. There is experimental evidence for this claim: The rate of microtubule disassembly in the presence of microtubule-associated proteins (MAPs), which co-purify with tubulin, is significantly slower than in the absence of MAPs.[14]

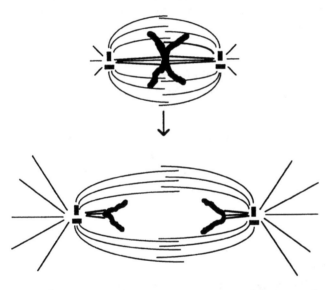

FIGURE 3 A "typical" mitotic spindle. Only major landmark features of metaphase (top) and anaphase (bottom) are shown. The basic shape of the spindle is determined by microtubules (lines) organized by centrosomes (small rectangles). A subset of microtubules, the kinetochore-to-spindle pole microtubules, grow directly inward from each centrosome towards the chromosomes (thick lines). During anaphase, chromosomes separate as the entire spindle lengthens due to growth of the spindle pole microtubules (thin lines). Chromosomes move towards each spindle pole as kinetochore-to-spindle pole microtubules shorten.

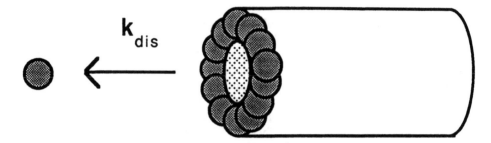

FIGURE 4   Mechanism of microtubule disassembly. The filled circles represent tubulin, the basic building blocks of microtubules. Tubulin assembles into microtubules by sequential addition to the end of a growing polymer. Likewise, a microtubule disassembles by sequential removal of subunits from the polymer end. Only this dissociation reaction is shown, and it is governed by a single first-order dissociation rate constant (k_{dis}).

Assuming 1500 tubulin subunits/micron, using the measured rate of microtubule disassembly (about 150 subunits/sec) and the dimensions of mitotic spindles in a *Drosophila* embryo (approximately 10 μm from spindle pole to kinetochore), we can calculate the expected minimum time of spindle movement to be 100 seconds, less than 2 minutes. *In vivo* measurements show that it takes about 3-5 minutes for the completion of mitosis in *Drosophila*. Thus the speed of chromosome separation in a *Drosophila* embryo is approaching a maximum set by the physicochemical properties of the molecular machinery.

DISCUSSION

Our analysis of developmental strategies has focussed on the unusually rapid pace of embryogenesis in the fruit fly *Drosophila melanogaster*. This analysis arose from a seemingly simple question: "How fast can a higher eukaryotic organism develop?" As is often the case, this initial question has led to related questions, about the strategies employed and the tradeoffs made as *Drosophila* evolved a life cycle highly specialized for speed. Interestingly, discussion of these questions has usually been restricted to ecological and behavioral considerations. For example, Stearns[27] discussed two concepts that characterize organisms subjected to disparate environmental pressures during evolution. These concepts, called r and K selection, describe strategies of adaptation for organisms evolving in low (r) or high (K) population densities. It is desirable to integrate these concepts with the ideas discussed in this chapter and with the results of laboratory experiments on selection for life history strategies, such as those of Mueller et al.[22] In this way it will be possible to understand how the processes within organisms meet the performance criteria associated with environmental selection pressures.

WHAT CONSTRAINTS WITHIN AN ORGANISM LIMIT ITS SPEED OF DEVELOPMENT?

Ultimately, intrinsic physicochemical properties of the molecular machinery must limit the speed of supramolecular processes. For example, apparently the speed of chromosome separation in *Drosophila* embryos approaches a limit set by the kinetics of microtubule dissociation. Such chemical kinetic limitations are among the generic constraints on properties of living systems that Newman[24] has discussed. However, at levels higher than an individual molecular interaction, various tactics for organizing networks of molecular processes can accelerate development. (These are genetic mechanisms, in Newman's[24] sense.) In gaining speed through these tactics, the embryo often pays a price in reliability, flexibility, or economy. We now review some of these tactics, and the tradeoffs associated with them, as introduced in the preceding survey.

NETWORK TACTICS: PARALLEL AND REDUNDANT PROCESSES

A process can be accelerated by performing parts of it in parallel. This tactic is evident in *Drosophila* in the increase in density of replication forks. An increase in parallelism requires extra copies of the macromolecular machinery performing the process—in this case, an increase in replication complexes. Economy is sacrificed, as more complexes are used. Furthermore, proper replication then requires a corresponding increase in other factors, such as gyrases. In particular, there must be an increase in the machinery for correcting errors during replication; otherwise, the reliability of replication will decrease. If the error rates in processes of molecular copying exceed threshold levels, the organism will become inviable.[8]

During evolution an organism may acquire redundant processes that increase the speed of development. For example, in *Drosophila* macromolecules are placed in the egg during oogenesis that might be generated by zygotic transcription in an embryo developing more slowly. Presumably this transfer of control from zygotic to maternal transcription requires the binding of transcription factors to additional cis-regulatory elements, both to activate maternal transcription and to suppress zygotic transcription. As another possible example of redundancy, the rapid generation of segments in long-germ development may require additional regulation to modify the slower sequential generation of segments in short-germ development. These processes would be redundant to the extent that the modifications for speed are added to, but do not replace, processes already available to generate an adult fruit fly more slowly. Redundant processes tax economy and reliability, in the same ways as does increased parallelism.

NETWORK TACTICS: DELETION OF PROCESSES

A process can be accelerated by deleting parts of it that occur sequentially. The omission of cytokinesis after mitosis in early cleavages of *Drosophila* is such a deletion. Lacking structures that would have been generated by the deleted processes, the embryo may economize but lose in reliability or flexibility. Specifically, with the deletion of cytokinesis, cell boundaries do not separate nuclei. The plasma membranes need not be synthesized, but the embryo relies more on intracellular mechanisms for localizing molecules near each nucleus.[12]

Another example of the deletion of an important cellular process is the absence of a heat shock response during the early cleavage stages in *Drosophila*.[19] The heat shock response is a ubiquitous reaction of cells to various environmental stresses, including heat and starvation. During stress a specific set of genes, the heat shock genes, are preferentially activated and transcribed at a high rate, producing a small set of proteins that may protect other proteins from thermal denaturation. At the same time general transcription is almost completely shut down. Before the blastoderm forms in *Drosophila*, a heat shock response does not occur; presumably the necessary transacting factors are not stored in the egg during oogenesis. The decrease in cellular activity that would accompany heat shock is clearly at odds with the rapid pace of early development. Apparently evolution has sacrificed an embryo's ability to deal with occasional external stresses for the sake of rapid development. Interestingly, heat shock genes are present in *Drosophila*; they become active only after the blastoderm has formed and zygotic gene expression begins in earnest.

The genome can be copied more rapidly if it is relatively small, as is the case in *Drosophila*. (Reduction in the number of bases to be copied is an example of accelerating a process by deleting sequential subprocesses.) However, reducing the genome size opposes parallelism and redundancy. As regards parallelism, a smaller genome is less likely to contain multiple copies of genes that could rapidly produce parallel-operating complexes. Similarly, a smaller genome has less coding capacity for redundant mechanisms that increase speed.

According to the exon hypothesis for the evolution of coding,[5] reproducing faster bacteria and viruses reduced the size of the genome by eliminating introns, the intervening sequences of DNA between the exons that encode peptides. *Drosophila* has retained introns, but used their potential coding capacity in some genes by using information in introns to generate more than one protein, through alternative splicing.[23] Alternative splicing allows an organism to make more proteins with a smaller genome that allows speedier development. If the sequence of bases in an intron is used to store more information, the intron is less available as a sequence-insensitive site for recombination, since recombination is likely to disrupt the sequence. Thus it might seem that using introns to encode alternative splicing would reduce the evolutionary flexibility gained through recombination. However, this loss of flexibility will be compensated, to a degree that is unclear, by the increased rate of recombination that accompanies acceleration of the life cycle in *Drosophila*.

WHAT PROGRAM OF RESEARCH FOLLOWS FROM THESE CONSIDERATIONS? EXPERIMENTS IN TIME—A GENETIC APPROACH

The preceding considerations suggest a number of experiments for which *Drosophila*, with its rich genetic heritage, is well suited. For example, can *Drosophila* be mutated so as to generate variants capable of either faster or slower development? What would be the phenotype of mutant "fast" or "slow" flies? Which genes would be affected? What kinds of mutations might we look for?

It is clear that some types of mutations are better candidates than others. For example, mutations that affect the rate of synthesis of intermediates are not good candidates because all intermediates must be available before substantial macromolecular synthesis can begin. The more promising candidates are likely to be those effecting mechanochemical reactions, such as the movement of chromosomes during mitosis or the rate of propagation of a replication fork. That is, mutations in elements involved in the regulation of large multicomponent structures may be uncovered with this simple screen. Formally, two general classes of mutants are envisioned:

1. Dominant, hypermorphic gain-of-function mutations, and
2. Recessive loss-of-function hypomorphs.

Dominant mutations have been extensively studied in *Drosophila*. They usually are associated with alterations either in structural genes or in regulatory regions of structural genes that result not in a loss of gene function, but rather in an increase or deregulation of gene activity. If viable, dominant mutations could accelerate development. Class 2 mutations are classical loss-of-function mutations that, if viable, should decelerate development. If *Drosophila* has achieved optimization with respect to speed, the majority of mutations would fall into class 2.

A recent technique for producing mutations in *Drosophila* has revolutionized our ability to generate and analyze alterations in the genome quickly and efficiently. This technique works by inducing specially designed mobile genetic elements (P-elements) to move to other sites within the genome. The gene disruption caused by the transposition of a P-element into the coding region would effectively result in a null mutation. Insertions into regions upstream of a structural gene (for example, encoding a protein involved in DNA synthesis), while having no direct effect on the structural gene, might alter the regulatory region, thus affecting the regulation of the gene.

CONCLUSION

In this inquiry we have explored a few threads in the fabric of processes that underlies the performance capabilities of *Drosophila*. Further explorations can use the rapidly growing body of information about the development of this organism to test hypotheses about its organization. A general hypothesis that emerged in the workshop was the idea that organisms often operate near a transition between distinct phases of performance. Our inquiry suggests this: Development in *Drosophila* may have accelerated to a threshold for reliability, beyond which a catastrophic increase in error rate would accompany a further increase in speed. Networks of processes in organisms may tend to evolve a connectivity such that they operate at the edge of chaos.[15,16] Oscillatory processes may be entrained at the edge of a domain of phase locking.[4] Such general hypotheses, more extensively developed and tested, may lead to a general theory for the development, physiology, and evolution of organisms.

ACKNOWLEDGMENT

We appreciate helpful discussion of this material with Arthur Baskin, and his comments on a draft of the manuscript.

REFERENCES

1. Alberts, B., D. Bray, J. Lewis, M. Raff, K. Roberts, and J. D. Watson. *Molecular Biology of the Cell*, 2nd Edition. New York: Garland, 1989.
2. Ashburner, M. *Handbook of Drosophila*. New York: Cold Spring Harbor, 1990.
3. Darnell, J., H. Lodish, and D. Baltimore. *Molecular Cell Biology*. New York: W. H. Freeman, 1986.
4. DeGuzman, C. G., and J. A. S. Kelso. "The Flexible Dynamics of Biological Coordination: Living in the Niche Between Order and Disorder." This volume.
5. Dorit, R. L., L. Schoenbach, and W. Gilbert. "How Big is the Universe of Exons?" *Science* **250** (1990): 1377–1382.

6. Edgar, B. A., and P. H. O'Farrell. "Genetic Control of Cell Division Patterns in the *Drosophila* Embryo." *Cell* **57** (1989): 177–187.
7. Edgar, B. A., and G. Schubiger. "Parameters Controlling Transcriptional Activation During Early *Drosophila* Development." *Cell* **44** (1986): 871–877.
8. Galas, D. J., T. B. L. Kirkwood, and R.F. Rosenberger. "An Introduction to the Problem of Accuracy." In *Accuracy in Molecular Processes*, edited by T. B. L. Kirkwood and R. F. Rosenberger. New York: Champman and Hall, 1986.
9. Gould, S. J. "Organic Wisdom, or Why Should a Fly Eat its Mother from Inside." In *Ever Since Darwin*. New York: W. W. Norton, 1977.
10. Huettner, A. F. "Maturation and Fertilization in *Drosophila Melanogaster.*" *J. Morphol.* **39** (1924): 249–265.
11. Karr, T. "Intracellular Sperm/Egg Interactions in *Drosophila*: A Three-Dimensional Structural Analysis of a Paternal Product in the Developing Egg." In *Mech. of Dev.* (1991).
12. Karr, T., and B. Alberts. "Organization of the Cytoskeleton in Early *Drosophila* Embryos." *J. Cell Biol.* **102** (1986): 1494–1509.
13. Karr, T. L., B. Drees, Z. Ali, and T. Kornberg. "The Engrailed Locus of *Drosophila Melanogaster* Encodes an Essential Function of Pre-Cellular Embryos." *Cell* **43** (1985): 591–601.
14. Karr, T. L., D. Kristofferson, D. L. Purich. "Mechanism of Microtubule Depolymerization: Correlation of Rapid Induced Disassembly Experiments with a Kinetic Model for Endwise Depolymerization." *J. Biol. Chem.* **255** (1980): 8567–8572.
15. Kauffman, S. A. "Antichaos and Adaptation." *Sci. Amer.* **265** (1991): 78–84.
16. Kauffman, S. A. "The Sciences of Complexity and 'Origins of Order.'" This volume.
17. Kornberg, A. *DNA Replication.* San Francisco: W. H. Freeman, 1980.
18. Laird, C. M. "DNA of *Drosophila* Chromosomes." *Ann. Rev. Genetics.* **7** (1973): 177–204.
19. Lindquist, S. L., and E. A. Craig. "The Heat Shock Proteins." *Ann. Rev. Genet.* **22** (1988): 631–677.
20. Micheli, G., C. T. Baldari, M. T. Carri, G. DiCello, and B. M. Nardelli. "An Electron Microscope Study of Chromosomal DNA Replication in Different Eukaryotic Systems." *Exp. Cell Res.* **137** (1982): 127–140.
21. Mittenthal, J. E. "Physical Aspects of the Organization of Development." In *Lectures in Complex Systems*, edited by D. Stein. Santa Fe Institute Studies in the Sciences of Complexity, Lectures Vol. I. Redwood City, CA: Addison-Wesley, 1989.
22. Mueller, L. D., P. Guo., and F. J. Ayala. "Density-Dependent Natural Selection and Trade-Offs in Life History Traits." *Science* **253** (1991): 433–435.
23. Nagoshi, R. N., M. McKeown, K. C. Burtis, J. M. Belote, and B.S. Baker. "The Control of Alternative Splicing at Genes Regulating Sexual Differentiation in *Drosophila Melanogaster.*" *Cell* **53** (1988): 229–236.

24. Newman, S. A. "Generic Physical Mechanisms of Morphogenesis and Pattern Formation as Determinants in the Evolution of Multicellular Organization." This volume.

25. Parker, G. A., and J. Maynard Smith. "Optimality Theory in Evolutionary Biology." *Nature* **348** (1990): 27–33.

26. Sander, K. "Specification of the Basic Body Pattern in Insect Embryogenesis." *Adv. in Insect Physiol.* **12** (1976): 125–238.

27. Stearns, S. C. "Life-History Tactics: A Review of the Ideas." *Quart. Rev. Biol.* **51** (1976): 3–47.

John Reinitz,* Eric Mjolsness,† and David H. Sharp‡
*Center for Medical Informatics, Department of Anesthesiology, Yale Medical School, 333 Cedar Street, New Haven, CT 06511; †Department of Computer Science, Yale University, P.O. Box 2158 Yale Station, New Haven, CT 06520-2158; and ‡Theoretical Division, Los Alamos National Laboratory, Los Alamos, NM 87545

A Connectionist Model of the *Drosophila* Blastoderm

We present a phenomenological modeling framework for development, and apply it to the network of segmentation genes operating in the blastoderm of *Drosophila*. Our purpose is to provide a systematic method for discovering and expressing correlations in experimental data on gene expression and other developmental processes. The modeling framework is based on a connectionist or "neural net" dynamics for biochemical regulators, coupled to "grammatical rules" which describe certain features of the birth, growth, and death of cells, synapses, and other biological entities. We present preliminary numerical results regarding regulatory interactions between the genes *Krüppel* and *knirps* that demonstrate the potential utility of the model.

1. INTRODUCTION

We sketch a modeling framework for development. Its purpose is to provide a systematic method for discovering and expressing correlations in experimental data on

Principles of Organization in Organisms,
SFI Studies in the Sciences of Complexity, Proc. Vol. XIII,
Eds. J. Mittenthal & A. Baskin, Addison-Wesley, 1992   **109**

gene expression and other developmental processes. In this report, we present preliminary results on the application of this modeling framework to the blastoderm of *Drosophila*. A further discussion of the underlying ideas is given in Mjolsness et al.[11]

2. BASIC APPROACH

The *configuration* of a developing embryo is specified by the number and internal state of its constituents. The constituents of an embryo include cells, cell nuclei, fibers, and synapses. The choice of state variables depends on the object to be described. For example, we describe the state of a cell nucleus in terms of concentrations of transcription factors, whereas a synapse could be described with membrane voltage, internal Ca^{++} concentration, and ionic conductance.

Developmental dynamics is modeled by a set of equations describing the transition between two configurations (state history). The model describes three fundamental dynamical processes.

Concentrations of regulatory molecules acting within and between cells and their organelles change in response to three factors: existing concentrations of the regulators, exchange of regulatory molecules among existing entities, and simple decay. With the exception of decay, which is represented by a simple exponential decay term, these processes are modeled with "connectionist" or "neural net" dynamical equations, which incorporate basic phenomenological features of the regulatory molecules. At almost all times, the developmental dynamics is represented by a system of coupled ordinary nonlinear differential equations, but, when the number or kind of fundamental entities changes (for example, when a cell divides), the system takes a discontinuous jump.

The continuous or discontinuous evolution of the system at a given moment of time is modeled by a set of "grammatical rules" (see, for example, Lindenmayer[10]). Rules are selected on an object-to-object basis, at each time. The rule selected by a particular object is determined by the state of that object, and of its neighbors. Each continuous dynamical process has an associated grammatical rule, in which there is no change in the type or number of fundamental entities. Changes in number and type of fundamental entities resulting from birth, growth, induction, and death processes are represented by grammatical rules that describe discontinuous dynamic processes. Finally, the model must describe the influence of spatial organization.

The combined action of continuous and discrete-time grammatical rules is illustrated in Figure 1. The figure shows a representative history of state variables v_i^a under a combination of continuous and discrete time neural net dynamics. v_i^a is the concentration of regulator a in object i. Time increases to the right; three generations of objects are shown. Continuous time dynamics is denoted by the stippled horizontal axes, and discrete time dynamics by the stippled ovals. The solid black curves are graphs of the functions $v_i^a(t)$. The rightmost oval denotes a type change;

the other two are mitoses. The level of a state variable may change under the action of a discrete time rule. Note, for example, that $v^l(t)$ is always above baseline before a mitosis and and below it afterwards.

There are different classes of rules to represent the different classes of events that occur during development. In general, a biological object such as a cell or synapse may undergo a variety of transformations. It may be born from a parent or it may die. Between birth and death its internal state will change as a result both of internal dynamics and by interactions with other objects of the same or different type. Birth and death processes are represented by discrete time rules only; changes in internal state and interactions with other objects may be represented by either continuous or discrete time grammar rules. These possibilities taken together amount to six classes of rules.

Figure 2 is a diagrammatic representation of the six classes of rules which we employ. In each diagram, the time axis runs in a horizontal direction, and a space axis in the vertical direction. The arrowheads on the solid lines point in the direction of increasing time. The dotted vertical arrows represent a spatial interaction: continuous time rules by three such arrows, discrete time by one. The dotted arrows point in the direction of the object which has chosen the illustrated rule. A discrete time rule is represented by a filled circle and a continuous time rule by a pair of arrowheads on a solid line without an intervening filled circle. The input

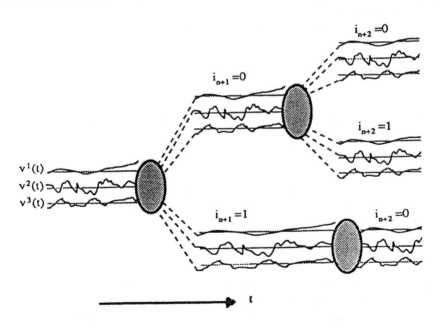

FIGURE 1 A schematic illustration of the history of the state variables v under a combination of continuous- and discrete-time grammatical rules.

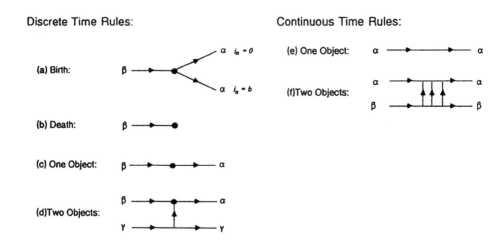

FIGURE 2   A diagrammatic representation of the six classes of grammatical rules.

and output object types are indicated on the left- and right-hand sides respectively of each diagram. In (a), more than two branches could occur; the branches have lineage indices i_n ranging from 0 to b.

The next two sections describe the application of this modeling framework to the *Drosophila* blastoderm.

3. APPLICATION TO THE *DROSOPHILA* BLASTODERM

Two features of the *Drosophila* blastoderm make it an especially suitable system for the initial application of the modeling framework outlined above. The first is that the regulatory dynamics of genes that lay out the basic body plan are dynamically separable from other aspects of the developmental dynamics during the blastoderm stage. The second is the availability of molecular probes for the products of these genes, which render the state variables v_i^a directly observable.

We review some elementary facts about the *Drosophila* blastoderm. Immediately following fertilization, the zygotic nuclei undergo a rapid series of mitoses without the formation of cells. After eight almost synchronous divisions, these nuclei migrate to the cortex of the egg, whereupon transcription of the zygotic genes begins. This stage is called the syncytial blastoderm, because no cells are present. After another five divisions, cell membranes are laid down and gastrulation begins.[4] The timing of these cell divisions is under the control of maternal gene products. The protein products of zygotic pattern-formation genes essential for laying down the basic body plan of the animal are expressed at this time in patterns that rapidly evolve from coarse- to fine-scale spatial resolution (reviewed in Akam[1] and Ingham[7]).

The *Drosophila* egg is approximately an ellipsoid, but asymmetries in its shape clearly define two axes, each with a polarity. These axes provide coordinates for the blastoderm as well. One axis runs in an anterior-posterior direction, and the other in a dorsal-ventral direction. The pattern-formation genes fall into two classes. To a reasonable degree of approximation, the level of expression of a member of the first class of genes is solely a function of location on the anterior-posterior axis; these genes are members of the anterior-posterior class. The expression level of a member of the second class of genes depends only on position along the dorsal-ventral axis; these genes belong to the dorsal-ventral class.

The separation of these two classes of genes by expression pattern carries over to their dynamical interactions. A member of one of these classes of genes does not regulate the expression of a member of the other class during the blastoderm stage, except perhaps in the region of the anterior or posterior pole. For the rest of this report, we focus on the zygotic anterior-posterior pattern-formation genes, often referred to as segmentation genes. The segmentation genes are dynamically coupled in a network of genetic regulation. A line of evidence leading to this conclusion is the observation that disabling one segmentation gene by mutation causes changes in the pattern of expression of many of the other segmentation genes.[2,3,5,8,12,13,14] The characterization of this regulatory network is one of the objectives of our modeling effort. A precise formulation of the regulatory network is required to interpret altered patterns of gene expression in terms of regulatory action.

The expression of the segmentation genes in the middle region of the blastoderm is approximately a function of anterior-posterior position only, so we model their dynamics of expression using a linear array of nuclei. At each time, a nucleus undergoes one of two types of transitions:

1. Mitosis: Replace each nucleus with a pair of daughter nuclei. Do not allow the synthesis of gene products. The mitoses are timed according to Foe and Alberts.[4]
2. Interphase: Allow protein concentrations to evolve by synthesis, exchange of material with neighboring nuclei, and by degradation. This process is described by the dynamical equations for interphase.

We consider N genes. During interphase the level of gene product a in nucleus i, denoted by v_i^a, is modeled by the the following connectionist equation[11]):

$$\frac{dv_i^a}{dt} = R_a g_a \left(\sum_{b=1}^{N} T^{ab} v_i^b + h^a \right) + D(n)[v_{i-1}^a - v_i^a) + (v_{i+1}^a - v_i^a)] - \lambda_a v_i^a . \quad (1)$$

The first term describes gene regulation and protein synthesis, the second term describes exchange of chemical products between neighboring nuclei, and the third term describes the decay of gene products.

In Eq. (1), g_a is a thresholding function of the form

$$g_a(u^a) = (1/2)((u/\sqrt{u^2 + 1}) + 1)$$

for all a, where $u^a = \sum_{b=1}^{N} T^{ab} v_i^b + h^a$. R_a is the maximum rate of synthesis from gene a. T^{ab} is a real number describing the influence of gene b on gene a. h^a is a threshold that is currently fixed at the same value of -10 for all genes. $D(n)$ depends on the number n of cell divisions that have taken place, such that $D(n) = 4D(n-1)$. λ_a is the decay rate of the product of gene a; the results given here assume a value of λ equivalent to a half-life of 30 minutes. The other parameters T^{ab}, R_a, and $D(0)$ are adjustable to fit data.

4. PRELIMINARY NUMERICAL RESULTS

The question we investigate first is the regulation of the central domain of *knirps* expression by other zygotic segmentation genes. Among the genes known to regulate *knirps* are *Krüppel, hunchback, giant,* and *tailless*. It is also known that *knirps* expression is reduced throughout the central domain in embryos mutant for *Krüppel*.[12] As a preliminary step, we have modeled the interaction of *Krüppel* and *knirps*. This truncated model may be valid in the posterior half of the *knirps* domain where

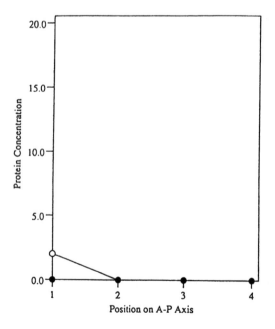

FIGURE 3   Distribution of the products of the genes *Krüppel* and *knirps* over a domain containing four nuclei at the beginning of cleavage cycle 13. The vertical axis is calibrated in arbitrary units of protein concentration; its scale was selected to be same as that used in Figures 4 and 5. The horizontal axis is calibrated in terms of individual nuclei in a line running along the anterior-posterior axis; nucleus 1 is the most anterior. Filled circles denote *knirps* concentrations; open circles denote *Krüppel* concentrations. Note that the only non-zero concentration is that of *Krüppel* in nucleus 1.

Krüppel is believed to be the major regulator. This domain is approximately eight nuclei in extent along the anterior-posterior axis at the end of the blastoderm stage.

For comparison of the model with data, we rely on double-labeling studies using fluorescence-tagged antibodies (unpublished data of R. Warrior and J. Reinitz). Some embryos were photographed under bright field optics as well as fluorescence in order to more accurately assess the developmental stage. Our model requires specification of initial conditions: these are illustrated in Figure 3, which shows the distributions of *Krüppel* and *knirps* proteins at the beginning of cleavage cycle 13. Given these initial conditions, we model the dynamics of gene products from that time until the onset of gastrulation, a period of about 70 minutes. During that period, one nuclear division takes place and the number of nuclei doubles.

To discover what values of T^{ab}, $D(0)$, and R^a best fit the data, we do a least-squares fit to the *trajectories* that the system follows. Given a set of fixed initial conditions (e.g., Figure 3), each v_i^a will follow a trajectory that depends on the values of the parameters in Eq. (1). Our aim is to find values of those parameters such that the trajectories given by Eq. (1) are as close as possible to the observed trajectories of changing protein concentrations in each cell. As an approximation to the observed trajectory, we compare the model to gene expression data at two

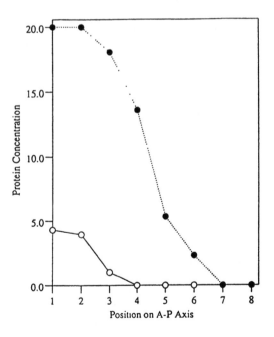

FIGURE 4 The graph shows our estimate of the relative levels of expression of *Krüppel* and *knirps* in a strip of eight nuclei running from the middle of the central *knirps* domain in a posterior direction. This estimate was made from photographs of embryos stained with the appropriate antisera (see text). The vertical axis is calibrated in arbitrary units of protein concentration. The horizontal axis is calibrated in terms of individual nuclei in a line running along the anterior-posterior axis; nucleus 1 is the most anterior. A cell division has occurred since the situation shown in Figure 3, so that nuclei 1 and 2 are daughters of nucleus 1 in Figure 3, and so on. Filled circles connected with dotted lines denote *knirps* concentrations; open circles connected with solid lines denote *Krüppel* concentrations.

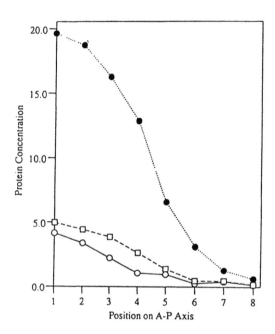

FIGURE 5 The graph shows the numerical output of the model for wild type and one mutant at the onset of gastrulation. The *Krüppel-Krüppel* connection strength was 5.1, the *Krüppel-knirps* connection strength was 3.9. The two corresponding connection strengths from *knirps* to *Krüppel* and itself were −.35 and 1.2, respectively. $R_{Krüppel}$ was .83, R_{knirps} was 3.4, and D(0) was .008. The axes and symbols are as described for Figure 4, with the addition that the open squares connected by dashed lines represent the distribution of *Krüppel* in an embryo numerically mutated for *knirps*; note that it extends more posteriorly than in wild type, in accordance with observations. Expression of *knirps* in this numerical mutant and of both *Krüppel* and *knirps* in a numerical mutant for *Krüppel* are not shown; all three were very close to zero in all nuclei.

times: mid cleavage cycle 14 and the onset of gastrulation. For example, the expression pattern just prior to gastrulation is shown in Figure 4. We measure the deviation between the data and the behavior of the model by taking the sum of squared differences between the observed protein concentration and that given by the model over each protein, cell, and time for which data exists. This deviation is then minimized using the method of simulated annealing.[9]

The behavior of the model after such a fit is shown in Figure 5. Results obtained using wild-type data allow us to predict the behavior of mutants without further experimental input. In particular, we find *knirps* expression greatly reduced in mutants for *Krüppel*; *Krüppel* expression extends more posteriorly in mutants for *knirps*. This is in qualitative agreement with experimental observations.[6] Although the numerical results reported here are quite preliminary, they indicate that the methods we use promise to be helpful in characterizing the network of genes that control pattern formation in *Drosophila* and other organisms. We are currently investigating a more comprehensive model of the blastoderm which includes the effect of other genes.

ACKNOWLEDGMENTS

Eric Mjolsness was supported in part by the Air Force Office of Scientific Research under grant AFOSR 88-0240. David H. Sharp's work was supported by the United States Department of Energy. John Reinitz is supported by grant 5-T15-LM07056 from the National Library of Medicine.

REFERENCES

1. Akam, M. "The Molecular Basis for Metameric Pattern in the *Drosophila* Embryo." *Development* **101** (1987): 1–22.
2. Carroll, S. B., and M. P. Scott. "Zygotically Active Genes that Affect the Spatial Expression of the *fushi tarazu* Segmentation Gene During Early *Drosophila* Embryogenesis." *Cell* **45** (1986): 113–126.
3. Carroll, S. B., A. Laughon, and B. S. Thalley. "Expression, Function, and Regulation of the *Hairy* Segmentation Protein in the *Drosophila* Embryo." *Genes & Development* **2** (1988): 883–890.
4. Foe, V. A., and B. M. Alberts. "Studies of Nuclear and Cytoplasmic Behavior During the Five Mitotic Cycles that Precede Gastrulation in *Drosophila* Embryogenesis." *J. Cell Sci.* **61** (1983): 31–70.
5. Frasch, M., and M. Levine. "Complementary Patterns of *Even-Skipped* and *fushi tarazu* Expression Involve their Differential Regulation by a Common Set of Segmentation Genes in *Drosophila*." *Genes & Development* **1** (1987): 981–995.
6. Harding, K., and M. Levine. "Gap Genes Define the Limits of Antennapedia and Bithorax Gene Expression During Early Development in *Drosophila*." *The EMBO J.* **7** (1988): 205–214.
7. Ingham, P. W. "The Molecular Genetics of Embryonic Pattern Formation in *Drosophila*." *Nature* **335** (1988.) :25–34
8. Jackle, H., D. Tautz, R. Schuh, E. Seifert, and R. Lehmann. "Cross-Regulatory Interactions Among the Gap Genes of *Drosophila*." *Nature* **324** (1986): 668–670.

9. Kirkpatrick, S., C. D. Gelatt, and M. P. Vecchi. "Optimization by Simulated Annealing." *Science* **220** (1983): 671–680.

10. Lindenmayer, A. "Mathematical Models for Cellular Interaction in Development." Parts I and II. *J. Theor. Biol.* **18** (1968): 280–315.

11. Mjolsness, E., D. H. Sharp, and J. Reinitz. "A Connectionist Model of Development." *J. Theor. Biol.*, submitted.

12. Pankratz, M. J., M. Hoch, E. Seifert, and H. Jackle. "*Krüppel* Requirement for *knirps* Enhancement Reflects Overlapping Gap Gene Activities in the *Drosophila* Embryo." *Nature* **341** (1989): 337–340.

13. Reinitz, J., and M. Levine. "Control of the Initiation of Homeotic Gene Expression by the Gap Genes Giant and Tailless in *Drosophila*." *Dev. Biol.* **140** (1990): 57–72,

14. Rushlow, C., K. Harding, and Mi. Levine. "Hierarchical Interactions Among Pattern-Forming Genes in *Drosophila*." In *Banbury Report 26: Developmental Toxicity: Mechanisms and Risk*. New York: Cold Spring Harbor Press, 1987.

Michael P. Stryker
W.M. Keck Foundation, Center for Integrative Neuroscience and Neuroscience Graduate Program, Department of Physiology, University of California, San Francisco, California 94143-0444

Activity-Dependent Reorganization of Afferents in the Developing Mammalian Visual System

Reprinted with permission from *Development of the Visual System*, edited by D. M. Lam and C. J. Shatz. The MIT Press, 1991.

The early experiments of Hubel and Wiesel[14] demonstrated that the neurons within radial columns in the primary visual cortex share many specific response properties. Each column is specific for *topography*, in that all neurons have their receptive fields in a particular portion of the visual field; for *ocular dominance*, in that all the neurons in a particular column will tend to respond more strongly to one eye than to the other; for *ON-* or *OFF-center types* in some species, in that the neurons within a column will tend to respond better to bright stimuli than to dark stimuli, or vice-versa; and for stimulus *orientation*, in that all the neurons within one column will respond selectively to bars or edges at one particular angle in the visual field. In the tangential dimension, the properties of neurons differ from one column to the next in a systematic fashion. Thus, the columns of the visual cortex are precisely organized with respect to three (or in many species, all four) of the response properties noted above. An understanding of this organization and its structural basis is essential before we can understand the mechanisms that give rise to this orderly arrangement of cortical response properties in development.

Principles of Organization in Organisms,
SFI Studies in the Sciences of Complexity, Proc. Vol. XIII,
Eds. J. Mittenthal & A. Baskin, Addison-Wesley, 1992   **119**

Experiments described below suggest that the major specific afferents to visual cortex are also precisely ordered in every respect in which the cortical neurons themselves are ordered. Such an orderly arrangement of afferents provides a structural basis for the tangential organization of cortical columns. The progressive reorganization of the afferent pathway in development may well be the mechanism by which the orderly arrangement of cortical columns is established. Further experiments described below indicate that relatively simple mechanisms of activity-dependent synapse rearrangement can account for the principal features of afferent organization and that the plasticity exhibited by the developing visual cortex is consistent with the existence of such mechanisms.

ORGANIZATION OF AFFERENTS IN RELATION TO CORTICAL COLUMNS

The first respect in which visual cortex is organized in the tangential domain is the map of the visual field onto the cortex, so that neighboring groups of columns represent neighboring points in the visual field.[19,20] The basic organization of this map was known from clinical neurology and anatomical studies and has been evident since the earliest physiological studies of Talbot and Marshall.[48] The structural basis of the topographic map is the orderly projection of neighboring retinal ganglion cells to neighboring points in the lateral geniculate nucleus, and the further orderly projection from geniculate to cortex.

A second feature of the organization of the visual cortex in the tangential domain is the system of cortical ocular dominance columns. These columns disrupt the larger-scale continuity of the map of the visual field by interdigitating regions in which responses favor one eye with those that favor the other eye at a scale of about half a millimeter. The structural basis of these ocular dominance columns was revealed in experiments in which the population of geniculocortical afferent terminals serving one eye was labelled by degeneration methods or autoradiographically, by transneuronal axonal transport of ^{3}H-sugars or amino acids injected into the vitreous humor of one eye.[18,41,53] In adult animals, the labelled afferent projection conveying information from one eye to the visual cortex takes the form of patches or stripes 350-500 μm wide, alternating with unlabelled patches of the same size that receive the other eye's projection.

Third, in several species, *ON*-center and *OFF*-center responses are also segregated, at a scale somewhat finer than that of the ocular dominance columns.[24] This pattern of organization also appears to have its structural basis in the organization of afferents to the cortex, as revealed by experiments in which geniculocortical afferents of the two center types were recorded in alternate patches of cortical layer

IV.[23,54] The borders between *ON-* and *OFF*-center patches were independent of those between the ocular dominance patches.

The orientation columns constitute a fourth type of columnar and tangential organization that was revealed in Hubel and Wiesel's[14] earliest studies of visual cortex and elaborated in later reports.[15,19] The structural basis of orientation selectivity has, however, been difficult to establish with confidence. The original model[14] proposed that neurons with simple-type receptive fields were endowed with orientation selectivity by virtue of the alignment in the visual field of the receptive fields of the lateral geniculate nucleus neurons from which the simple cell received its input. In this case, the arrangement of afferents could constitute a structural basis for the orientation columns just as they do for the other sorts of cortical columns. Later models[2] proposed that orientation selectivity was produced largely or completely by intracortical circuitry. In this latter case, the arrangement of afferents might have nothing to do with cortical orientation columns, making them different from the other sorts of cortical columns.

AFFERENTS TO SINGLE ORIENTATION COLUMNS

We have now investigated the arrangement of the geniculocortical afferents that provide the thalamic input to orientation columns by recording from afferent terminal arbors in the major input layer, layer IV, of the cortex.[5,6] As originally suggested by Helen Sherk, the factor that prevents one from discriminating the electrical activity produced by the many afferent terminal arbors through which a microelectrode must pass on its way through the cortex is not the small size of their extracellularly recorded action potentials in comparison to the electrical noise of the microelectrode, since the microelectrode noise can be as little as 5 μV while afferent spikes are 10 to 100 μV. Instead, it is the ongoing discharge of cortical cells, which produce spikes of some 100's of μV, that occludes recognition of most of the signals from afferents. By silencing the cortical discharge, it becomes routine to record and plot 10–40 afferent receptive fields on a single vertical penetration through the visual cortex.

The design of our experiment was to align a microelectrode so that it passed down a single orientation column in the primary visual cortex of the ferret. Recordings made at a series of cortical depths guaranteed this alignment and allowed us to determine the preferred orientation of the cortical cells before withdrawing the electrode to a position in layer III just above the major input layer. In the earlier experiments, cortical cells were then silenced by killing them, using superfusion of the excitotoxin kainic acid.[54] In later experiments, the cortical cells were silenced more quickly and with less damage by superfusing them with muscimol, a potent analog of the inhibitory neurotransmitter GABA that acts on the postsynaptic GABA$_A$ receptors. Once the cortical cells were silent, the microelectrode was advanced again slowly into and through layer IV, where the action potentials of many afferent single units then became individually discriminable. These units had visual response properties identical to those of their parent cell bodies recorded in the lateral geniculate nucleus (LGN), and their responses to electrical stimulation of the LGN confirmed that they were the terminals of geniculate cells. In most experiments, the use of a blind procedure ensured that the plotting of the afferent receptive fields was not influenced by knowledge of the prior results from the cortical cells.

Figure 1 shows the results of this experiment for three cases. In the cases illustrated, the afferent receptive fields, shown as ellipses, were disposed about an elongated region of the visual field, and the axes of elongation matched the preferred orientations of the cortical cells. Both of these findings were generally true. In 16 out of 18 such experiments, the afferent receptive fields were found to be elongated with better than 90% confidence (and with at least 99.99% confidence in 13 of 16 cases). The agreement between cortical orientation selectivity and the elongation of the collection of afferent receptive fields is illustrated for these 16 cases in Figure 2. While there are three cases of mismatch by as much 25 principal axis of degrees, overall the match is good, as indicated by the proximity of the data points to the line drawn to the prediction of perfect agreement between the orientation of the collection afferent receptive fields and that of the cortical cells.

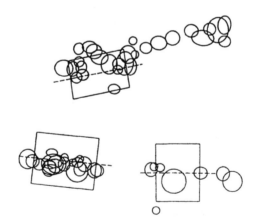

FIGURE 1   Receptive fields of a layer III cortical cell and the collection of geniculocortical afferents recorded in layer IV immediately below the cortical cell in three separate experiments. The cortical receptive fields are drawn as rectangles, and the dashed lines indicate the preferred orientations for the cortical cells as determined from orientation tuning histograms. Each ellipse is the receptive-field center of a single geniculocortical afferent terminal. Note that the principal axis of elongation of the collection of afferent receptive fields matches the preferred orientation of the cortical neuron in each case. Data from Chapman et al.[5,6]

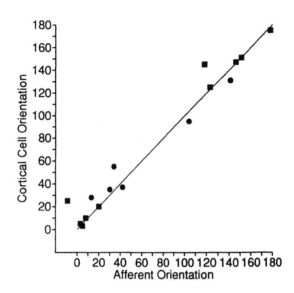

FIGURE 2 Scatter plot of the relationship between the preferred orientation of the cortical neuron (ordinate) and the principal axis of elongation of the collection of afferent receptive fields (abscissa) in 16 of 18 experiments at which the collection of afferent receptive fields was significantly elongated. A line is drawn to indicate the predicted result of a perfect match between afferent and cortical orientations. The squares plot experiments done using a "blind" procedure in which the experimenter plotting the afferent receptive fields was unaware of the results from the cortical neuron recordings. Note that the match between cortical and afferent orientations is generally good, and that there is no case of large mismatch. Data from Chapman et al.[5,6]

The findings of this experiment are exactly as would be predicted by the Hubel and Weisel[14] model. These findings would appear to be surprising, at least at first sight, if orientation selectivity were produced by purely intracortical mechanisms. This experiment thus provides strong evidence that, at least in adult animals, orientation columns are similar to the other cortical columns in that their arrangements correspond to, and may be determined by, the arrangement of their geniculocortical afferent inputs. As discussed below, however, such an arrangement of input might also be expected if orientation selectivity were produced initially by some intracortical mechanism, following which afferent terminals were allowed to refine or stabilize in a manner that depends on the correlation between cortical and afferent activity. Thus, such findings in adult animals do not answer the chicken and egg question of who organizes whom—do the afferents come first, and organize the cortical columns, or vice-versa? For that we need to turn to studies of development.

HOW DO GENICULOCORTICAL AFFERENTS COME TO BE ORGANIZED IN DEVELOPMENT? THE OCULAR DOMINANCE COLUMNS

We know little with certainty about the mechanisms of thalamocortical afferent organization in development. A large body of evidence in many systems indicates that the generation of appropriate numbers of target neurons, their migration to appropriate positions, the outgrowth of axons, their navigation along appropriate pathways, their recognition of the target structure, and their formation of at least coarsely topographic maps are all governed by molecular mechanisms of specificity, and all take place normally in the absence of neuronal activity.[11]

At least a coarse topographic specificity appears to be present in the growth of afferent arbors into the cortical plate from the sub-plate zone, within which they may have become organized during a waiting period of as long as several weeks.[42] In several respects, however, the geniculocortical afferents appear to exhibit little specificity in their initial growth into the cortical plate, and a number of results suggest that afferents organize under the influence of patterns of neural activity. In particular, the initial growth of eye-specific inputs into the visual cortex does not take place in the form of ocular dominance patches. Instead, geniculocortical afferents serving the two eyes initially make connections to the cortex in a completely overlapping pattern.[21,26,27,36] Ocular dominance patches then develop by the progressive segregation of these initially overlapping inputs. This development was most clearly revealed by the progressive changes in the transneuronal labelling pattern of visual cortex following an injection into one eye of animals at different ages, as shown in Figure 3.

Nearly all of our work on the development of geniculocortical afferent specificity has focused on the ocular dominance columns. The profound influence of neural activity on their development and plasticity makes them an excellent model system for studies of the organization of neural connections. This influence was noted long ago in the clinic, where surgical removal of cataracts that had been acquired in adulthood "miraculously" restored sight. In contrast, when patients with cataracts that had occluded vision in one eye from the time of birth were treated by similar surgery, useful vision was not restored, in spite of the fact that no serious histological damage was evident in the retina or in visual structures in the brain.[40]

Changes in the developing visual system in experimental animals can explain such clinical findings. Most neurons in the cat's visual cortex ordinarily respond to stimulation through either eye.[14] Such binocular responses in the visual cortex are unaffected by even years of monocular visual deprivation in adult animals, but as little as a few days or weeks of monocular visual experience during a sensitive period in early life leaves most cortical neurons unresponsive to the eye whose vision had been occluded.[17,50] In young monocularly deprived animals, the two eyes were entirely normal, and neurons driven by the deprived eye in the lateral geniculate

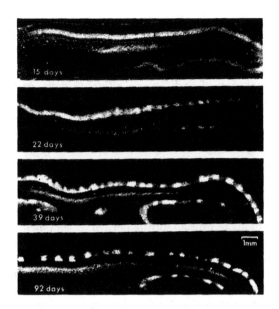

FIGURE 3 Development of ocular dominance in the cat. Progressive changes in the pattern of transneuronal labelling of the projection from one eye to the visual cortex of the cat. The ipsilateral eye of each kitten was injected with [^{3}H]-proline approximately one week before perfusion at the ages indicated. Sections of the visual cortex were exposed for autoradiography and were photographed in dark field, making the labelled regions bright. Note that labelling is uniform along the major input layer of cortex at 15 days, indicating that afferents serving the labelled eye spread over the entire tangential extent of the layer and were completely intermingled with those serving the other eye. Following this time, labelling becomes increasingly patchy, as the afferent terminals serving the two eyes segregate. Data from experiments of LeVay, Stryker, and Shatz.[26]

nucleus, which is the major source of input to visual cortex, appeared to be nearly normal.[51] Thus, neonatal monocular visual deprivation produces a change in the visual cortex, where inputs from the two eyes first have the opportunity to interact on single neurons, rather than at some more peripheral stage of the visual system.

Binocular visual deprivation for a similar period in early life produces no ill effects, suggesting that the changes produced by unilateral visual deprivation are due to a competitive interaction between the geniculocortical afferents serving the two eyes rather than merely to disuse of the occluded eye's afferents.[52] This conclusion is reinforced by failure of monocular deprivation to produce changes either in the most peripheral portion of the visual field, which is viewed through only one eye, or in a region of LGN and visual cortex in which input from the seeing eye was experimentally removed.[10,45] The changes of binocular connections in the developing visual cortex also do not depend on light deprivation, since effects similar to those of monocular occlusion were produced when the image of one eye was merely blurred.[50] Instead they appear to be due entirely to alterations of the spatial and temporal patterns of neural discharge in geniculocortical afferents. Perhaps the most striking finding is that equal amounts of neural activity presented asynchronously to the two eyes, by occlusion of each eye on alternate days, or by surgically or optically misaligning images in the two eyes, cause the partial segregation of visual responses into ocular dominance columns to become nearly absolute.[16,49] In such a cortex each cortical column contains cells driven exclusively by one eye or the other. This phenomenon strongly suggested that the relative timing of neural activity in the two eyes played an important role in preserving some binocular connections in development.

The sensitive period for these deprivation effects is at its height at the time at which the geniculocortical afferent terminals are rearranging from their initial projection pattern of complete overlap during normal development. Thus, the plasticity produced by monocular deprivation may represent merely the outcome of the normal developmental process in the presence of abnormal patterns of activity.

The preceding experiments demonstrate that alterations in visual experience can alter the course of geniculocortical afferent segregation and cause ocular dominance columns to form abnormally. Ocular dominance columns do not form at all, however, when neural activity is blocked. Stryker and Harris[47] stopped all neural activity in the two eyes by repeatedly injecting tetrodotoxin (TTX), the

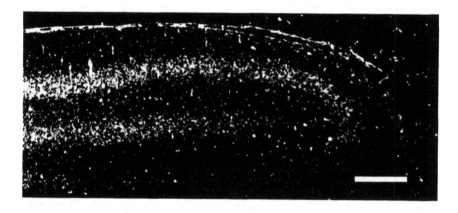

FIGURE 4 Transneuronal labelling pattern produced as described for Figure 3 in the visual cortex of a cat subjected to bilateral retinal activity blockade beginning at 14 days of age and continuing until the time of perfusion at 45 days of age. Note that unlike normal cats illustrated in Figure 3, the labelling pattern is uniform, indicating complete overlap between afferents serving the two eyes. Scale = 1 mm. Unpublished data from experiments of Stryker and Harris.[47]

voltage-sensitive sodium channel blocker, during the period in which ocular dominancecolumns normally develop. This treatment also dramatically reduced neural activity in LGN and visual cortex. The effect of the treatment was to cause geniculocortical afferents to remain in their infantile state of complete overlap. Figure 4 shows the uniform transneuronal labelling pattern following injection of [^{3}H]-proline into one eye of a 45-day-old kitten in which neural activity in both eyes had been blocked continuously beginning at 14 days of age, prior to the time that ocular dominance columns begin to segregate in normal development. Compare this labelling pattern to the clearly patchy, segregated pattern observed in the normal 39-day-old kitten shown in the next to last panel of Figure 3. This failure of eye-specific segregation was apparent physiologically as well, in that nearly all neurons in the cortex were driven well through both eyes, in contrast to the situation in normal animals in which many neurons are strongly dominated or driven exclusively by one eye or the other. These experiments suggest that the normal developmental rearrangement of geniculocortical synaptic connections to form ocular dominance columns requires neural activity. Since ocular dominance columns form, to a considerable extent, *in utero* in the monkey,[8,27,36] and in cats reared with bilateral lid suture or in total darkness, it appears that the maintained activity of retinal ganglion cells in darkness is sufficient for segregation, and that visually driven activity is not required.

Is there information in the pattern of maintained activity of retinal ganglion and geniculate cells in darkness? In adult cats, neighboring ganglion cells of the same center type tend to fire together over time periods of a millisecond to a few tens of milliseconds, and this correlation of activity decreases with increasing distance across the retina.[31] Activity correlated over longer time scales is also present.[28,39] Even before the retinal circuitry has developed *in utero*, ganglion cells have rhythmic activity and the activities of neighboring neurons may be correlated.[29] Such correlated activity within one eye, and its absence between the two eyes, could be the source of the information used by the developing visual system to distinguish the afferents serving one eye from those serving the other.

Stent[46] and Changeux and Danchin[3] proposed mechanisms to account for the effects of visual deprivation during early life. These mechanisms were formally similar to the rule described by Hebb,[12] which postulates that synapses are strengthened to the extent that the activities of pre- and postsynaptic neurons are correlated and that synapses are weakened otherwise. A Hebb rule for the adjustment of geniculocortical synaptic strengths would be expected to allow the geniculocortical afferents serving each eye to remain together in normal development, since their correlated activities would allow them to cooperate in activating the cortical cells to which they provided input. The absence of correlation between activity in the two eyes would not allow cooperative activation of cortical cells, and would therefore cause the two eyes' afferents to segregate from one another. It was suggested that such a rule could also explain the effects of early monocular and binocular visual deprivation, alternating monocular occlusion and experimental strabismus, and the effects of binocular activity blockade.

A MATHEMATICAL MODEL OF OCULAR DOMINANCE COLUMN DEVELOPMENT

By explaining, at a qualitative level, how a simple neural mechanism could produce precise patterns of connections in development, the Hebb rule was tremendously appealing. Quantitatively, however, it was not clear whether such an explanation would work with realistic elements. It was also not clear what degree or extent of correlated activity in the retina or LGN was necessary for such an explanation to work, what pattern of initial connections was compatible with such an explanation, and what was the role of intracortical interconnections in the process. Finally, a genuine model of development should allow one to predict the widths of the ocular dominance columns from the input parameters.

Miller, Keller, and Stryker[33] constructed and analyzed a mathematical model of the development of ocular dominance columns capable of addressing quantitative questions. The model, illustrated in Figure 5, incorporates a minimal set of features consistent with the experiments above. First, there are two sets of afferents in the model, corresponding to the two eyes or to the layers of the LGN that serve the two eyes, and these afferents initially make widespread overlapping connections, some of which become ineffective or are removed in development. The extent of synapses between these afferent arbors and the cortical cells is described by the "arbor function" A in the model. Second, correlated activity among afferents serving one eye, and the absence of correlation between the eyes, plays an important role. The correlation functions between the afferents serving the left eye is described by the CLL function in the model, and that between the two eyes by the CLR function. Third, postsynaptic activity in the cortex is communicated via intracortical synaptic connections. These pathways by which cortical cells influence one another's activity are described by the I (corticortical "interaction") function in the model. Finally the change in strength S of each synaptic connection between the afferents and the cortical cells was hypothesized to change by a Hebb rule, as described in the differential equation at the bottom of Figure 5, and the model was carried forward in time from its initial state.

This model was studied mathematically by linear stability analysis, and the evolution of its neural connections was simulated in the computer. The model robustly reproduces many of the biological phenomena described above. Figure 6 shows the similarity between real ocular dominance columns on the left and those produced by the model. Ocular dominance columns formed with a characteristic spacing in the presence of activity, and the model reproduced the known effects of monocular deprivation on column size and spacing. Receptive fields refined during development, and afferent arbors broke up into patches resembling those observed anatomically. All of these similarities between the model and biological development indicate that a simple rule for synaptic plasticity in a system with initial connectivity like that of the developing visual cortex can at least in principle account for the rich structure observed biologically.

A new insight obtained from the mathematical analysis was that the spacing of the ocular dominance columns was determined by the corticocortical interaction I function if that function selected a spacing small enough to contain the initial afferent arbor A. If the corticocortical interaction function I selected for a spacing that was too large, then the spacing would be constrained by the maximum that could be sustained by the arbor function A. A sufficient spread of the correlation function C^{LL} was important for allowing monocular cortical neurons to develop at all, but beyond that its role was purely permissive, and it played no role in setting the spacing of ocular dominance patches.

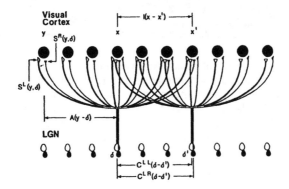

$$\partial_t S^L(x,\delta,t) = \lambda A(x-\delta)\sum_{y,\beta} I(x-y)\left[C^{LL}(\delta-\beta)S^L(y,\beta,t)+C^{LR}(\delta-\beta)S^R(y,\beta,t)\right]-\text{DECAY}$$

$$\partial_t S^R(x,\delta,t) = \lambda A(x-\delta)\sum_{y,\beta} I(x-y)\left[C^{RR}(\delta-\beta)S^R(y,\beta,t)+C^{RL}(\delta-\beta)S^L(y,\beta,t)\right]-\text{DECAY}$$

FIGURE 5   Cartoon illustrating elements of the model of Miller et al.[33] (1) Afferents from the lateral geniculate nucleus ("LGN" in figure) project to the visual cortex. Afferents (open and filled ellipses) serving each of the two eyes make equivalent initial projections to the cortex. Synaptic interconnections among cortical cells (filled circles) may be either excitatory or inhibitory. (2) The afferents project to all cortical cells in a compact region, making a terminal arborization; the strength of the connection between a cortical point y and a geniculate point δ is given by the arbor function $A(y-\delta)$, which is zero outside the arbor radius. (3) The degree of correlation in firing among incoming afferents from retinotopic positions δ and δ' is represented by the correlation functions $C^{LL}(\delta-\delta')$, $C^{RR}(\delta-\delta')$ (not illustrated), and $C^{LR}(\delta-\delta')$ gives the correlation between a left eye afferent from δ and a right eye afferent from δ', etc. (4) Each synapse has a physiological strength, which varies with time during development. This is illustrated by the functions $S^L(y,\delta)$ and $S^R(y,\delta)$. (5) Finally, there is some influence of activity at a cortical point x' on the strength of synapses at a cortical point x. This spread of influence, as a function of distance, is summarized in the corticocortical interaction function $I(x-x')$, which may be both excitatory and inhibitory at different distances. See text for discussion.

(a)   (b)

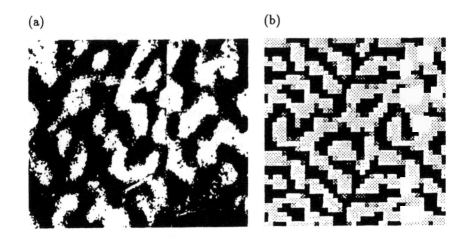

FIGURE 6   (a) Tangential view of ocular dominance columns labelled in flat-mounted section through layer IV of cat visual cortex by transneuronal transport of wheat-germ agglutinin conjugated to horseradish peroxidase injected into one eye. Data from Anderson et al.[1] (b) Simulated ocular dominance columns from model of Miller et al.[33] shown in a similar view. Note similarity between the tangential arrangements of ocular dominance columns in the experimental observations and model.

An important feature of the model is that each of the parameters can be, and has been to a limited extent, measured experimentally. The correlation functions C^{LL} and C^{LR} may be measured by straightforward cross-correlation studies in the LGN. To date, such data is available in quantity only for the retina in adult animals, but there is no technical barrier to obtaining it from the LGN in kittens at the ages at which ocular dominance columns begin to form. The arbor function A may be measured from anatomical reconstructions of geniculocortical afferent arbors like the one illustrated in LeVay and Stryker[25] or others labelled using more modern techniques. An exact measurement of A would also take the spread of the dendrites of postsynaptic cortical cells into account as well; while this would be straightforward to do, it may well be that those dendrites are quite short at the relevant time in development. The corticocortical interaction function I may be measured by experiments like those of Hess, Negishi, and Creutzfeldt,[13] which plot the effect, as a function of distance, of pharmacologically exciting a group of distant neurons on the visual responses of a local neuron. Miller and Stryker[34] and Miller[32] discuss the rather good agreement between the column spacings observed experimentally and the current experimental estimates of the values of the model parameters.

The mathematical model was formulated to incorporate a generic Hebb-rule synapse, in which the correlation between the electrical activities of the presynaptic terminal and the postsynaptic cortical cell-controlled changes in synaptic efficacy. The agreement between the predictions of the model and the results of a wide variety of experiments on the normal and experimentally perturbed development of ocular dominance columns led us to ask whether such a mechanism was the only one that could work so satisfactorily. Could other proposed mechanisms of plasticity be excluded on the grounds that they would necessarily fail to reproduce the biological phenomena of interest? Miller's analysis showed that all of the biologically plausible mechanisms of synaptic plasticity that we knew had been proposed in this system could be described in the same mathematical framework that we had used to analyze the Hebb synapse model. As one extreme example, the model was applied to a hypothetical mechanism in which the afferent terminals interact with one another through diffusible tropic or trophic substances and the postsynaptic cells just sit there like potatoes.[33] If one assumes such a non-Hebbian mechanism, the activity of postsynaptic cells plays no role whatever in synaptic plasticity. None of the mathematical behavior differed as our model was applied to different biological mechanisms; what did differ was the biological interpretation of the model parameters. For example, in the original model, the corticocortical interaction function I represented the net synaptic interaction among cortical cells as a function of their separation. In the mathematically identical presynaptic trophic substance model, the I function represented the release, diffusion, degradation, and uptake of the hypothetical trophic factor or factors. In either case, the mathematical model tells us what the spatial extent of the net interactions between synapses on different cells had to be in order to produce ocular dominance columns of the experimentally observed spacing. But it does not tell us the biological mechanism by which such interactions are effected.

The *quantitative* answer given by the mathematical model is, however, enormously helpful to the biologist engaged in the search for the answer to the normal *qualitative* question of what mechanism of synaptic plasticity is responsible for some feature of development, for the model allows one to use measurements of, for example, the diffusion constant of a hypothetical trophic substance, to completely exclude it as an explanation for the phenomenon. If one puts forward a number of alternative hypotheses about the biological mechanisms involved in ocular dominance column formation, one can then measure the real values of the biological features that correspond to the model parameters under the assumptions implicit in each mechanism. If a proposed mechanism is not operative, it is unlikely that the measured values would agree with the ones required by the model except by chance, and if they do not agree, the mechanism simply cannot be the correct one. It is thus true that, with appropriate choices of values for the parameters, very many biological mechanisms could produce the same behavior in the model, and the eager modeler could produce computer pictures that beautifully mimic real development under the assumption of whichever mechanism he chose. The model does not tell us the answer—it is no substitute for doing biology. But the model does tell us what to measure to see whether a proposed mechanism is consistent with the phenomenon it hopes to explain.

For the reasons above, biologists and modelers should not be satisfied by computer pictures that merely resemble real development. Given the different backgrounds of most modelers and most biologists, such pictures may constitute the only language common to the two camps, and the model certainly should be capable of producing such pictures; but more is needed. For productive interaction with biology, it is necessary in addition that the elements of a model correspond in some fairly direct way to the elements of the biological system, and the model parameters must, at least in principle, be susceptible to straightforward measurement from the biological system. Only then can one test the model, make quantitative predictions, and refine it to account for further features of the biology.

DOES CORTICAL ACTIVITY PLAY A ROLE IN THE FORMATION OF OCULAR DOMINANCE COLUMNS?

Despite the success of the Hebb synapse model in reproducing the development of ocular dominance columns, we have seen above the need to gather more evidence that such a mechanism is actually operative in normal development. Meaningful or behaviorally significant vision was found not to be necessary for ocular dominance plasticity,[4] consistent with a Hebb synapse explanation of development, in which the statistics of neural activity are sufficient to account for ocular dominance plasticity. By introducing controlled patterns of activity into the two optic nerves using electrical stimulation, Stryker[47] showed that ocular dominance columns did not form when activity in the two eyes was simultaneous, but that an equal amount of activity delivered alternately to the two eyes did allow ocular dominance segregation. These experiments were consistent with the Hebb synapse prediction that development and plasticity was controlled by the timing of neural activity. Another simple prediction of a Hebb synapse mechanism is that the neural activity relevant to ocular dominance development and plasticity is the activity in the cortex involving cortical cells and their geniculocortical afferent inputs. The experiments described above had all interfered with activity at earlier stages of the visual system as well. Reiter, Waitzman, and Stryker[47] tested this prediction by infusing TTX into a region of cortex to block the discharge of cortical cells and their geniculo-cortical afferent terminals and then instituting a period of monocular deprivation to study whether the deprivation would cause a shift in ocular dominance. Consistent with the prediction of a Hebb synapse model, the cortical activity blockade completely prevented plasticity.

The experiments above have located the neural activity relevant to ocular dominance plasticity in the cortex, but they do not reveal whether it is the presynaptic activity, the postsynaptic activity, or both (as postulated by a Hebb synapse mechanism) that are important. In many earlier experiments in which the responses of cortical cells were perturbed by substances infused into the cortex, ocular dominance plasticity was disrupted to a greater or lesser extent, consistent with a role for postsynaptic activity, but these substances appeared likely to have presynaptic effects on afferent terminals as well. Reiter and Stryker[37] selectively blocked postsynaptic activity during a period of monocular deprivation by infusing the GABA$_A$

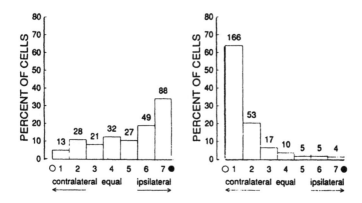

FIGURE 7 Ocular dominance histograms compiled from single unit responses in area 17. Monocular eyelid closures were performed in different animals either ipsilateral or contralateral to the muscimol-infused hemisphere. Results are plotted as if the eyelid sutured was always ipsilateral to the treated hemisphere, and control recordings were obtained from unaffected regions of that same hemisphere. That is, responses from single cells were plotted such that an ocular dominance of 1 indicates a cell driven exclusively by the open eye; 4, a cell driven equally by the two eyes; and 7, a cell driven exclusively by the occluded eye. All animals received intracortical muscimol infusions for 8–10 days and were monocularly deprived for 5–7 days beginning three days after the onset of the muscimol infusion. The direction of ocular dominance shift in favor of the occluded eye within the area blocked by the muscimol infusion was the same in all animals tested and opposite to the direction of shift in control areas outside of the blockade. (a) Ocular dominance distribution of 258 visually responsive units recorded within the region of cortex in which discharges had been blocked by muscimol infusion during the period of monocular deprivation. (b) Ocular dominance distribution of 260 visually responsive units recorded in regions of cortex outside of the muscimol-induced blockade, including contralateral control hemisphere as well as unblocked areas anterior to the blocked region. Data from experiments of Reiter and Stryker.[37]

agonist muscimol into the visual cortex, a substance that, as described in the first section of this paper, powerfully inhibits all cortical cells but appears to have no effect on activation of or synaptic release from afferent terminals. In the region of cortex in which cortical discharge was completely inhibited, not only was the normal synaptic plasticity prevented, but inputs from the less active, occluded eye came to dominate over those from the more active, non-deprived eye, as shown in Figure 7. This form of synaptic plasticity in the reverse direction from normal is exactly what would be predicted by the Hebb synapse model. In this case, the activity of the less active, occluded eye is better correlated with that of the inhibited postsynaptic cortical cells than is the activity of the more active open eye. In adjacent regions of cortex, in which the cells were able to respond to their inputs, ocular dominance plasticity in the normal direction was evident. Since identical patterns of afferent activity produced opposite types of plasticity, depending on whether the postsynaptic cortical cells were able to respond to their inputs, the role of the postsynaptic cells is clearly crucial, as postulated by the Hebb synapse model. This experiment further suggests that a process coupled to postsynaptic membrane voltage or conductance controls the direction of synaptic plasticity, which favors more active inputs when the postsynaptic cell can be depolarized by them but less active inputs when the postsynaptic cell is inhibited. Finally, local responses rather than action potentials in the postsynaptic cells appear to be responsible at least for the reverse plasticity, since it took place while spikes were blocked in the cortical cells.

FUTURE DIRECTIONS

We have seen above that ocular dominance columns may result from activity-dependent reorganization of the geniculocortical afferents serving the left and right eyes using a mechanism involving Hebb-rule synapses. To date, this hypothesis has passed all of the tests, several of them quite formidable, to which it has been subjected. Although we will not have conclusive evidence for this hypothesis until we understand the mechanisms of plasticity in more detail than we do at present, we may nevertheless design experiments that would be difficult to reconcile with other mechanisms. For example, if we could create ocular dominance columns with dramatically abnormal spacing by selectively perturbing corticocortical synaptic interactions, we would have compelling evidence that the columns did emerge by a self-organizing process like the one we have modeled.

Directly interfering, at a molecular level, with proposed mechanisms of plasticity is another approach of potentially great value. But it is an approach that is also fraught with difficulty because some of the molecular machinery responsible for plasticity may contribute significant neuronal activity as well, and alterations of activity may affect plasticity by any of a variety of mechanisms. Recent work on blocking the NMDA receptor (which currently appears to be the most likely molecular candidate for the correlation detector required by a Hebb synapse) has illustrated these difficulties (compare the interpretations of Kleinschmidt, Bear, and Singer,[22] with those of Miller, Chapman, and Stryker,[33]). Ideally we should eventually have the molecular tools to interfere with plasticity at a stage beyond that involved with transmembrane currents; such tools could alter plasticity without affecting neural activity.

Could explanations along similar lines also account for aspects of afferent and cortical organization other than the ocular dominance columns? What about the refinement of topographic maps, the formation in some species of ON/OFF patches, and the organization of ocular dominance columns? Malsburg,[30] Fraser,[9] and others have modeled map refinement and binocular segregation in the retinotectal system using similar principles, and the experiments of Constantine-Paton and her associates (reviewed in Constantine-Paton, Kline, and Debski[7]) provide strong support for the operation of similar principles in that system. In Miller's recent models and in preparation,[33] interactions among cortical simple cells and between them and their *ON*-center and *OFF*-center inputs can give rise to an arrangement of orientation columns and a partial segregation of *ON* and *OFF* afferents similar to the arrangement observed experimentally. Finally, Shatz and Stryker[44] have presented evidence that neural activity is essential for the segregation of the two eyes' inputs to form different layers of the lateral geniculate nucleus.

Each of these phenomena appears individually to be explicable in terms of an activity-dependent reorganization of afferents. One might well imagine that the different patterns of activity present at different stages of development could use similar mechanisms of synaptic plasticity to give rise to these different forms of organization, each building on organization generated previously or at prior stages of the visual system. At early stages, and certainly *in utero* and before eye-opening, only intrinsic activity would be present, but at later times, visual inputs might stimulate the patterns of activity relevant to normal development and plasticity. Such mechanisms might be widespread in the developing central nervous system. To devise and test a comprehensive explanation of how all of these sorts of reorganization take place in the same population of afferents is a challenge for the future.

ACKNOWLEDGMENT

The studies reviewed here were carried out principally by my students, Barbara Chapman, Ken Miller, Holger Reiter, and Kathleen Zahs, and constituted parts of their Ph.D. dissertations. I am grateful to them for their creativity and energy.

REFERENCES

1. Anderson, P.A., J. Olavarria, and R. C. Van Sluyters. "The Overall Pattern of Ocular Dominance Bands in Cat Visual Cortex." *J. Neurosci.* **8** (1988): 2183–2200.
2. Benevento, L. A., O. D. Creutzfeldt, and U. Kuhnt. "Significance of Intracortical Inhibition in the Visual Cortex." *Nature* **238** (1972): 124–126.
3. Changeux, J. P., and A. Danchin. "Selective Stabilization of Developing Synapses as a Mechanism for the Specification of Neuronal Networks." *Nature* **264** (1976): 705–712.
4. Chapman, B., M. D. Jacobson, H. Reiter, and M. P. Stryker. "Ocular Dominance Shift in Kitten Visual Cortex Caused by Imbalance in Retinal Electrical Activity." *Nature* **324** (1986): 154–156.

5. Chapman, B., K. R. Zahs, and M. P. Stryker. "Receptive Field of Geniculo-cortical Afferents Tend to be Aligned Along Preferred Orientation of Cortical Cells." *Soc. Neurosci. Abstr.* **15** (1989): l055.
6. Chapman, B., K. R. Zahs, and M. P. Stryker. "Relation of Cortical Cell Orientation Selectivity to Alignment of Receptive Fields of Geniculocortical Afferents that Arborize within a Single Orientation Column in Ferret Visual Cortex." *J. Neurosci.* **11**, in press.
7. Constantine-Paton, M., H. T. Cline, and E. Debski. "Patterned Activity, Synaptic Convergence, and the NMDA Receptor in Developing Visual Pathways." *Ann. Rev. Neurosci.* **13** (1990): 129–154.
8. Des Rosiers, M. H., O. Sakurada, J. Jehle, M. Shinohara, C. Kennedy, and L. Sokoloff. "Demonstration of Functional Plasticity in the Immature Striate Cortex of the Monkey by Means of the ^{14}C-Deoxyglucose Method." *Science* **200** (1978): 447–449.
9. Fraser, S. E. "Differential Adhesion Approach to the Patterning of Nerve Connections." *Dev. Biol.* **79** (1980): 117–130.
10. Guillery, R. W. "Bionocular Competition in the Control of Geniculate Cell Growth." *J. Comp. Neurol.* **144** (1972): 117–130.
11. Harris, W. A., and C. E. Holt. "Early Events in the Embryogenesis of the Vertebrate Visual System: Cellular Determination and Pathfinding." *Ann. Rev. Neurosci.* **13** (1990): 155–169.
12. Hebb, D. O. *The Organization of Behavior.* New York: Wiley, 1949.
13. Hess, R., K. Negishi, and O. Creutzfeldt. "The Horizontal Spread of Intracortical Inhibition in the Visual Cortex." *Exp. Brain Res.* **22** (1975): 415–419.
14. Hubel, D. H., and T. N. Wiesel. "Receptive Fields, Binocular Interaction and Functional Architecture in the Cat's Visual Cortex." *J. Physiol.* **160** (1962): 106–154.
15. Hubel, D. H., and T. N. Wiesel. "Shape and Arrangement of Columns in Cat's Striate Cortex." *J. Physiol.* **165** (1963): 559–568.
16. Hubel, D. H., and T. N. Wiesel. "Binocular Interaction in Striate Cortex of Kittens Reared with Artificial Squint." *J. Physiol.* **165** (1965): 1041–1059.
17. Hubel, D. H., and T. N. Wiesel. "The Period of Susceptibility to the Physiological Effects of Unilateral Eye Closure in Kittens." *J. Neurophysiol.* **206** (1970): 419–436.
18. Hubel, D. H., and T. N. Wiesel. "Luminar and Columnar Distribution of Geniculocortical Fibers in the Macaque Monkey." *J. Neurophysiol.* **146** (1972): 421–450.
19. Hubel, D. H., and T. N. Wiesel. "Sequence Regularity and Geometry of Orientation Columns in the Monkey Striate Cortex." *J. Comp. Neurol.* **158** (1974): 267–294.
20. Hubel, D. H., and T. N. Wiesel. "Uniformity of Monkey Striate Cortex: A Parallel Relationship Between Field Size, Scatter, and Magnification Factor." *J. Comp. Neurol.* **158** (1974): 295–306.

21. Hubel, D. H., T. N. Wiesel, and S. LeVay. "Plasticity of Ocular Dominance Columns in Monkey Striate Cortex." *Phil. Trans. Roy. Soc. London (B)* **278** (1974): 377–409.
22. Kleinschmidt, A., M. F. Bear, and W. Singer. "Blockade of NMDA Receptors Disrupts Experience-Dependent Modifications of Kitten Striate Cortex." *Science* **238** (1987): 355–358.
23. LeVay, S., and S. K. McConnell. "ON and OFF Layers in the Lateral Geniculate Nucleus of the Mink." *Nature* **300** (1982): 422–441.
24. LeVay, S., S. K. McConnell, and M. B. Luskin. "Functional Organization of Primary Visual Cortex in the Mink (Mustela Vision), and a Comparison with the Cat." *J. Comp. Neurol.* **257** (1987): 422–441.
25. LeVay, S., and M. P. Stryker. "The Development of Ocular Dominance Columns in the Cat." In *Aspects of Developmental Neurobiology*, edited by J. A. Ferrendelli, 83–98. Bethesda: Society for Neuroscience, 1979.
26. LeVay, S., M. P. Stryker, and C. J. Shatz. "Ocular Dominace Columns and Their Development in Layer IV of the Cat's Visual Cortex: A Quantitative Study." *J. Comp. Neurol.* **179** (1978): 223–244.
27. LeVay, S., T. N. Wiesel, and D. H. Hubel. "The Development of Ocular Dominance Columns in Normal and Visually Deprived Monkeys." *J. Comp. Neurol.* **191** (1980): 1–51.
28. Levick, W. R., and W. O. Williams. "Maintained Activity of Lateral Geniculate Neurones in Darkness." *J. Physiol.* **170** (1964): 582–597.
29. Maffei, L., and L. Galli-Resta. "Correlation in the Discharges of Neighboring Rat Retinal Ganglion Cells During Prenatal Life." *Proc. Natl. Acad. Sci. USA* **87** (1990): 2861–2864.
30. Malsburg, C. von der. "Development of Ocularity Domains and Growth Behavior of Axon Terminals." *Biol. Cybernetics* **32** (1979): 49–62.
31. Mastronarde, D. N. "Correlated Firing of Cat Retinal Gaglion Cells: I. Spontaneously Active Inputs to X and Y Cells." *J. Neurophysiol.* **49** (1983): 303–324.
32. Miller, K. D. "Correlation-Based Models of Neural Development." In *Neuroscience and Connectionist Theory*, edited by M. A. Gluck and D. E. Rumelhart, 267–353. Hillsboro, NJ: Lawrence Erlsbaum, 1983.
33. Miller, K. D. "Orientation-Selective Cells Can Emerge from a Hebbian Mechanism Through Interactions Between ON- and OFF-Center Inputs." *Soc. Neurosci. Abs.* **15** (1989): 794.
34. Miller, K. D., B. Chapman, and M. P. Stryker. "Visual Responses in Adult Cat Visual Cortex Depend on N-Methyl-D-Asparate Receptors." *Proc. Natl. Acad. Sci. USA* **856** (1989): 5183–5187.
35. Miller, K. D., J. B. Keller, and M. P. Stryker. "Ocular Dominance Column Development: Analysis and Simulation." *Science* **245** (1989): 605–615.
36. Rakic, P. "Prenatal Development of the Visual System in the Rhesus Monkey." *Phil. Trans. Roy. Soc. London (B)* **278** (1977): 245–260.
37. Reiter, H. O., and M. P. Stryker. "Neural Plasticity Without Postsynaptic Action Potentials: Less-Active Inputs Become Dominant When Kitten Visual

Cortical Cells are Pharmacologically Inhibited." *Proc. Nat. Acad. Sci. USA* **85** (1988): 3623–3627.

38. Reiter, H. O., D. M. Waitzman, and M. P. Stryker. "Cortical Activity Blockade Prevents Ocular Dominance Plasticity in the Kitten Visual Cortex." *Exp. Brain Res.* **65** (1986): 182–188.

39. Rodieck, R. W., and P. S. Smith. "Slow Dark Discharge Rhythms of Cat Retinal Ganglion Cells." *J. Neurophysiol.* **29** (1966): 942–953.

40. Senden, Marjus von. *Space and Slight: The Perception of Space and Shape in the Congenitally Blind Before and After the Operation*, translation by P. Heath. Glencoe, IL: Free Press, 1960.

41. Shatz, C. J., S. Lindstrom, and T. N. Wiesel. "The Distribution of Afferents Representing the Right and Left Eyes in the Cat's Visual Cortex." *Brain Res.* **131** (1977): 103–116.

42. Shatz, C. J., and M. B. Luskin. "The Relationship Between the Geniculocortical Afferents and Their Cortical Target Cells During Development of the Cat's Primary Visual Cortex." *J. Neurosci.* **6** (1986): 3655–3668.

43. Shatz, C. J., and M. P. Stryker. "Ocular Dominance in Layer IV of the Cat's Visual Cortex and the Effects of Monocular Deprivation." *J. Physiol.* **281** (1978): 267–283.

44. Shatz, C. J., and M. P. Stryker. "Prenatal Tetrodotoxin Infusion Blocks Segregation of Retinogeniculate Afferents." *Science* **242** (1988): 87–89.

45. Sherman, S. M., R. W. Guillery, J. H. Kaas, and K. J. Sanderson. "Behavioral, Electrophysiological, and Morphological Studies of Binocular Competition in the Development of the Geniculocortical Pathways of Cats." *J. Comp. Neurol.* **158** (1974): 1–18.

46. Stent, G. S. "A Physiological Mechanism of Hebb's Postulate of Learning." *Proc. Nat. Acad. Sci. USA* **70** (1973): 997–1001.

47. Stryker, M. P., and W. Harris. "Binocular Impulse Blockade Prevents the Formation of Ocular Dominance Columns in Cat Visual Cortex." *J. Neurosci.* **6** (1986): 2117–2133.

48. Talbot, S. A., and W. H. Marshall. "Physiological Studies on Neural Mechanisms of Visual Localization and Discrimination." *Amer. J. Ophthalmol.* **24** (1941): 1255–1263.

49. Van Sluyters, R. C., and F. B. Levitt. "Experimental Strabismus in the Kitten." *J. Neurophysiol.* **43** (1980): 686–699.

50. Wiesel, T. N., and D. H. Hubel. "Single-Cell Responses in Striate Cortex of Kittens Deprived of Vision in One Eye." *J. Neurophysiol.* **26** (1963): 1003–1017.

51. Wiesel, T. N., and D. H. Hubel. "Effects of Visual Deprivation on Morphology and Physiology of Cells in the Cat's Lateral Geniculate Body." *J. Neurophysiol.* **26** (1963): 978–993.

52. Wiesel, T. N., and D. H. Hubel. "Comparison of the Effects of Unilatral and Bilateral Eye Closure on Cortical Unit Responses in Kittens." *J. Neurophysiol.* **28** (1965): 1029–1040.

53. Wiesel, T. N., D. H. Hubel, and D. M. K. Lam. "Autoradiographic Demonstration of Ocular Dominance Columns in the Monkey Striate Cortex by Means of Transneuronal Transport." *Brain Res.* **79** (1974): 273–279.
54. Zahs, K. R., and M. P. Stryker. "Segregation of ON and OFF Afferents to Ferret Visual Cortex." *J. Neurophysiol.* **59** (1988): 1410–1429.

K. Obermayer, H. Ritter,* and K. Schulten
Beckman Institute and Department of Physics, University of Illinois at Urbana-Champaign, Urbana, IL 61801 U.S.A.; *new address: Technische Fakultät, Universität Bielefeld, W-4800 Bielefeld, Germany

A Model for the Development of the Spatial Structure of Retinotopic Maps and Orientation Columns

1. INTRODUCTION

Topographic maps are one of the major architectural features of the cortex. Their main characteristic is a macroscopic, spatial pattern of the tuning properties of (at least a subset of) the neurons within some cortical region. It is generally accepted that for the most part these patterns of tuning properties are not established genetically, but instead evolve during ontogenesis in a self-organizing process[1,12] and show in many cases a considerable degree of adaptability even in adult life.[26,27]

A considerable amount of research has been directed at unraveling the precise nature of the underlying adaptive processes. As a result we now know that the formation of topographic maps does not rely on a single mechanism. Instead, several mechanisms seem to be at work supporting each other and, thereby, providing a significant degree of redundancy.

There is evidence that both chemical markers and neural activity are important factors for correct map formation. The former factor seems to predominantly guide the formation of the coarse structure of the connectivity between afferents and

Principles of Organization in Organisms,
SFI Studies in the Sciences of Complexity, Proc. Vol. XIII,
Eds. J. Mittenthal & A. Baskin, Addison-Wesley, 1992   **141**

target neurons that is required for a topographic map,[4,18] while the latter factor is largely responsible for a subsequent refinement of the topographic structure, which will not occur if electrical activity is blocked.[38,42,46]

In the present contribution we will report results on the capability and the limits of a model for topographic map formation that regards the activity-based mechanisms only. Although an ultimate, detailed description of map formation will have to include further mechanisms, there are several reasons to study simplified models based on a single mechanism only.

First, there is the question if, although being the result of the complex interaction of several different mechanisms, the process of map formation admits a simple, phenomenological description, based on a small set of simple rules. In the case of a positive answer, these rules would provide a useful, intermediate-level description of an important type of structure formation in organisms.

Second, even if such a description should turn out to be unfeasible and a composite model involving several mechanisms to be the only viable alternative, a necessary first step towards such a model is a thorough understanding of the properties of candidates for its constituent parts. A model considering only activity-based processes could, therefore, be considered as a "module" for the larger process (cf. J. Mittenthal et al.[28]).

Third, if part of the reason for having several mechanisms at work is the implementation of functional redundancy, the effects of different mechanisms should be functionally similar. Then the study of a single mechanism can already provide a very representative case.

Finally, restricted aspects of the spatial structure of a map may be entirely due to a single process and, therefore, may be explained by a simpler model. For instance, activity-based mechanisms may be the main force behind the formation of a spatial organization of more subtle tuning properties, such as orientation selectivity. Such properties require more intricate connectivity patterns that are harder to realize by diffusion controlled processes, by fiber-fiber interactions, or by adhesion-type processes than those required for more elementary tuning properties, such as receptive field position.

The previous issues provided the motivation for exploration into the capabilities, consequences, and limits of a simple, activity-based model for explaining the observed structures of topographic maps in various cortical areas.

In previous reports we have investigated simple maps of purely spatial stimulus features, such as the somatotopic map in somatosensory cortex, where a distorted image of the two-dimensional body surface is formed.[30,32] In the present contribution we want to discuss to what extent the same model can account for the structure of more complex maps, such as the hierarchical feature maps in the visual cortex, where the stimulus features "retinal position" and "orientation" are represented simultaneously.

Previous approaches to this question were either limited by considering only a subset of all features,[17,21,24,47,49,52] or they were based on replacing the stimuli by prespecified parameter vectors.[6,47] Both restrictions are overcome in the present approach (cf. Obermayer et al.[33]); input patterns are directly defined in terms of retinal intensity distributions, and cell specificity for both retinal position and orientation develops simultaneously and despite the fact that the encoding of position and orientation is only implicit in the patterns and overlapping on the same shared set of receptor locations. The merit of the present approach is that it provides a basis for assessing the degree of approximation entailed by more simplified models, such as those used in earlier approaches, or a low-dimensional version of the presented model itself. With regard to the latter, a comparison of simulation results shows a very satisfactory agreement in the structure of maps generated through prespecified parameter vectors and the retinal intensity distributions. Therefore, we can conclude that results of a recent mathematical analysis of the simplified version of the model[39,40] can be used to relate statistical properties of the input patterns with the hierarchical structure of feature maps in the visual cortex.

2. TOPOGRAPHIC MAPS AND DIMENSION REDUCTION

Frequently the spatial order of the tuning properties of neurons within a topographic map reflects just the spatial origin of the afferent signals, such as the location of a tactile stimulus on the skin. However, there are also maps where further stimulus properties, such as orientation in the case of visual stimuli, become expressed in a spatial pattern. Even in a case where there are more than two stimulus features, the spatial pattern can be characterized by a smooth variation of stimulus features at most points in the map. Such distribution of stimulus specificity represents a dimension-reducing projection from a higher-dimensional "feature space" of stimulus properties onto the two-dimensional cortical sheet. A mathematical consequence of any such projection would be either to suppress some of the additional feature dimensions, or to exhibit discontinuities which appear as interspersed "jumps" of tuning properties in the topographic map.

In the visual cortex both alternatives seem to be realized. Retinotopic location defines the "primary" feature that is mapped smoothly across visual cortex. The spatial variation of additional "secondary" features, such as orientation preference, specificity, and ocular dominance is smooth only within small local domains, which are separated by boundaries where discontinuous jumps occur. In the case of orientation, there are also "foci" where orientation selectivity is zero; i.e., in these locations the stimulus feature "orientation" is suppressed under the projection.

Color Plate 1 shows an example of such an orientation pattern, in this case obtained from the cortex in area 17 of a macaque monkey, using the voltage-sensitive dye technique.[3] Color is used to encode different values of the stimulus feature "orientation" across the cortical sheet. Besides exhibiting the discontinuities described above, the map also exhibits a hierarchical structure; the variation of the stimulus feature "orientation" is highly repetitive across the primary map of retinal location, giving rise to a large number of small maps, each containing a complete topographic representation of the full range of the secondary feature "orientation." Although the precise details of the spatial structure vary between species, a similar structure can be found in several higher mammals, e.g., macaque monkey,[3,11,13,14,50] cat,[22,48] tree shrew,[45] and ferret.[37]

It, therefore, seems that topographic maps reflect some of the information processing strategies realized in the brain to match abstract feature spaces onto the spatial structure of its "parallel hardware."[29] This makes it an intriguing question also from the viewpoint of neural computation to consider to what extent such structures can be understood in terms of simple pattern-formation processes within the underlying substrate, or, stated in a more abstract way, can be understood in terms of rules for elementary local adaptation steps.

In the following, we will consider a specific model that is able to generate many of the features of these maps on the basis of a simple adaptive process, the so-called "self-organizing feature map" algorithm. However, before we turn to the formulation of the model (section 5), the next two sections provide a discussion of a low-dimensional "caricature." The value of this simplified version lies in the fact that it is possible to derive analytical results relating the structure of the map with statistical properties of the stimuli. Simulating the full model then leads to the satisfying result that the maps obtained with both models are similar to some extent, thus showing that the simpler model is well justified in many situations.

3. "LOW-DIMENSIONAL" MODEL

The self-organizing feature map algorithm (SFM) is a neural network algorithm for the formation of a topographic representation of a set of patterns (given as vectors in some "input" or "feature"-space V) on a discrete set, A, of points ("cells") endowed with a topology.

Figure 1(a) shows a schematic drawing of the model. The cells $\vec{r} \in A$ are arranged on a two-dimensional lattice (the network layer) to match the topology of the cortical layer containing the feature map. The cells should not be identified with single neurons, but rather with groups of neurons or with small patches of tissue, where neurons with common response properties are located. Periodic boundary conditions were chosen for the network layer as well as for the position coordinates in feature space.

The receptive field properties of the model cells are characterized by a feature vector $\vec{w}_{\vec{r}}$, which is associated with each cell $\vec{r}$ and whose components $(\vec{w}_{\vec{r}})_k$ are interpreted as receptive field properties of the cell. The feature vectors, $\vec{w}_{\vec{r}}$, as a function of the cell locations, $\vec{r}$, describe the spatial distribution of selectivity of cells over the cortical layer.

In the following we will consider these receptive field properties: position of the receptive field centers in visual space $(x_{\vec{r}}, y_{\vec{r}})$, orientation preference $(\phi_{\vec{r}})$, and a quantity, which qualitatively can be interpreted as orientation specificity $(q_{\vec{r}})$. If $q_{\vec{r}}$ is zero, then the cells respond in an unspecific manner. The larger $q_{\vec{r}}$ becomes,

Color Plates

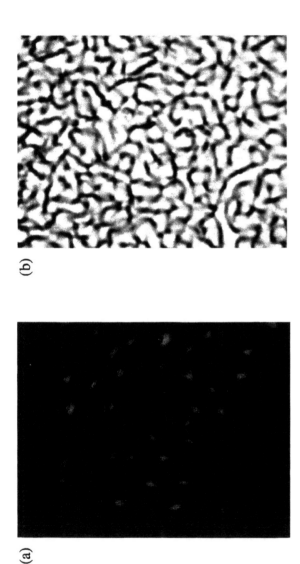

(a)

(b)

PLATE 1 Color-coded representation of orientation preference in monkey striate cortex. (a) Distribution of orientation preference obtained from a 6mm × 8mm patch of the striate cortex of an adult macaque (*macaca nemestrina*) using voltage sensitive dyes (see Blasdel and Salama, *Nature* **321** (1986): 579–586). The region is located near the border to area 18 close to midline. Orientation preferences are coded by color, a complete cycle of from 0° to 180° being mapped onto a color circle. (b) Magnitude of the gradient of the orientation values presented in (a) to visualize the distribution of continuous regions and sudden breaks (fractures). Black regions denote a high magnitude of the orientation gradient. The data was provided by Dr. G. Blasdel.

PLATE 2 Spatial distribution of orientation preferences (color) and orientation specificity (brightness) for a (model)-orientation column system generated by the SFM algorithm (low-dimensional version) using an isotropic neighborhood function. Each image pixel corresponds to one cell in the network layer. Orientation preference is indicated by color (light blue→green→orange→purple→blue correspond to angles of $0° \rightarrow 45° \rightarrow 90° \rightarrow 135° \rightarrow 180°$ relative to the vertical axis). The values of orientation specificity are normalized and indicated by brightness (black: zero, bright: one). The network layer contains 65,536 cells (arranged on a 256×256 square lattice with periodic boundary conditions). The initial state of the network was "topographic." The parameters of the simulation were: $q_{pat} = 12$, $\sigma_h = 5$, $\epsilon = 0.01$ and 10^7 iterations.

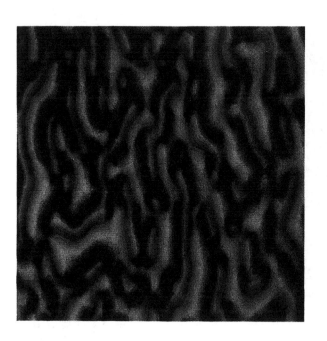

PLATE 3 Spatial distribution of orientation preferences and specificity for a (model)-orientation column system generated by the SFM algorithm (low-dimensional version) using all anisotropic neighborhood function. Color and brightness indicate orientation preference and selectivity, as described in color plate 2. The network layer contains 65,536 cells (arranged on a 256 × 256 square lattice with periodic boundary conditions). The initial state of the network was "topographic." The parameters of the simulation were: $q_{pat} = 14$, $\sigma_{h1} = 6$, $\sigma_{h2} = 4$, $\epsilon = 0.01$ and 10^7 iterations.

PLATE 4 Spatial distribution of orientation preferences and specificity for a (model)-orientation column system generated by the SFM algorithm (low-dimensional version) using an isotropic neighborhood function. The pattern distribution was biased towards the orientation values indicated by blue. Color and brightness indicate orientation preference and selectivity as described in color plate 2. The network layer contains 16,384 cells (arranged on a 128×128 square lattice with periodic boundary conditions). The parameters of the simulation were: $q_{pat} = 6$, $\sigma_{h1} = 5$, $\epsilon = 0.01$ and 10^7 iterations. The pattern probability distribution $P(\vec{v})$ was characterized by $a = 0.7$, $b = 0.3$.

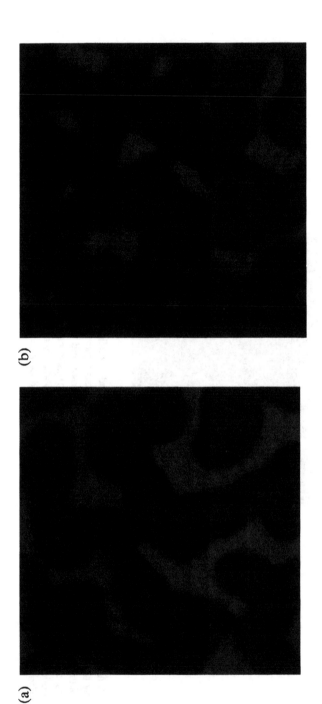

(a)

(b)

PLATE 5 Spatial distribution of orientation preferences and specificity for a (model)–orientation column system generated by the SFM algorithm (low-dimensional version) using an isotropic neighborhood function. The pattern distribution was restricted to two orthogonal orientations ("blue" and "orange"). Color and brightness indicate orientation preference and selectivity as described in color plate 2. The network layer contains 16,384 cells (arranged on a 128 × 128 square lattice with periodic boundary conditions). The parameters of the simulation were: $q_{pat} = 6$, $\sigma_{h1} = 5$, $\epsilon = 0.01$ and 10^7 iterations. The pattern probability distribution $P(\vec{v})$ was characterized by (a) $a = 0.5$, $b = 0.5$ and (b) $a = 0.65$, $b = 0.35$.

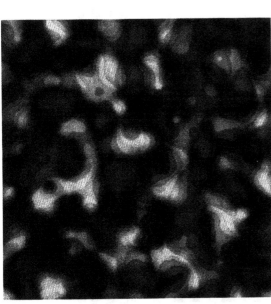

(b)

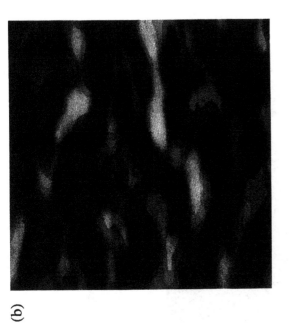

(a)

PLATE 6 Spatial distribution of orientation preferences and specificity for a (model)-orientation column system generated by the SFM algorithm (high-dimensional version) for an (a) isotropic and (b) anisotropic neighborhood function after 30,000 adaptive steps. The network layer contains 65,536 cells (arranged on a 256 × 256 square lattice with periodic boundary conditions), and the input layer contains 900 randomly distributed cells. The initial connection strengths were chosen randomly from the interval $[0, 1]$ and normalized to unity. The parameters of the simulation were (a) $\epsilon_i = 0.09, \epsilon_f = 0.02, \sigma_1 = 0.23$ and $\sigma_2 = 0.09$ and $\sigma_h(0) = 240, \sigma_h(15,000) = 60, \sigma_h(30,000) = 2$ and (b) $\sigma_{h1}(0) = 300, \sigma_h(15,000) = 80, \sigma_h(30,000) = 3$. The arguments denote the number of adaptation steps. Each image pixel corresponds to one cell in the network layer. The angle ϕ_F of the major principal axis of the receptive fields to some reference axis ("preferred orientation") is indicated by color as described in color plate 2. The ratio of the width of the receptive field along its major and minor principal axis ("orientation specificity") is indicated by brightness (dark: unspecific; bright: specific).

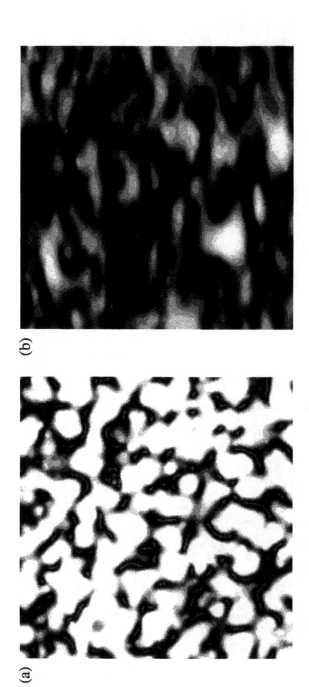

PLATE 7 Correlation between orientation specificity and the magnitude of the orientation gradient for the simulation results shown in (a) color plate 6(a) and (b) color plate 6(b). Orientation specificity is indicated by brightness (dark: unspecific; bright: specific). Regions of high magnitude of the orientation gradient ("fractures" and "foci") are marked red. Note the excellent correlation between the "fractures," "foci," and the regions of low orientation specificity.

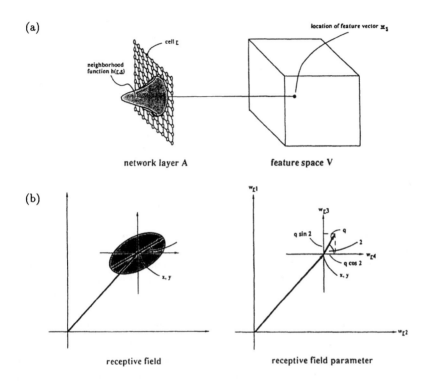

FIGURE 1 The "low-dimensional" network model. (a) The model consists of a set of cells which are arranged on a square lattice (network layer A). The receptive field properties of each cell are described by a feature vector, which is an element of a four-dimensional feature space V. The neighborhood function $h(\vec{r}, \vec{s})$ implements cooperative learning between neighboring cells during map formation (see Eqs. 4 and 5). (b) Construction of the components of the feature vector (left) for an elongated receptive field (right). In order to visualize the feature vector, its components can be rearranged (left), such that each receptive field (right) is characterized by a two-dimensional "orientation"-vector located at the position $x_{\vec{r}}$, $y_{\vec{r}}$ of the center of the receptive field. Preferred orientation $\phi_{\vec{r}}$ is encoded by half the angle of the vector to the $(w_{\vec{r}})_3$ axis, and orientation specificity by its length $q_{\vec{r}}$.

the sharper the cells are tuned to the preferred orientation. These properties are combined to a four-dimensional feature vector:

$$\vec{w}_{\vec{r}} = (x_{\vec{r}},\ y_{\vec{r}},\ q_{\vec{r}}\cos(2\phi_{\vec{r}}),\ q_{\vec{r}}\sin(2\phi_{\vec{r}})).\tag{1}$$

Figure 1(b) illustrates how this feature vector is constructed for the case of an elongated receptive field. The position of the receptive field is given by the coordinates $(x_{\vec{r}},\ y_{\vec{r}})$ of its centroid, orientation preference by the orientation $\phi_{\vec{r}}$ of the receptive fields major axis, and orientation specificity by its elongation. Follow-ing Swindale[47] orientation preference and specificity are treated in their cartesian forms, $q_{\vec{r}}\cos(2\phi_{\vec{r}})$ and $q_{\vec{r}}\sin(2\phi_{\vec{r}})$, where a factor 2 is introduced because preferred orientation is π-periodic. Therefore, each four-dimensional feature vector (1) may be visualized as a two-dimensional orientation vector (Figure 1(b), right) located at the position $x_{\vec{r}}$, $y_{\vec{r}}$ of the receptive field centroid. Its direction and its length describe preferred orientation and orientation specificity, respectively.

The input to the network layer consists of localized and oriented stimuli. They are described by a feature vector also, which is of the same dimensionality as $\vec{w}_{\vec{r}}$. It is constructed from a stimulus intensity distribution in the same way as the feature vector is constructed from the receptive fields. Its components,

$$\vec{v} = (x,\ y,\ q\cos(2\phi),\ q\sin(2\phi))\,, \tag{2}$$

correspond to the stimulus properties position in the visual field $(x,\ y)$, orientation ϕ of its major principal axis, and elongation q.

The next step is to introduce a measure of similarity between stimuli as well as between the feature vectors describing the receptive field properties of the cells. There arises a problem, which all models using preprocessed feature vectors (e.g., the models of Durbin and Mitchinson[6] and Kohonen[17]) have to face, since there is an ambiguity in comparing magnitudes of change of different stimulus features. In the following we will define a distance measure $d(\vec{v}, \vec{w}_{\vec{r}})$ via the Euclidean norm. But note that this definition is arbitrary, and can be justified only by comparing the model predictions with experiments. The quantities q and $q_{\vec{r}}$ now become parame-ters which can only be interpreted qualitatively, and whose values ultimately have to be determined by fitting the resulting maps to experimental data. In section 5, therefore, we will introduce a version of the SFM model, where the preprocessed fea-ture vectors are replaced by the stimuli themselves, and a distance between stimuli can be defined using the natural metric of the vector space spanned by the activity patterns.

A set of stimuli (described by the stimulus vectors vectors $\vec{v} \in V$) drives the model to adapt its feature vectors $\vec{w}_{\vec{r}}$ such that (i) the variation of its components $(w_{\vec{r}k})$ with cell position $\vec{r}$ is as continous as possible, and (ii) the resulting vectors $\vec{w}_{\vec{r}}$ span the range over which the feature combinations vary in the set of input patterns. These are two complementary requirements: (i) favors "uniformity," while (ii) demands "diversity" for the feature vectors $w_{\vec{r}}$.

To satisfy (i) and (ii), we use the SFM algorithm.[15,16,17] Map formation pro-ceeds by an iterative sequence of steps: At the beginning of each step, a feature vector $\vec{v}$ is chosen at random according to a probability distribution $P(\vec{v})$. Then the cell $\vec{s}$, whose stimulus vector $\vec{w}_{\vec{s}}$ is closest to $\vec{v}$, is selected:

$$\vec{s} = \min_{\vec{r}} d(\vec{v}, \vec{w}_{\vec{r}}) \tag{3}$$

and the attached feature vectors are updated according to the SFM update law.

$$\vec{w}_{\vec{r}}(t+1) = \vec{w}_{\vec{r}}(t) + \varepsilon(t)h(\vec{r}, \vec{s}, t)(\vec{v} - \vec{w}_{\vec{r}}(t)) \tag{4}$$

where $h(\vec{r}, \vec{s}, t)$ is given by:

$$h(\vec{r}, \vec{s}, t) = \exp\left(-\frac{(r_1 - s_1)^2}{\sigma_{h1}^2(t)} - \frac{(r_2 - s_2)^2}{\sigma_{h2}^2(t)}\right). \tag{5}$$

Equation (3) introduces an element of competition, whereas $h(\vec{r}, \vec{s}, t)$ in Eq. (4) ensures local cooperativity of cells in the learning process. The width of the Gaussian function (5) may be interpreted as the length scale below which the response properties of neurons are kept correlated within a cortical layer.[17,24] Its shape is parametrized by the quantities σ_{h1} and σ_{h2}.

4. RESULTS FOR THE LOW-DIMENSIONAL MODEL

Too little is known about the statistical properties of the afferent patterns driving map formation, such that neither a set of patterns nor their probability of occurance could be derived from experimental data. We, therefore, decided for an "unbiased" probability distribution $P(\vec{v})$ and drew the patterns with equal probability from the surface of a cylinder in the four-dimensional feature space spanned by the coordinates x, y, $q\cos(2\phi)$, and $q\sin(2\phi)$. $P(\vec{v})$ is given by

$$P(\vec{v}) = \begin{cases} \mathcal{N}\delta(q - q_{pat}) & \text{if } x, y \in [0, d]; \\ 0 & \text{else.} \end{cases} \tag{6}$$

Figure 2 illustrates this pattern distribution. In order to allow a graphical representation, the second "position" dimension in feature space and the second dimension of the network layer is omitted. The goal of the map formation process is to represent the cylindrical pattern manifold in four-dimensional space by the (discrete) cells of the two-dimensional network layer, under the constraint that neighboring cells in the network represent neighboring regions on the pattern manifold.

Let's define the state of the network layer by the set $\{\vec{w}_r\}$ of all feature vectors $\vec{w}_r$. Then the expectation value $E(\Delta\vec{w}_{\vec{r}}|\{\vec{w}_{\vec{r}}\})$ for a change $\Delta\vec{w}_{\vec{r}}$ of the feature vectors $\vec{w}_{\vec{r}}$ during one iteration of the SFM-algorithm is given by:

$$E(\Delta\vec{w}_{\vec{r}}|\{\vec{w}_{\vec{r}}\}) = \int (\vec{v} - \vec{w}_{\vec{s}(\vec{v})})P(\vec{v})d\vec{v}. \tag{7}$$

The quantity $\vec{s}(\vec{v})$ denotes the cell $\vec{s}$ whose label fulfills Eq. (3), i.e., whose feature vector is next to the input vector. For any given state $\{\vec{w}_{\vec{s}}\}$, the sets

$$\Gamma(\vec{s}) = \{\vec{v} \mid |\vec{v} - \vec{w}_{\vec{s}}| \leq |\vec{v} - \vec{w}_{\vec{r}}| \ \forall \vec{r}\} \tag{8}$$

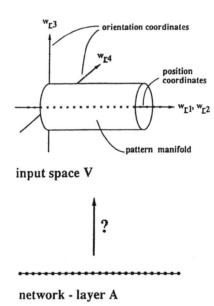

input space V

?

network - layer A

FIGURE 2 Dimension-reducing mapping of a cylindrical manifold in a four-dimensional feature space V onto a two-dimensional network layer A. For the sake of illustration, one "position" feature dimension and one network dimension is suppressed. The location of the feature vectors $w_{\vec{r}}$ in feature space is indicated by small crosses for the regime $q_{pat} < q_{thres}$ (see text).

render a tesselation of the pattern manifold. Each set $\Gamma(\vec{s})$ is called the *tesselation cell* of unit $\vec{s}$, and it corresponds to the set of all $\vec{v}$ for which the unit $\vec{s}$ is the winner unit according to Eq. (3). Equation (7) then can be split into a sum of integrals, each performed over one tesselation cell:

$$E(\Delta \vec{w}_{\vec{r}} | \{\vec{w}_{\vec{r}}\}) = \sum_{s} h(\vec{r}, \vec{s}) \int_{\Gamma(\vec{s})} (\vec{v} - \vec{w}_{\vec{s}}) P(\vec{v}) d\vec{v}. \tag{9}$$

For fixed number N^2 of cells in the network layer and for a fixed length d, the final map crucially depends on the variance $\int v_{3,4}^2 P(\vec{v}) dv_{3,4} = q_{pat}^2$ of the pattern distribution given by Eq. (6) along the feature dimensions describing orientation. If q_{pat} is smaller than a certain threshold q_{thres}, then the stationary states $\{\tilde{\vec{w}}_{\vec{r}}\}$ of Eqs. (3) and (4) defined by Ritter et al.[40]

$$E(\Delta \vec{w}_{\vec{r}} | \{\tilde{\vec{w}}_{\vec{r}}\}) = 0 \tag{10}$$

are given by

$$\tilde{x}_{\vec{r}} = \frac{d}{N} r_{1,2} + \alpha \text{ or } x_{\vec{r}} = \frac{d}{N}(N - r_{1,2}) + \alpha$$

$$\tilde{y}_{\vec{r}} = \frac{d}{N} r_{2,1} + \alpha \text{ or } x_{\vec{r}} = \frac{d}{N}(N - r_{2,1}) + \alpha \tag{11}$$

$$\tilde{q}_{\vec{r}} \cos 2\tilde{\phi}_{\vec{r}} = 0$$

$$\tilde{q}_{\vec{r}} \sin 2\tilde{\phi}_{\vec{r}} = 0$$

where α can assume any value between 0 and d, and d/N is the magnification factor of the mapping. A more detailed analysis shows that these stationary states are stable. For the cylindrical pattern distribution shown in Figure 2, the feature vectors of the network cells (small crosses) would, in this case, all be located on the x, y-axis in the center of the cylindrical pattern manifold. The resulting map is a topographic representation of visual space, but since $(w_{\vec{r}})_3$, $(w_{\vec{r}})_4$, and therefore $\tilde{q}_{\vec{r}}$ are zero, the feature orientation is not represented.

In order to form orientation selective cells, i.e., cells $\vec{r}$ for which the third and fourth component of $\vec{w}_{\vec{r}}$, i.e., $(w_{\vec{r}})_3$ and $(w_{\vec{r}})_4$, are different from zero, q_{pat} must exceed a threshold, q_{thres}, above which the "purely retinotopic" solution (11) of Eqs. (3) and (4) becomes unstable. This threshold can be calculated following the approach of Ritter and Schulten[39] (details will be published elsewhere) and is given by

$$q_{thres} = \sqrt{\frac{e}{2}} \frac{d}{N} \sigma_h \tag{12}$$

where $\sigma_h = \min(\sigma_{h1}, \sigma_{h2})$ and e is the Euler constant. Therefore, solution (11) becomes unstable if the range σ_h of the neighborhood function, projected back to the feature space V, falls below q_{pat}, the standard deviation of the set of patterns along the orientation feature dimension. Let $\vec{u}_{\vec{r}} = \vec{w}_{\vec{r}} - \tilde{\vec{w}}_{\vec{r}}$ denote the deviation of the feature map from its stable states given by Eq. (11). These deviations result from the stochastic nature of the adaptation process (3)–(5) and their size is proportional to the learning step ϵ. At the threshold q_{thres} a set of modes

$$\hat{\vec{u}}_{\vec{k}} = \frac{1}{N} \sum_{\vec{r}} e^{i\vec{k}\vec{r}} \vec{u}_{\vec{r}} \tag{13}$$

characterized by

$$\hat{\vec{u}}_{\vec{k}} \perp \vec{k} \tag{14}$$

and

$$\begin{aligned}|\vec{k}| &= 2/\sigma_h & \text{if } \sigma_{h1} = \sigma_{h2} \\ k_x &= \pm 2/\sigma_{h1} \\ k_y &= 0 \end{aligned} \Bigg\} \quad \text{if } \sigma_{h1} < \sigma_{h2} \tag{15}$$

becomes unstable and for $q_{pat} > q_{thres}$ orientation, selective cells form.

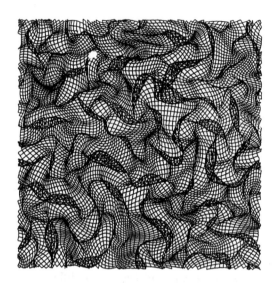

FIGURE 3 Topographic representation of visual space generated by the "low-dimensional" model above threshold after 10^7 iterations using an isotropic neighborhood function ($\sigma_{h1} = \sigma_{h2} = \sigma_h$). The network layer contains $65,536$ cells (arranged on a 256×256 square lattice with periodic boundary conditions). The initial state of the network was "topographic," i.e., $(W\vec{r})_3 \equiv (W\vec{r})_4 \equiv 0$. The parameters of the simulation were: $q_{pat} = 12$, $\sigma_h = 5$, and $\epsilon = 0.01$. The position $(x_{\vec{r}},\ y_{\vec{r}})$ of every second cell $\vec{r}$ in visual space is indicated by a dot. Points $(x_{\vec{r}''}, y_{\vec{r}''})$, $(x_{\vec{r}'}, y_{\vec{r}'})$ belonging to cells $\vec{r}$, $\vec{r}'$ that are nearest neighbors in the network layer A are connected by lines. The value of x increases to the right, the value of y to the top.

Color Plate 2 shows the final distribution of orientation preference $\phi_{\vec{r}}$ (color) and selectivity $q_{\vec{r}}$ (saturation) along the network layer for a map generated in the regime above threshold with an isotropic neighborhood function ($\sigma_{h1} = \sigma_{h2}$). The presence of only very small black areas indicate that almost all cells have become orientation specific. The cells form domains of continuously changing orientation, in which iso-orientation regions are organized as parallel slabs. The slabs start and end at vortices containing orientationally unspecific cells (dark spots). Orientation preference changes by $180°$ in a clockwise or counterclockwise fashion around these foci. Neighboring domains have similar slab orientations but, on a larger length scale, the directions of the domains are distributed isotropically. The multiple representations of a complete cycle of preferred orientation indicate that orientation plays the role of a "secondary" feature.

Figure 3 illustrates the topographic representation of visual space generated by the SFM algorithm.[15,16,17] The diagram presents the locations $(x_{\vec{r}},\ y_{\vec{r}})$ of receptive field centers in visual space for all cells in the network layer. Receptive field centers of neighboring cells were connected by lines. An ideal topographic projection of visual space to the network layer would give rise to a square lattice with equal mesh size in Figure 3, since the receptive field centroids of neighboring units are equally spaced in this case. The overall preservation of the lattice topology in Figure 3 and the absence of any major distortions demonstrat that, indeed, "position" varies in a topographic fashion across the cell layer on a large length scale. On a small length scale, however, numerous distortions are visible. These distortions are the result of the constraint to map a two dimensional feature space on a two-dimensional cortical surface, such that the response properties of the cells vary smoothly over the cortical surface. Since visual space is mapped only once along the network layer, "position" plays the role of the primary stimulus variable.

Figure 4 shows the absolute values of the two-dimensional complex Fourier transform:

$$\hat{w}_{\vec{k}} = \sum_{\vec{r}} e^{i\vec{k}\vec{r}} q_{\vec{r}}(\cos(2\phi_{\vec{r}}) + i\sin(2\phi_{\vec{r}})) \tag{16}$$

of the orientation coordinates in the $\vec{k}$-plane. The origin of the $\vec{k}$-plane is located in the center of the image. Each pixel corresponds to one mode and its brightness indicates the square amplitude of the mode. The fact that, for an isotropic neighborhood function $h(\vec{r}, \vec{s})$, a continous ring of modes becomes unstable is reflected in the Fourier spectrum. The orientation map is characterized by wave vectors from a ring-shaped region in the Fourier domain.

Figure 5 shows the two-point autocorrelation functions

$$S_{ij}(\vec{s}) = \langle (w_{\vec{r}-\vec{s}})_i (w_{\vec{r}})_j \rangle_{\vec{r}} \quad (i, j \in 3, 4). \tag{17}$$

Figure 5(a) shows the autocorrelation function S_{33} as a function of the distance $\vec{s}$ between cells in the network layer. The origin of the $\vec{s}$-plane is located in the center

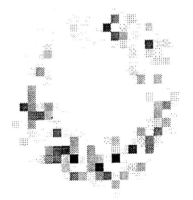

FIGURE 4   Discrete Fourier spectrum of the orientation map shown in Color Plate 2. The origin of the $\vec{k}$-plane is in the center of the ring; k_x increases to the right, k_y to the top. Each pixel corresponds to one mode of the discrete spectrum. The absolute values of the amplitudes are normalized to one, and their value is indicated by brightness (white: zero, black: one).

(a)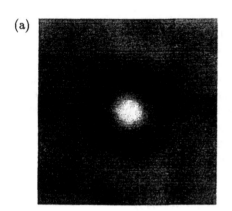

(b)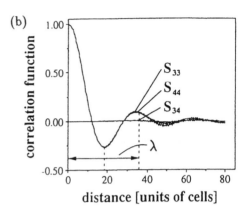

distance [units of cells]

FIGURE 5   Two-point correlation functions $S_{ij} = \langle (w_{\vec{r}-\vec{s}})_i (w_{\vec{r}})_j \rangle^{\vec{r}}$ of the "orientation" components $(w_{\vec{r}})_3$ and $(w_{\vec{r}})_4$ for the orientation map shown in Color Plate 2. (a) Autocorrelation function S_{33}. The origin of the $\vec{s}$-plane (see text) is in the center; s_x increases to the right, s_y to the top. Each pixel corresponds to a distance between cells in the network layer. Zero correlation is indicated by medium gray, negative correlations by dark, and positive correlations by bright values. (b) Correlation functions S_{33}, S_{44}, and S_{34} as a function of the absolute value $|\vec{s}|$ of cell distance. Since the correlation functions are rotationally symmetric, they were averaged over all directions. The slight anisotropy visible in Figure 5(a) is the result of statistical fluctuations. H will vanish in the limit of infinitely large maps. The arrow indicates the wavelength λ associated with the Fourier modes located on the ring in Figure 4.

of the image and the brightness indicates a positive (white), zero (medium gray), or negative (black) value of S_{33}. The autocorrelation function depends only on the absolute value $|\vec{s}|$. Figure 5(b) presents the correlation functions S_{33}, S_{44}, and S_{34} as a function of the absolute value $|\vec{s}|$ of cell distance averaged over all directions. The arrow in Figure 5(b) indicates the wavelength λ associated with the wavenumber $|\vec{k}|$ of modes from the ring-shaped region in the Fourier domain. The correlation function has a Mexican hat form: neighboring cells prefer correlated orientations, with the degree of correlation decreasing with distance. At a separation of $\vec{s} = \lambda/2$, the preferred orientation between cells is anti-correlated; i.e., preferred orientations are more likely to be orthogonal. If cells are separated by a distance larger than λ, the preferred orientations are uncorrelated, as one would expect from an isotropic arrangement of iso-orientation slabs over the cortical surface. The cross-correlation function $S_{34}(\vec{s})$ was found to be very small.

Color Plate 3 and Figures 6-8 show the the distribution of orientation preference and selectivity (Color Plate 3), the "position" map (Figure 6), the Fourier spectrum (Figure 7), and the correlation functions (Figure 8) for maps generated with an anisotropic neighborhood function ($\sigma_{h1} \neq \sigma_{h2}$) in the regime q_{pat} above threshold q_{thres}. In contrast to the isotropic case, most of the iso-orientation slabs and distortions in the "position" map run parallel to the major axis of the neighborhood function $h(\vec{r}, \vec{s})$. The fact that for an anisotropic neighborhood function, only two modes become unstable at the threshold is reflected in the Fourier spectrum (Figure 7) of the maps. The spectrum consists of only two groups of modes which differ in the sign of $\vec{k}$. Consequently, the two-point autocorrelation function (shown in Figure 8) has no longer rotational symmetry. It essentially consists of a central positive "bar" accompanied by two negative "lobes." The autocorrelation as a function of cell distance $\vec{s}$ orthogonal to the major axis of $h(\vec{r}, \vec{s})$ has pronounced Mexican hat character, which is no longer true for values of $\vec{s}$ parallel to the major axis. The cross-correlation S_{34} between both "orientation" coordinates is, again, small.

The corresponding orientation map is shown in Color Plate 2.

What happens if the probability distribution $P(\vec{v})$ is biased towards a certain subset of patterns? Such a situation is artificially created in "deprivation experiments," where an animal is exposed predominantly to patterns of a single orientation. For this purpose, a simulation using only stimuli representing a narrow

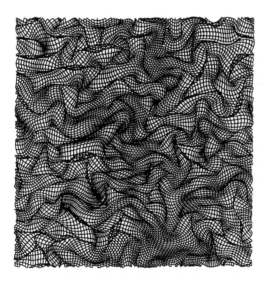

FIGURE 6 Topographic representation of visual space generated by the "low-dimensional" model above threshold after 10^7 iterations using an anisotropic neighborhood function ($\sigma_{h1}/\sigma_{h2} = 1.5$). The parameters of the simulation were: $q_{pat} = 14$, $\sigma_{h1} = 6$, and $\epsilon = 0.01$. Visualization of data is described in the legend of Figure 3.

FIGURE 7 Discrete fourier spectrum of the orientation map shown in Color Plate 3. For explanation see legend of Figure 4.

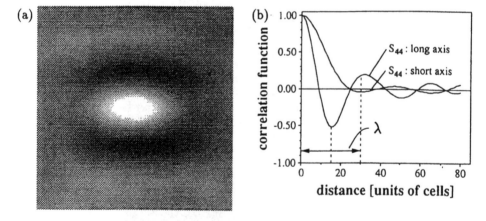

FIGURE 8 Two-dimensional correlation functions of the components $(w_{\vec{r}})_3$ and $(w_{\vec{r}})_4$ for the orientation map shown in Color plate 3. (a) Autocorrelation function S_{33}. Display of data is described in Figure 5(a). (b) Autocorrelation function S_{44} as a function of cell distance parallel (s_x) and orthogonal (s_y) to the major axis of the neighborhood function $h(\vec{r}, \vec{s})$. The arrow indicates the wavelength λ associated with the Fourier modes located in the center of the group of modes with high energy shown in Figure 7.

distribution of orientations has been carried out. The "biased" pattern distribution $P(\vec{v})$ yields a final map that exhibits larger regions of cells selective to the predominant orientation.

Color Plate 4 shows the final distribution of orientation preference and selectivity for a map generated by a probability distribution $P(\vec{v})$ "biased" to pattern orientations between $\phi = 0$ and $\phi = \pi/4$, given by

$$P(\vec{v}) = \begin{cases} a\mathcal{N}\delta(q - q_{pat}) & \phi \in [0, \frac{1}{4}\pi], x, y \in [0, d]; \\ b\mathcal{N}\delta(q - q_{pat}) & \phi \in [\frac{1}{4}\pi, \pi], x, y \in [0, d]; \\ 0 & \text{else}; \end{cases} \tag{18}$$

with $a = 0.7$, $b = 0.3$. Regions of cells selective to the orientation of the predominant patterns form a network, which contains "blobs" of cells with other orientation preferences. The larger the value of a compared to b, the larger the regions responding to the predominant orientation. The map still has a hierarchical structure, but the number of foci in the map is found to be smaller than in the "non-deprived" case.

If stimulus patterns are restricted to two orthogonal orientations only, given by a probability distribution:

$$P(\vec{v}) = \begin{cases} a\mathcal{N} & \phi = 0, \quad x, y \in [0, d]; \\ b\mathcal{N} & \phi = \pi/2, \quad x, y \in [0, d]; \\ 0 & \text{else }; \end{cases} \tag{19}$$

the final distribution of orientation preference (Color Plate 5) shows alternating stripes or blobs, which are separated by cells unspecific for orientation (dark bands). The width of the stripes depends on the relative probabilities of occurrence: the higher the probability for a pattern the larger the corresponding region on the map. The resulting maps very much resemble the spatial structure of an ocular dominance column system. This is not surprising, because restricting the stimulus distribution in feature space to two orthogonal orientations is equivalent to reducing the pattern manifold shown in Figure 2 to the intersection of the (hyper-)cylinder with any (hyper-)plane, which includes the $w_{\neq 1,2}$-(hyper-)axis, hence reducing the dimensionality of the feature space from four to three.[1] The new pattern manifold consists of two lines[2] on opposite sides of the $w_{\neq 1,2}$-(hyper-)axis. If the feature, which is now coded by the distance between these planes, is re-interpreted as the degree of correlation in the activity between both eyes[10] (large values stand for small correlation in activity), and the corresponding receptive field property as the amount a particular neuron is driven by a particular eye, the same model describes the development of retinotopic maps and ocular dominance stripes.[34,35]

---

[1] The axis to the right in Figure 2 represents two coordinates.

[2] Actually two planes, since the $w_{\neq 1,2}$-axis represents a plane in the four-dimensional feature space.

5. "HIGH-DIMENSIONAL" MODEL

The use of feature vectors in models of development entails several problems: (i) the relevant features describing the presented patterns as well as the receptive field properties of the cells must be specified in advance, (ii) an appropriate coding of features, like the representation of preferred orientation in cartesian coordinates, has to be found, (iii) there is an ambiguity in comparing differences in magnitude between features of different quality, and (iv) the models cannot be extended to an arbitrary number of features in a natural way.

To overcome these difficulties, the preprocessed feature vectors were replaced by activity patterns themselves. The activity values v_k at certain points k ("input-cells," "receptors") of an "input-layer" (Figure 9) were combined to pattern vectors $\vec{v} = (v_1, v_2, \ldots, v_d)$ of a high-dimensional input space V of dimension d. The components $(w_{\vec{r}})_k$ of the vectors $\vec{w}_{\vec{r}}$ describing the cell's response properties can now be interpreted as weights or connection strengths of afferents connecting input cells k with network cells $\vec{r}$.

The weight values are iteratively refined using a variant of the feature map algorithm described above. For each step, a pattern vector $\vec{v}$ is selected at random from a given ensemble described by $P(\vec{v})$ and the weights are updated according to:

$$(w_{\vec{r}})_k(t+1) = \frac{((w_{\vec{r}})_k(t) + \varepsilon(t)h(\vec{r}, \vec{s}, t) \cdot v_k)}{\sqrt{\sum_i ((w_{\vec{r}})_i(t) + \varepsilon(t)h(\vec{r}, \vec{s}, t) \cdot v_i)^2}} \tag{20}$$

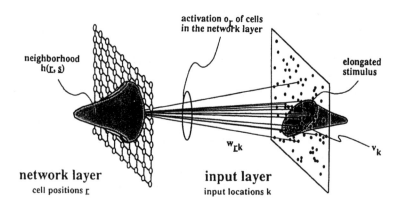

FIGURE 9 The "high-dimensional" network model. The model consists of an input layer containing randomly distributed input cells and a network layer consisting of feature selective cells arranged in a square lattice. Each cell in the network layer is connected to every cell in the input layer by modifiable links. The stimulus is given by the spatial activity distribution over the input layer. The neighborhood function $h(\vec{r}, \vec{s})$ describes the cooperation between neighboring cells during learning.

where the index $\vec{s}$ denotes the network cell whose input $o_{\vec{s}} = \vec{w}_{\vec{s}} \cdot \vec{v}$ is a global maximum. The neighborhood function $h(\vec{r}, \vec{s}, t)$ is again given by Eq. (5).

---

6. RESULTS FOR THE "HIGH-DIMENSIONAL" MODEL

Very little is known about the statistical properties of the activity profile, which drives the formation of feature maps in the cortical layer. Before birth (or eye opening), activity is generated spontaneously in subcortical structures and seems to be spatially correlated.[9,23] After eye-opening, stimulus-induced activity as well as spontanous activity is present.[25] Unfortunately, there is insufficient experimental data to derive a set $\{\vec{v}\}$ of patterns together with their probability $P(\vec{v})$ of occurrence. Therefore, we constructed sets of patterns on the basis of three general assumptions, namely, (i) that pattern activity should be locally correlated (artificially introducing globally synchronized activity patterns seems to hinder the refinement of topographic projections[5]), (ii) that only features to be represented in the map are encoded in the presented patterns, and (iii) that each feature combination is generated with equal probability (under "undeprived" conditions). A "minimal" choice is a set of patterns with elliptic shape, defined by the three random numbers x, y (stimulus center) and $\alpha \in [0, 180^\circ]$ (stimulus orientation) chosen from a uniform distribution. Cell activities v_k were then defined by:

$$
v_k = \exp\left[-\frac{1}{\sigma_1^2}((x_k - x)\cos\alpha - (y_k - y)\sin\alpha)^2 \right.
$$
$$
\left. -\frac{1}{\sigma_2^2}((x_k - x)\sin\alpha + (y_k - y)\cos\alpha)^2 \right],
$$

(21)

where (x_k, y_k) denotes the spatial location of cell k in the input layer. Parameters σ_1 and σ_2 are fixed and specify the length of the major and the minor axis of the ellipsoidal intensity distribution.

Color Plate 6 shows the final distribution of orientation preference (color) and selectivity (saturation) along the network layer for cells initially unspecific both to stimulus orientation and position. Most of the cells have developed elongated receptive fields, whose aspect ratio approximately matches the aspect ratio of the stimuli presented to the receptor surface. These cells became selective for stimuli, whose orientation matches the orientation of their receptive fields. The high degree of orientation specificity is indicated by the bright colors in Color Plate 6. The map on the left (Color Plate 6(a)) was generated using an isotropic, the map on the right (Color Plate 6(b)) using an anisotropic neighborhood function. The spatial structure of orientation preference is more irregular compared to maps generated by the "low-dimensional" model (Color Plates 2 and 3 and Figures 3 and 6). While in some regions the slab-like structure dominates, there are other regions displaying more patchy domains.

A gradient filter applied to the orientation values (Figure 10) reveals that vortices with high orientation changes are often connected by bands of rapid orientation shifts, where the orientation gradient can become very large ("fractures"). These areas with rapid orientation change coincide with regions containing cells that did not become tuned to stimuli of a particular orientation. These cells keep almost circular receptive fields and, consequently, respond equally well to stimuli of all orientations. These correlations between the magnitude of the orientation gradient and the orientation specificity are displayed in Color Plate 7. They result from the property of the SFM algorithm to generate maps, where receptive field properties change smoothly along the network layer.

For almost circular receptive fields, small changes in the shape of the receptive fields are sufficient to greatly change the direction of their major principal axis. Since even for such receptive fields, intracortical mechanisms are likely to greatly enhance the otherwise low orientation specificity of the associated cells,[43] regions containing this type of cells may develop into the "vortice" and the "fractures" between orientation domains.

The shape of the receptive field of orientation selective cells in the model generally reflects the shape of the presented stimuli. Cells located within the foci, however, have circular-shaped receptive fields, although circular stimuli have not been presented to the network. Their presence ensures a smooth progression of the shape of the receptive fields also between cells of orthogonal orientation preference and, therefore, contributes essentially to the spatial continuity of the cell's response properties over the cortical surface.

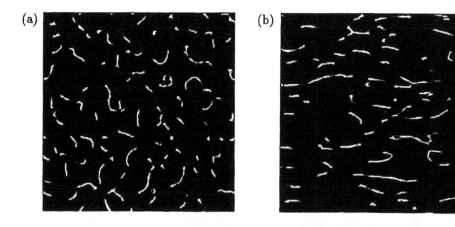

FIGURE 10   Magnitude of the gradient $|\nabla_{\vec{r}}\phi_{\vec{r}}|$ of the orientation values $\phi_{\vec{r}}$ (orientation of the major principal axis of the receptive fields) for the maps shown in Color Plate 6(a) and Color Plate 6(b). Bright and dark regions indicate areas of rapid and smooth change in orientation preference, respectively.

The final size of the receptive fields depends on the size of the presented patterns as well as on the range σ_h of the neighborhood function. In the case of non-oriented patterns ($\sigma_1 = \sigma_2 = \sigma$), the size of the cell's receptive field can be calculated analytically.[31] Let $G_{\vec{r}}$ denote the mean square radius of the receptive field:

$$G_{\vec{r}} = \frac{\sum_k (\vec{x}_k - \vec{X}_{\vec{r}})^2 w_{\vec{r}k}}{\sum_i w_{\vec{r}i}} \qquad (22)$$

where $\vec{X}_{\vec{r}}$ denotes the center of gravity of the receptive field cell $\vec{r}$. Then the final value of $G_{\vec{r}}$ is given by:

$$G_{\vec{r}} \approx \Gamma + M_{\vec{r}}^{-1} \sigma_h^2 \qquad (23)$$

where Γ is the mean square radius of the presented patterns (which is equal to $\sigma_{1,2}$ for Gaussian-shaped stimuli (21)), and $M_{\vec{r}}$ is the local magnification factor of the mapping from stimulus centroids (x, y) to neuron coordinates $\vec{r}$. Equation (23) shows that the cells develop receptive fields whose area (which is proportional to $G_{\vec{r}}$) is the sum of two terms: the first term is essentially the area of a typical stimulus ($\propto \Gamma$) and the second term is essentially the area $\propto \sigma_h^2$ in the adjustment zone in the network layer, but "projected back" (inverse magnification factor M^{-1}) onto the input sheet. The size of the cells' receptive fields, therefore, reflects the size of the region within which the response properties of the cells are kept correlated.

As in the "low-dimensional" case, the overall conservation of lattice topology and the absence of any larger discontinuities in the mapping of the stimulus dimensions "position," the x,y-coordinates, demonstrates that the position of the

(a)   (b)

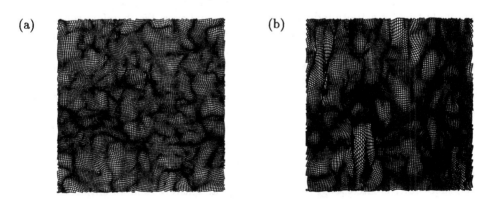

FIGURE 11 "Position" map corresponding to the orientation maps shown in Color Plate 6(a) and Color Plate 6(b). Every second cell $\vec{r}$ of the network layer is "projected" onto the center of its receptive field center in the input layer. Points belonging to cells that are nearest neighbors in the network layer are connected by lines.

receptive field centroids varies in a topographic fashion across the network layer (see Figure 11). On a small scale, however, numerous local distortions are again visible, similar to the results obtained with the more simplified low-dimensional model.

7. DISCUSSION

7.1 BIOLOGICAL RELEVANCE

The model presented above describes the formation of cortical feature maps. Biologically plausible principles (activity-dependent synaptic plasticity, correlated response properties of neighboring cells, and a competitive learning describing an activity-based developmental process) are formulated in a mathematically simple form. This process generates a representation of a given stimulus distribution based on the similarity between stimuli and on the probability of their occurence. The observed model dynamics does not necessarily mirror the actual sequence of events in the formation of cortical maps. The emphasis of this model (and of similar models[6]) lies on the role of basic principles and on an analysis of their capacity to yield the observed organization of cortical maps. We mainly focused on two aspects: (i) the degree of similarity between observed and artificially created maps, and (ii) the role of the stimulus distribution and the spatial range of the local adaptation rules on which the model is based.

Very little is known about the statistical properties of the afferent activity profile of the cortical layers containing the feature map. Spontaneous retinal activity is spatially (locally) correlated and retinal ganglion cells maintain substantial discharge in the absence of visual stimulation[25] and before birth.[9,23] The same is probably true for activity induced by natural visual cues. We therefore constructed sets of patterns with localized activity profiles, based on very general assumptions, or we restricted simulations to more abstract feature vectors. Still, different sets of patterns exist, which are compatible with these assumptions. For the given map, the SFM model may act as a tool to predict the statistical properties a set of patterns must possess in order to shape the cortical maps by an input-driven self-organizing process.

A further important issue is the time course of the map formation process. The exclusion of marker-based processes that create much of the coarse initial order in the map poses a severe limitation for the present model to account for these aspects. Although simulations show that the purely activity-based model can succesfully create topographic maps from an entirely random initial structure, and that these maps are not significantly different from maps obtained after some coarse topographic prestructuring,[3] such extrapolation of the model is likely to be biologically inadequate. In particular, the formation of correct maps from entirely random initial connectivity requires a range of the local adaptation steps that varies

---

[3]This is true as long as pattern distribution by itself leads to a hierarchical representation of features.

during the process, since, otherwise, convergence time increases drastically and only partially ordered maps are generated. At present, it is unclear to what extent these questions are relevant to biological systems. Some results on these issues will be reported elsewhere.[7]

7.2 SPATIAL STRUCTURE OF THE MAPS

Cortical maps in the primary visual cortex of various species typically exhibit[3,11,13,14,22,48,50]: a hierarchical mapping of position (primary stimulus variable) and orientation (secondary variable), a patch- or slab-like shape of iso-orientation domains, foci, and "fractures". Maps exhibiting these features can be generated by the SFM algorithm for a broad range of model parameters and independently of the initial state. The model predicts some constraints regarding the set of stimulus patterns. If the formation of the visual map employs externally or internally generated activity patterns driving a process governed by Eqs. (3)–(5) or (20), and if the configuration space of the network is not confined by some prestructuring, then the eccentricity of the patterns would have to be within a certain range, large enough to exceed the threshold required to express orientation in the map, but still sufficiently small not to destroy the observed feature hierarchy.

The constraint of dimension reduction leads to a trade-off between the representation of orientation and the degree of retinotopy. One consequence of this trade-off is visible in the position map: the simultaneous representation of orientation above threshold causes multiple distortions in the position map. These distortions are separated by $\lambda/2$ (see Figure 5(b)) from each other on the average and are linked to the spatial pattern of orientation patches. At the fractions and singularities, the rate of change of retinotopic location with cortical distance is, on the average, reduced compared to its change with cortical distance within orientation patches. This gives rise to an anticorrelation, . . .the resolution of the cortical representation between the features retinotopic location and orientation. The resulting modulation of the position map should be experimentally detectable in a high-resolution study of receptive field centroids.

As a first step towards a more quantitative comparison, spectra and correlation functions have been calculated for the SFM maps. The two-dimensional Fourier spectrum of the orientation values for the isotropic neighborhood function exhibit discrete modes located on a ring around the origin in the $\vec{k}$-plane; for the anisotropic neighborhood function, it consists of two localized groups of modes, which differ in the sign of $\vec{k}$. These findings seem to be in agreement with data from area 17 at the macaque[53] and area 18 in the cat.[48] The spatial autocorrelation functions are short ranged and have Mexican hat shape similar to the orientation-autocorrelation function found for the orientation column system of cat's area 18.[48] Note that there is almost no correlation between the two orientation coordinates $(w_{\vec{r}})_3$ and $(w_{\vec{r}})_4$. The finite range of the spatial autocorrelation functions reflects the finite length of the neighborhood function at the threshold q_{pat} spectum and correlation functions indicate that a phenomenological description of the structure of the orientation column system in terms of a filter process acting on spatial random noise[41] might capture essential features of the system.

7.3 INHOMOGENEOUS PATTERN DISTRIBUTION AND DEPRIVATION EXPERIMENTS

In deprivation experiments the statistical properties of the afferent activity pattern are changed (i) by artificially changing sensory input (e.g., by rearing in darkness, under selective exposure or under stroboscopic illumination), or (ii) by directly changing the firing pattern of cells in the subcortical structures (e.g., blocking activity by chemicals like TTX or αBTX).

Elements of the orientation column system in V1 of the macaque monkey are present at birth, but the pattern still undergoes some refinement afterwards.[51] Its development, if input driven, must to some extent depend on internally generated activity pattern. In the cat, however, a large number of cells, especially located in the upper layers, develop orientation selectivity during a critical period after birth, and artificially changing afferent inputs during this period of time lead to changes in the numbers and the spatial distribution of feature-selective cells.[12] The effects of prolonged dark rearing suggest that first-order area 17 cells (located in layer IV and VI) develop orientation selectivity, while the other cells depend on external visual input.[20] Exposure to a limited range of stimulus orientations leads to an increase in the proportion of orientation, selective cells,[1] and the range of orientations is restricted to the range of orientations to which the animal is exposed.[2] A columnar system seems still to be maintained in layer IV with columns of the experienced orientation being larger, while deprived columns fail to extend into non-granular layers.[44]

In order to compare maps generated by the SFM algorithm with the spatial structure of maps developed under deprived conditions, the effect of sensory deprivation was imitated by an appropriate inhomogeneous probability distribution $P(\vec{v})$ of patterns. Model maps were compared under the assumption that maps emerging during the critical period correspond to an "optimal" state of the network. Since the "optimal" states of the network depend on the probability distribution $P(\vec{v})$, (i) the essential features of the final map should depend on the stimulus distribution presented last (and for a sufficiently long time), and (ii) should not depend on the initial state.

If patterns are restricted to a single orientation, the SFM algorithm generates maps where deprived columns are no longer present, but if "deprived" patterns are generated with some probability, then a column system emerges, where regions corresponding to the experienced orientation are larger than the regions corresponding to "deprived" orientation.

It was mentioned in section 6 that if case $P(\vec{v})$ is restricted to two orientations, the SFM model can be re-interpreted to apply to the formation of ocular dominance columns. The stripe width depends on the correlation structure of the patterns and increases monotonically with decreasing correlation. With proper interpretation the results shown in Color Plate 5 are in qualitative agreement with results from monocular and reverse suture experiments[19] as well as impulse blockade.[46] In the case of "artificial strabismus" conditions[8] (zero activity correlation between both eyes), the SFM model wrongly predicts a reversal of feature hierarchy,[34] ocular dominance now being the primary feature instead of "position." In order to explain the experimental findings, one has to assume a two-stage process, where the first stage leads to a topographic prestructuring for the second stage, the input-driven process. If a rough topographic prewiring is assumed, which constrains the configuration space of the network, a stripe-like pattern emerges again.

ACKNOWLEDGMENTS

The authors would like to thank Dr. G. Blasdel for providing data on the orientation column system in monkey striate cortex. The help and support of R. Brady and R. Kufrin in all technical matters concerning the use of the Connection Machine system, and the financial support of one of the authors (K. Obermayer) by a scholarship of the Boehringer-Ingelheim Fonds is greatly acknowledged. This research has been supported by the National Science Foundation (grant number 9017051). Computer time on the Connection Machine CM-2 has been made available by the National Centers for Supercomputer Applications at Urbana-Champaign and Pittsburgh supported by the National Science Foundation.

REFERENCES

1. Blakemore, C., and R. C. Van Sluyters. "Innate and Environmental Factors in the Development of the Kitten's Visual Cortex." *J. Physiol. (London)* **248** (1975): 663–716.
2. Blasdel, G. G., D. E. Mitchell, D. W. Muir, and J. D. Pettigrew. "A Physiological and Behavioural Study in Cats of the Effects of Early Visual Experience with Contours of a Single Orientation." *J. Physiol. (London)* **265** (1977): 615–636.
3. Blasdel, G. G., and G. Salama. "Voltage Sensitive Dyes Reveal a Modular Organization in Monkey Striate Cortex." *Nature* **321** (1986): 579–585.
4. Bolz, J., N. Novak, M. Götz, and T. Bonhoeffer. "Formation of Target-Specific Neuronal Projections in Organotypic Slice Cultures from Rat Visual Cortex." *Nature* **346** (1990): 359–362.
5. Cook, J. E., and E. C. C. Rankin. "Impaired Refinement of the Regenerated Retinotectal Projection of the Goldfish in Stroboscopic Light: A Qualitative WGA-HRP Study." *Exp. Brain Res.* **63** (1986): 421–430.
6. Durbin, R., and M. Mitchinson. "A Dimension Reduction Framework for Understanding Cortical Maps." *Nature* **343** (1990): 644–647.

7. Erwin, E., K. Obermayer, and K. Schulten. "Convergence Properties of Self-Organizing Maps." *Artificial Neural Networks I*, edited by T. Kohonen et al., 409-414. North Holland, 1991.

8. Freeman, R. D., G. Sclar, and I. Ohzawa. "Cortical Binocularity is Disrupted by Strabismus More Slowly than by Monocular Deprivation." *Dev. Brain Res.* 3 (1982): 311-316.

9. Galli, L., and L. Maffei. "Spontaneous Impulse Activity of Rat Retinal Ganglion Cells in Prenatal Life." *Science* 242 (1988): 90-91.

10. Goodhill, G. J., and D.J. Willshaw. "Application of the Elastic Net Algorithm to the Formation of Ocular Dominance Stripes." *Network* 1 (1990): 41-59.

11. Grinvald, A., E. Lieke, R. D. Frostig, C. Gilbert, and T. M. Wiesel. "Functional Architecture of Cortex Revealed by Optical Imaging Technique." *Nature* 324 (1986): 361-364.

12. Hirsch, H. V. B. "The Role of Visual Experience in the Development of Cat Striate Cortex." *Cell. Mol. Neurobiol.* 5 (1985): 103-121.

13. Hubel, D. H., and T. N. Wiesel. "Sequence Regularity and Geometry of Orientation Columns in the Monkey Striate Cortex." *J. Comp. Neurol.* 158 (1974): 267-294.

14. Hubel, D. H., D. N. Wiesel, and P. N. Stryker. "Anatomical Demonstration of Orientation Columns in Macaque Monkey." *J. Comp. Neurol.* 177 (1978): 361-380.

15. Kohonen, T. "Self-Organized Formation of Topologically Correct Feature Maps." *Biol. Cybern.* 43 (1982): 59-69.

16. Kohonen, T. "Analysis of a Simple Self-Organizing Process." *Biol. Cybern.* 44 (1982b): 135-140.

17. Kohonen, T. *Self-Organization and Associative Memory.* New York: Springer-Verlag, 1983.

18. Kuljis, R. O., and P. Rakic. "Hypercolumns in Primate Visual Cortex Can Develop in the Absence of Cues from Photoreceptors" *Proc. Natl. Acad. Sci. USA* 87 (1990): 5303-5306.

19. LeVay, S., T. N. Wiesel, and D. H. Hubel. "The Development of Ocular Dominance Columns in Normal and Visual Deprived Monkeys." *J. Comp. Neurol.* 191 (1980): 1-51.

20. Leventhal, A. G., and H. V. B. Hirsch. "Receptive Field Properties of Different Classes of Neurons in the Visual Cortex of Normal and Dark-Reared Cats." *J. Neurophysiol.* 43 (1980): 1111-1132.

21. Linsker, R. "From Basic Network Principles to Neural Architecture: Emergence of Orientation Columns." *Proc. Natl. Acad. Sci. USA* 83 (1986): 8779-8783.

22. Löwel, S., B. Freeman, and W. Singer. "Topographic Organization of the Orientation Column System in Large Flat-Mounts of the Cat Visual Cortex: A 2-deoxyglucose Study." *J. Comp. Neurol.* 255 (1987): 401-415.

23. Maffei, L., and L. Galli-Resta. "Correlations in the Discharges of Neighboring Rat Retinal Ganglion Cells During Prenatal Life." *Proc. Natl. Acad. Sci. USA* **87** (1990): 2861–2864.
24. von der Malsburg, C. "Self-Organization of Orientation Sensitive Cells in the Striata Cortex." *Kybernetik* **14** (1973): 85–100.
25. Mastronarde, D. N. "Correlated Firing of Retinal Ganglion Cells." *TINS* **12** (1989): 75–80.
26. Merzenich, M. M., J. H. Kaas, J. T. Wall, M. Sur, R. J. Nelson, and D. J. Felleman. "Progession of Change Following Median Nerve Section in the Cortical Representation of the Hand in Areas 3b and 1 in Adult Owl and Squirrel Monkeys." *Neurosci.* **10** (1983): 639–665.
27. Merzenich, M. M., R. J. Nelson, M. P. Stryker, M. S Cynader., A. Schoppman, and J. M. Zook. "Somatosensory Cortical Map Changes Following Digit Amputation in Adult Monkeys." *J. Comp. Neurol.* **224** (1984): 591–605.
28. Mittenthal, J. E., A. B. Baskin, and R. E. Reinke. "Patterns of Structure and Their Evolution in the Organization of Organisms: Modules, Matching, and Compaction." This volume.
29. Nelson, M. E., and J. M. Bower. "Brain Maps and Parallel Computers." *TINS* **13** (1990): 401–406.
30. Obermayer, K., H. Ritter, and K. Schulten. "Large-Scale Simulation of a Self-Organizing Neural Network: Formation of a Somatotopic Map." *Parallel Processing in Neural Systems and Computers*, edited by R. Eckmiller, G. Hartmann, and G. Hauske, 71–74. North Holland, 1990.
31. Obermayer, K., H. Ritter, and K. Schulten. "A Neural Network Model for the Formation of Topographic Maps in the CNS: Development of Receptive Fields." *Proceedings of the IJCNN 1990*, Vol. II, 423–429. IEEE Computer Society Press, 1990.
32. Obermayer, K., H. Ritter, and K. Schulten. "Large-Scale Simulations of Self-Organizing Neural Networks on Parallel Computers: Application to Biological Modelling." *Parallel Computing* **14** (1990): 381–404.
33. Obermayer, K., H. Ritter, and K. Schulten. "A Principle for the Formation of the Spatial Structure of Cortical Feature Maps." *Proc. Natl. Acad. Sci. USA* **87** (1990): 8345–8349.
34. Obermayer, K., H. Ritter, and K. Schulten. "Development and Spatial Structure of Cortical Feature Maps: A Model Study." *Advances in Neural Information Processing Systems 3*, edited by D. S. Touretzky and R. Lippman, 11-17. Morgan Kaufmann, 1991.
35. Obermayer, K., G. G. Blasdel, and K. Schulten. "A Neural Network Model for the Formation of the Spatial Structure of Retinotopic Maps, Orientation- and Ocular Dominance Columns." *Artificial Neural Networks I*, edited by T. Kohonen et al., 505–511. North Holland, 1991.
36. Orban, G. *Neuronal Operations in the Visual Cortex.* Berlin: Springer-Verlag, 1984.

37. Redies, C., M. Diksic, and H. Riml. "Functional Organization in the Ferret Visual Cortex: A Double-Label 2-deoxyglucose Study." *J. Neurosci.* **10** (1990): 2791–2803.

38. Reh, T. A., and M. Constantine-Paton. "Eye-Specific Segregation Requires Neural Activity in Three-Eyed Rana Pipiens." *J. Neurosci.* **5** (1985): 1132–1143.

39. Ritter, H., and K.Schulten. "Convergence Properties of Kohonen's Topology Conserving Maps: Fluctuations, Stability and Dimension Selection." *Biol. Cybern.* **60** (1988): 59–71.

40. Ritter, H., K. Obermayer, K. Schulten, and J. Rubner. "Self-Organizing Maps and Adaptive Filters." In *Physics of Neural Networks,* edited by E. Domani, J. L. van Hemmen, and K. Schulten, 281–306. New York: Springer-Verlag, 1991.

41. Rojer, A. S., and E. L. Schwarz "Cat and Monkey Cortical Columnar Patterns Modeled by Bbandpass-Filtered 2d White Noise." *Biol. Cybern.* **62** (1990): 381–391.

42. Schmidt, J. "Formation of Retinotopic Connections: Selective Stabilization by an Activity-Dependent Mechanism." *Cell. Mol. Neurobiol.* **5** (1985): 65–84.

43. Silito, A. M. In *Cerebral Cortex,* edited by A. Peters and E. G. Jones, Vol. 2, 91–117. New York: Plenum Press, 1984.

44. Singer, W., B. Freeman, and J. Rauschecker. "Restriction of Visual Experience to a Single Orientation Affects the Organization of Orientation Columns in Cat Visual Cortex." *Exp. Brain Res.* **41** (1981): 199–215.

45. Skeen, L. C., A. L. Humphrey, T. T. Norton, and W. C. Hall. "Deoxyglucose Mapping of the Orientation Column System in the Striate Cortex of the Tree Shrew Tupaia Glis." *Brain Res.* **142** (1978): 538–545.

46. Stryker, M. P., and W. Harris. "Binocular Impulse Blockade Prevents the Formation of Ocular Dominance Columns in Cat Visual Cortex." *J. Neurosci.* **6** (1986): 2117–2133.

47. Swindale, N. V. "A Model for the Formation of Orientation Columns." *Proc. R. Soc. Lond.* **B215** (1982): 211–230.

48. Swindale, N. V., J. A. Matsubara, and M. S. Cynader. "Surface Organization of Orientation and Direction Selectivity in Cat Area 18." *J. Neurosci.* **7** (1987): 1414–1427.

49. Takeuchi, A., and S. Amari. "Topographic Organization of Nerve Fields." *Biol. Cybern.* **35** (1979): 63–72.

50. Ts'o, D. Y., R. D. Frostig, E. Lieke, and A. Grinvald. "Functional Organization of Primate Visual Cortex Revealed by High Resolution Optical Imaging." *Science* **249** (1990): 417–420.

51. Wiesel, T. N., and D. H. Hubel. "Ordered Arrangement of Orientation Columns in Monkeys Lacking Visual Experience." *J. Comp. Neur.* **158** (1974): 307–318.

52. Willshaw, D. J., and C. von der Malsburg. "How Patterned Neural Connections Can Be Set Up by Self-Organization." *Proc. R. Soc. Lond.* *B* **194** (1976): 431–445.

Physiology and Development: Commentary

The preceding chapters continue to explore, in the context of development, issues previously raised in the context of physiology—the generative and design approaches to discovering principles of organization, and relations of models and data in these approaches. The environment influences intrinsic processes in development as well as in physiology. However, whereas physiological changes in activity are often reversible, development can change the morphology and behavior of an organism irreversibly. Hence a discussion of development must deal with the influence of antecedent conditions—that is, with history.

APPROACHES THROUGH GENERATION AND DESIGN

As in physiology, to characterize a developmental process one should specify its dynamic and several types of parameters—conditions for starting and stopping the process, boundary conditions, and parameters adjusting the rate of performance.[1] The parameters may be specified in diverse ways. For example, as Nelson points out, generation of the lung is constrained by boundary conditions from the rib cage and the diaphragm. An endpoint of its development, the shape of the lung, apparently is fixed by these boundary conditions rather than by its intrinsic generative processes. This example illustrates for development the complementarity of generative and design approaches, discussed in the essay Physiology: Commentary. The design

Principles of Organization in Organisms,
SFI Studies in the Sciences of Complexity, Proc. Vol. XIII,
Eds. J. Mittenthal & A. Baskin, Addison-Wesley, 1992   **167**

criterion that the lungs should conform to the shape of the thoracic cavity, despite the imprecise specification of this shape, is met by the process that generates a fractal structure with a shape adjusted through interaction with the boundary. Another example of this complementarity occurs in neural development: Obermayer et al. provide a generative approach to the development of topographic mappings from a many-dimensional feature space into two dimensions, while Durbin and Mitchison[2] argued that the same characteristics of mappings can be inferred from a design argument.

The chapters explore further a major theme of the book, the use of models to understand macro-micro relations. Stryker's chapter, in particular, shows that a model can give focus and direction to the collection of micro-level data in order to understand macro-micro relations. Models can also provide level- insensitive descriptions of these relations; for example, a network model can describe the development of genetic and neural networks. However, many biological mechanisms may be compatible with a given mathematical model. As Stryker emphasizes, to resolve controversies about qualitatively different mechanisms that can be subsumed under one quantitative model, experiments are necessary.

THE INTERFACES OF INTRINSIC PROCESSES WITH HISTORY AND ENVIRONMENT DURING DEVELOPMENT

What are the initial conditions for a period of development, and how important are they? Karr and Mittenthal argue that in *Drosophila* conditions set up during oogenesis and fertilization play a central role in the rapidity of early development.

The chapters show that in studying development it is necessary to distinguish whether one process provides initial conditions for another or they operate concurrently, interacting within a single dynamical system. These alternatives are evident in the formation of spatial patterns: Gene activity might provide proteins that would implement a formshaping process without feedback to the genetic network. Alternatively, formshaping might proceed through a continual dialogue between network and material processes. These alternatives also appear in the development of behavior: Development might generate a newborn organism that was a neurological blank slate, leaving the structuring of neural organization to proceed concomitantly with experience. Or (as seems the case), intrinsic processes might generate most of the neural organization, leaving sharply limited aspects plastic for response to the environment.

When processes interact in development, both may be intrinsic to the organism, or one may be intrinsic and the other extrinsic. For example, intrinsic processes and experience contribute to neural organization. The responsiveness of development in a region of an embryo to input from the environment of the region, within the embryo or from its surroundings, varies during development. A period of maximal integration throughout the embryo may generate the major structures characteristic of a large group of organisms—a phylotypic stage.[3] Later morphogenesis depends less on global integration, though inputs to a region from other regions and from the external world can modify development, as Stryker and Obermayer et al. show.

REFERENCES

1. Alberch, P., S. J. Gould, G. F. Oster, and D. B. Wake. "Size and Shape in Ontogeny and Phylogeny." *Paleobiology* **5** (1979): 296–317.
2. Durbin, R., and G. Mitchison. "A Dimension-Reduction Framework for Understanding Cortical Maps." *Nature* **343** (1990): 644–647.
3. Wray, G. A., and R. A. Raff. "Pattern and Process Heterochronies in the Early Development of Sea Urchins." *Seminars in Developmental Biology* **1** (1990): 245–251.

Physiology, Development, Evolution, and Their Evolution

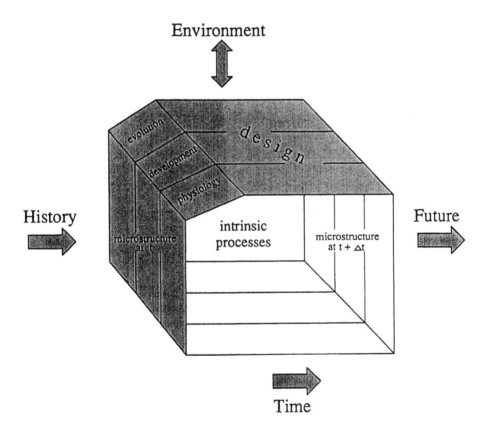

The preceding sections offer an expanding vista of time scales and approaches to understanding structure. The first section dealt with problems in physiology, and emphasized approaches through design and through generative dynamics at macro- and micro-scales. To these approaches the second section added a short-range historical perspective needed to analyze development and its physiology. The chapters in the first half of this section widen the inquiry further, asking how the dynamics of evolution generate patterns of constancy and change in the processes of physiology and development. The remaining chapters suggest broad principles that may be valid at many time scales and levels of organization. Here we introduce the first group of chapters in this section.

Patterns of Constancy and Change: Introduction

Through evolution the processes of physiology and development show patterns of constancy and change. Constancy may obtain at levels from the molecular to the organismal. For example, some species have retained a nearly constant morphology for a hundred million years; this is morphological stasis. Recurrent patterns of change are known. For example, in homoplasy diverse branches of a lineage converge toward a common morphology. During evolution the patterns of constancy and change have themselves changed. One long-term change is the evolution of higher levels of organization; cellular and multicellular organizations have evolved from the molecular level at which life originated.

Organisms and their environment evolve as a composite dynamical system, from which such patterns of constancy and change must emerge. To analyze the dynamics of this system, it is desirable to view it as an association of interacting processes. Material and network processes interact with each other and with extrinsic processes. Presumably in this way patterns of constancy and change evolve.

In the first chapter, Wake points out a dilemma in analyzing patterns such as stasis and homoplasy: Such phenomena may occur because intrinsic processes with generic outcomes dominate morphogenesis, or as a result of particular regimes of natural selection. He argues that neither of these extremes is likely to be the case. Rather, a conjunction of functionalist (externalist, selection) and structuralist (internalist, self-organization) viewpoints is essential for understanding the evolution of structure. Neither of these viewpoints is primary, and the appropriate mix of them

Principles of Organization in Organisms,
SFI Studies in the Sciences of Complexity, Proc. Vol. XIII,
Eds. J. Mittenthal & A. Baskin, Addison-Wesley, 1992   **173**

is context dependent. To analyze the composite system of organisms and environment as an association of interacting processes, Wake considers relations among processes at different levels of organization that contribute to the morphology of salamanders.

To understand patterns of constancy and change, it is desirable to characterize limits on the transitions among structures that are likely to occur. To some extent these limits can be inferred from sequences of morphologies and of ecological constraints. In an empirical approach such as Wake's, these sequences must be inferred from close comparative study of related species. Stork shows an alternative approach: Modelling the course of evolution can show how the structure of an organism manifests the sequence of constraints through which its lineage has evolved. Stork presents a model for the evolution of behaviors involving abdominal flexion in crayfish—swimming and flipping. The analysis pits processes at vastly different scales against each other. It shows that a neural network can include connections that were useful in mediating an earlier evolved behavior but that are irrelevant for a later evolved one.

Mandell and Selz propose molecular interpretations for macro-level phenomena of evolution. They suggest that punctuated equilibrium, an alternation between morphological stasis and rapid transition to new morphologies, may be causally related to recurrent phases in the kinetics of allosteric enzymes. They propose that the action of neurohormones throughout an organism may modulate its physiology but also modify its gametes, thereby allowing the inheritance of acquired characteristics.

Although there are diverse mixtures of self-organization and selection, a long-term trend must obtain: Because the consequences of selection are cumulative, as evolution proceeds, selection must make an increasing contribution to the structure of organisms. Specifically, Newman suggests that early in evolution, embryos developed mainly through processes generic to the materials composing them. These materials resemble viscous fluids—the cytoplasm of a single cell, or an aggregate of cells that are motile and adhesive. They are also excitable media, in which cells can respond actively to chemical, mechanical, and electrical inputs by producing chemicals, exerting forces, and generating potentials. During evolution the networks of genes that generate cells and tissues were modified under selection, in ways that altered the conditions in which a material process is used (heterostasy), the time and place of its use (heterochrony, heterotopy), the structures it generates, or its stability.

Goodwin and Kauffman suggest that a generic material process became subject to refined genetic control during the evolution of segmentation in *Drosophila*. That is, a reaction-diffusion-like process that could generate stripes may have come under regulation by a network of genes that can produce or suppress individual stripes. Goodwin and Kauffman show that a description based on the generic features of a gene-regulated stripe-forming system can account for many aberrations of segmentation.

David B. Wake
Museum of Vertebrate Zoology and Department of Integrative Biology, University of California, Berkeley, CA 94720, U.S.A.

Homoplasy: The Result of Natural Selection, or Evidence of Design Limitations?

Reprinted with permission from *The American Naturalist* **138**(8) (1991): 543–567.

Similarity in morphological form may arise from common ancestry (failure to evolve), from parallel evolution, from convergence, or from reversal to an apparently ancestral condition. Homoplasy from convergence, parallelism, and reversal is common, and its ubiquity creates difficulties in phylogenetic analysis. Convergent evolution often is considered to be one of the most powerful lines of evidence for adaptive evolution. But an alternative explanation for convergence and other evolved similarities is that limited developmental and structural options exist. Identical forms can be obtained when particular developmental phenomena are triggered by very different kinds of stimuli, or when constraints exist that shape external form or limit morphological expression to a few options. Examples from plethodontid salamanders are used to illustrate an approach combining internalist and externalist analytical methods. In order to understand how morphologies evolve in lineages, both functionalist and structuralist approaches are necessary, combined in a context in which phylogenetic hypotheses and their tests are continuously pursued. When homoplasy is rampant, as in

Principles of Organization in Organisms,
SFI Studies in the Sciences of Complexity, Proc. Vol. XIII,
Eds. J. Mittenthal & A. Baskin, Addison-Wesley, 1992   **175**

salamanders, we can expect discordance with phylogenetic analyses based on nonmorphological data sets.

Recently Allan Larson and I argued that an understanding of the evolution of biological form—morphology—was unlikely unless one combined two distinct and independent approaches: neo-Darwinian functionalism and biological structuralism, in a context of rigorous phylogenetic analysis.[74] In this Presidential Address to the American Society of Naturalists, an organization dedicated to the conceptual unification of biological knowledge, I expand on this theme.

Our approach requires that structuralism, with its focus on organismal-level phenomena such as the generation, self-organization, and transformation of specific form,[15,40,45,79] and neo-Darwinism, with its focus on population-level phenomena and historical contingency, must be used concurrently, even while retaining their differences and separate domains of explanation. In this way conflicts between the different modes of explanation are highlighted and interpreted, not simply argued away or ignored by default.

My central theme is the phenomenon of non-divergent evolutionary change among lineages, including convergent morphological evolution, parallelism, and some kinds of reversal—in other words, what phylogeneticists term homoplasy.[25,38,56,80] Convergence and parallelism (for a useful discussion of the distinction between these terms, see Patterson[39]) often are considered to constitute strong evidence of the functioning of natural selection. Patterson[39] states that "the general explanation for convergence is functional adaptation to similar environments," but I will argue that alternatives must always be considered. In recent years increasing attention has been given to the possibility that parallelism is a manifestation of internal design constraints (e.g., Alberch[2]), and so both functionalist and structuralist constructs predict its occurrence.

The ubiquity of homoplasy is a major concern in phylogenetic analysis (cf. Rieppel[45] and Sanderson and Donoghue[51]), which ironically is its primary means of detection. One assumes that homoplasy is sufficiently rare that when parsimony is used in interpreting results of the application of cladistic methods to large data sets, an accurate assessment of genealogical relationships is obtained. But the level of homoplasy varies among taxa, and it is of interest to inquire why and to investigate causative factors. If homoplasy is rampant, existing cladistic methods fail.

To discuss homoplasy implies that one understands homology, one of the most ancient and difficult of biological topics. Homoplasy is essentially false homology, and Patterson[38] proposes three tests for its detection, similarity, congruence (essentially phylogenetic analysis in the hands of most systematists), and conjunction. Parallelism fails only the congruence test, while convergence fails both the similarity (only "superficial" similarity, see below) and the congruence tests. Patterson does not specifically deal with reversal, but it seems likely that it would be considered a kind of parallelism in his scheme.

The taxon I study, the salamander family Plethodontidae, displays substantial homoplasy. The family includes well over 200 species, and the level of homoplasy is known to increase as the number of taxa included in a study increases.[51] I will argue that both functionalist and structuralist explanations are necessary to understand the biological bases and evolutionary significance of the morphological homoplasy that is routinely encountered by me and other systematists.

My research focus is the Family Plethodontidae, the Lungless Salamanders, and my examples all come from this group. Isolated cases lack the impact that arises when one must confront the often conflicting lines of evidence that arise from a long-term focus on the evolution of a diverse monophyletic taxon. My specific examples range from ecomorphology at the level of the whole organism, to complex integrated systems such as feeding, and even include alternative states of a single osteological character. I purposely take a broad perspective on biological form. For a morphologist such as myself, homoplasy is fundamentally a form concept—two distantly related organisms are convergent if they have evolved similar features from non-homologous ancestral states. The convergent forms may function similarly, but similarity in physiological performance, or ethological or ecological biological role, is neither assumed nor expected.[67]

Several themes, not necessarily independent, emerge:

1. *The Bauplan.* There is a place in modern biology for the concept of the bauplan (cf. Rieppel[44] and Wagner[62]), which is manifest in design parameters that set limits to morphological variation, make reversion to developmental default states expected, and lead to evolutionary stopping points. The bauplan promotes parallelism, as well as reversal to ground plans.

2. *Heterochrony.* Evolutionary changes in the timing of development among taxa can strongly influence form generation, and these become manifest in the morphological expressions of paedomorphosis and peramorphosis.[5] The evolutionary significance of paedomorphosis is debated, but salamander biologists, confronted by vivid examples such as the axolotl, are convinced of its importance. Permanently larval forms have evolved repeatedly within the order Caudata, and some of the most persistent phylogenetic questions concern the phylogenetic relationships of such genera as *Siren, Necturus,* and *Proteus.*[24] These larval forms represent parallel truncation of ontogeny by paedomorphosis. However, heterochrony is important in less dramatic, but more pervasive and ultimately more significant ways. The most insidious effects are: (1) paedomorphic homoplasy, the phylogenetic reversal of individual morphological traits to apparently ancestral states (from "backing down" ontogenetic trajectories in cases of prior recapitulatory evolution), or the parallel evolution of similar juvenile traits; and (2) peramorphic homoplasy, parallel phylogenetic extensions of ontogenies.[37,77] Parallelism is promoted when multiple features of organisms are affected simultaneously in species with direct development, which accordingly lack the tell-tale persistent larval state.[1]

3. *Genomic constraints on morphogenesis.* Cell and molecular phenomena have major implications for organismal and evolutionary biology in unexpected ways. Specifically, the enormous genome sizes of salamanders impose on them large cells and slow cell cycles, with many implications for biological form and evolutionary pattern. Plethodontid genomes range in size from about 12 to nearly

80 pg DNA per haploid genome,[53] and are the largest of all terrestrial vertebrates, exceeded among vertebrates only by lungfishes. Phenomena at the molecular level, such as increased genome size, can generate homoplasy at higher levels of organization.

4. *Adaptation and its consequences.* Ecological specialization driven by natural selection can lead to severe tests of design limitations, and can lead to the dissolution of constraints on features quite unrelated to the specialization in question. Ecological specialization leads to convergence, and in the context of developmental constraints, to parallelism. Specifically, I will examine microhabitat specialization, locomotion, miniaturization, age at maturity, and modifications of ancestral biphasic life cycles, all of which relate directly to different kinds of homoplasy.

5. *Interaction among constraints at different hierarchical levels.* Hierarchical organization can have profound implications for homoplasy, especially in situations in which downward causation (such as population ecological limitations on body size) and upward causation (such as cell size) simultaneously impact a focal level (in this case, the individual organism).

THE PLETHODONTID SALAMANDERS

Herpetologists generally have considered the lungless salamanders of the family Plethodontidae to be the most highly derived of the nine or ten families of salamanders (e.g., Duellman and Trueb[11]). About two-thirds of living species of salamanders are included in the family. Plethodontids have a wide range of life histories and ecologies, and occupy a full array of habitats, from strictly aquatic to fully terrestrial.[64] The relatively small subfamily Desmognathinae is restricted to eastern North America, but the large subfamily Plethodontinae, with three tribes, is more widespread.

Plethodontids are the only salamanders that have been successful in the tropics, in the sense that they are widespread (latitudinally, altitudinally, and ecologically) and speciose. All tropical species (salamanders occur only in the New World tropics) are members of a phylogenetic "twig," the supergenus *Bolitoglossa* of the tribe Bolitoglossini. This single lineage has undergone an extraordinary radiation, centered in Middle America. The supergenus includes about 45% of all species of salamanders, and their evolution has featured extreme specialization and adaptive radiation of a highly derived ancestral stock.[65,74,76] The tropical salamanders have evolved with substantial homoplasy, and my attempts to solve the formidable phylogenetic problems that result have been a driving force in the development of ideas presented in this paper. Progress is being made in defining monophyletic taxa among the approximately 150 species,[64,72,77] but problems remain.

Within the tropics, ecological generalists and more ancestral lineages are found at middle and high elevations; ecological specialists occur in specific microhabitats and elevational zones, usually associated with cloud forest bromeliads, lowland fossorial habitats, and lowland arboreal habitats. Because so little was known concerning these organisms, and in order to pursue the study of adaptation, morphological evolution and phylogenetics outlined in the introduction, I established a series of elevational transects between Veracruz, Mexico, and western Panama.[74,76,78] As many as 23 species (the highest number anywhere in the world) are distributed along a single long transect. Morphological homoplasy is common in phylogenetically independent lineages occupying different regions (for example, convergence in body form among small bromeliad-dwelling members of the genera *Chiropterotriton* in eastern Mexico, *Dendrotriton* in Nuclear Central America, and *Nototriton* in Talamancan Central America), and even within regions (as in the evolution of interdigital webbing in species of different groups of *Bolitoglossa*).[74] Some general features of the tropical salamanders, in addition to their relative morphological simplification, include specialized visual systems featuring large, frontally oriented eyes,[46] a highly projectile tongue used for feeding on terrestrial arthropods,[35] and direct development in which eggs are laid in terrestrial sites and the aquatic larval stage is bypassed. With one exception (an undescribed aquatic species from Mexico), all of the species appear to be strictly terrestrial throughout life.

The following examples of homoplasy in plethodontids exemplify my analytical approach. All examples arise from problems initially pursued in the bolitoglossines, but in every instance a broader frame of reference is required.

DESIGN LIMITATIONS AND HOMOPLASY
THE PREMAXILLARY BONE

My first example is a relatively well understood, but very complex, case—the alternative forms of the premaxillary bone in plethodontid salamanders.[64,77,78] In most nonplethodontid salamanders the bone arises as a paired structure at the front of the skull and remains paired throughout life. In some taxa the bones fuse during ontogeny. Parallelism and reversal have characterized the history of fusion and separation of the premaxilla in plethodontids.

A caenogenetic novelty inferred to have been inserted into the larval stage of a plethodontid ancestor suggests a functional interpretation; fusion associated with improved feeding performance in larvae became a fixed feature of the normal ontogeny and now constitutes a synapomorphy of the family. The plesiomorphous ontogeny within the family is for the larval or embryonic bone to separate at metamorphosis or, in direct developing forms, at about the time of hatching. This restores the plesiomorphous state (two separated bones) for the order Caudata—the bones essentially revert to the bauplan.

This sequence of events persists even in the face of evolution of direct development and loss of the larval stage; this is significant, for it means that the ontogenetic trajectory stabilized once it was modified by caenogenesis. Subsequently, evolution of a single bone in adults has proceeded in parallel in different plethodontid lineages by paedomorphosis (retention of the larval or embryonic state, generally but not necessarily nonadaptive as far as the premaxilla is concerned) or by peramorphic fusion (adaptive evolution associated with strengthening of the jaws).

This example illustrates the phylogenentic persistence of bauplans, and the existence of a default state associated with metamorphosis. The point is that an analysis of the potentialities and the limits of morphogenetic systems that characterize lineages may provide a unique perspective on how morphological evolution proceeds, and why patterns of change are so restricted and stable. It is not the adapted larval skull which influences adult skull form, but rather a default to ground plan, which restores an ancestral pattern. This example also illustrates the need for an appreciation of the whole ontogeny of a character, its holomorphology, as opposed to its instantaneous manifestation, say, in an adult (cf. de Queiroz[10]).

NUMBERS OF DIGITS

The ancestral condition for salamanders, found in the vast majority of living species and in fossil out-groups, is four digits in the fore limb and five in the hind limb.[55] However, species with four digits (toes) on the hind limb are found in several different families (e.g., Plethodontidae, Salamandridae, Hynobiidae). Figure 1 is a "scenariogram" (see Wake and Larson[74]), a collapsed cladogram (based on phylogenetic hypotheses discussed in Lombard and Wake[81]) in which only relevant taxa and otherwise most inclusive taxa are illustrated. Within the Plethodontidae four toes occur in widely dispersed lineages—the genera *Batrachoseps* and *Hemidactylium*, and one species in the genus *Eurycea*.[55,74] Factors controlling limb and digit development have been studied extensively in salamanders, and many aspects are well understood (see reviews in Shubin and Alberch[55]; Bryant et al.[8]; Sessions and Larson[53]; Oster et al.[41]). In contrast to the situation in lizards, where digital and limb reduction has evidently been a gradual process (Lande[28]; Lande also defines conditions under which limb loss in amniotes could occur rapidly), evolutionary modifications of developmental events in the limb bud of amphibians result in the production of entire (rather than partial) digits in the postaxial portion of the limb; experimental disruption of developmental regulation (for example, with the mitotic inhibitor colchicine) mimics patterns seen in nature.[4] While cellular-level controlling factors remain to be discovered, there are important correlations between the number of digits in amphibians and such phenomena as cell size (directly related to genome size), cell number, and the number and size of cells in limb buds of different size.

The four-digited feet of unrelated salamander taxa are nearly identical in structure, and they are produced independently by the operation of one of the two developmental options, which produce either four or five toes. This is a direct example of design limitations, in which alternative states are sharply defined. One might predict that four-toed variants would appear in species with very small limb buds, in species with very large genomes (and hence relatively few cells in limb buds), or both. The miniaturized species *Parvimolge townsendi* of eastern Mexico has five toes, but individuals with four have been found. The outer (postaxial) toe of the four-toed individuals is larger than the fourth toe of five-toed individuals. This toe contains either one or two skeletons (Figure 2), direct evidence of the existence of design limitations. Alberch and Gale[4] found a similar phenomenon in a large species of *Bolitoglossa* which has very large genome size.

A possible reverse example is a case involving *Batrachoseps wrighti*, a member of a genus which normally has four toes. A single individual more than 10% larger than any other member of the species so far reported has one hind limb bearing five toes.[6] However, there are only three, instead of the normal four metapodial elements; two of these are much larger than normal, and each bifurcated so that five digits appear distally (Figure 2). Only experimental studies will determine

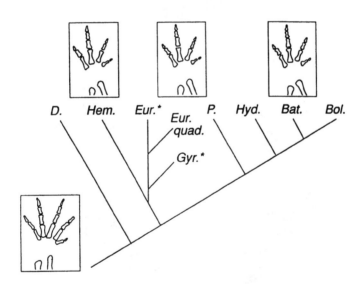

FIGURE 1   A "scenariogram" (see text) illustrating digital loss in the family Plethodontidae. The distal ends of the tibia and fibula, the metatarsals, and the phalanges are drawn; the tarsals and joints are cartilaginous in these species. The independent evolution of four toes from an ancestral five-toed state has taken place in three separate lineages. This is a collapsed cladogram (based on Lombard and Wake[36]), in which only the most inclusive relevant taxa are shown. D. = subfamily Desmognathinae; P. = tribe Plethodontini of the subfamily Plethodontidae; Hem. = *Hemidactylium*; Eur.* = all members of the genus *Eurycea* except *Eur. quad.* (= *E. quadramaculatus*); Gyr* = all members of the tribe Hemidactyliini except *Hemidactylium* and *Eurycea* (Gyr* is probably paraphyletic); Hyd. = *Hydromantes*; Bat. = *Batrachoseps*; Bol. = supergenus *Bolitoglossa*.

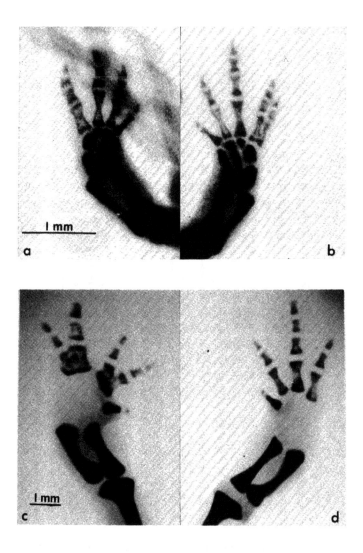

FIGURE 2 Hind feet of abnormal individuals of two species of plethodontid salamanders, printed directly from radiographs. Above, a specimen of *Parvimolge townsendi*, a normally five-toed species from Mexico, which has only four digits and four skeletons (the last of abnormally large diameter) on the left (a), but five skeletons within a single external fourth digit on the right (b, the tarsals are often ossified in this species, as in the specimen radiographed). Below. a specimen of *Batrachoseps wrighti*, a normally four-toed species from Oregon, with the normal pattern on the right (d) and an abnormal pattern on the left that shows a bifurcation of two metapodial elements resulting in five digits (c).

if abnormalities such as this are more likely in relatively larger individuals (with correspondingly different balances between limb bud dimensions and cell numbers), as I suspect. Alternatively, this may be a case of abnormal regeneration following injury (although no other evidence of injury is present).

These examples shows that homoplasy, in this case parallelism, can be a manifestation of design limitations in the form of developmental constraints, which are only indirectly related to adaptive processes. Thus, selection for very small size may have, as an incidental side effect, the loss of a toe. A functionalist interpretation would view this as a correlated response to selection, while a structuralist would focus on design limitations.

ORGANIZATION OF WRISTS AND ANKLES

The wrist and ankle (mesopodia) of plethodontids consist of a group of cartilages that arise during development by processes of condensation, segmentation, and bifurcation of blastemal masses.[55,41] There are many conceivable organizations for this complex set of elements (primitively eight in the wrist and nine in the ankle), but variation is highly ordered and only a few of the possible variant conditions are found, even in cases in which higher than expected amounts of variation are encountered.[17,18,21,22] The ordering is sufficiently great that comparative anatomists have named each element and treated the named units as homologues, a procedure that has been challenged by some developmental biologists (e.g., Goodwin and Trainor[16]). Relatively few rearrangements of the ancestral patterns are found in plethodontids, but all that characterize species or genera have arisen more than once.[74]

Functionally significant rearrangements associated with climbing have arisen independently in the genera *Aneides*[63] and *Chiropterotriton*.[64] I restrict the treatment to the distal postaxial portion of the tarsus, in which the ancestral condition finds a large distal tarsal 4 (dt 4) contacting the fibulare, thereby preventing the small dt 5 from articulating with the centrale. In the derived condition dt 5 is large and articulates with the centrale, thus excluding the relatively small dt 4 from articulating with the fibulare (Figure 3). The discovery of apparently atavistic variants in different species suggests a possible morphogenetic mechanism. In one variant (a specimen of *Pseudoeurycea anitae*) an extra element ("m") is present and dt 4 and 5 are of approximately equivalent size (Figure 3). The tissue which in the variant forms a separate element is incorporated into dt 4 in the ancestral condition (and other individuals of this species), but a kind of developmental switch permits it to become associated with dt 5 to produce the functionally significant rearrangement. Recently Neil Shubin and I discovered the alternative variant in an adult *Aneides flavipunctatus*. In this specimen we also found element "m," in this case disconnected from distal tarsal 5. This entire area remains undivided in a number of bolitoglossine genera, all of which either have very large genomes, very small size, or both, and here the element has been termed distal tarsal 4-5 (but it also incorporates "m"); it is unclear as to how many times this derived failure to separate has evolved.[72]

The point of this example is that adaptively significant changes can arise from alternative developmental states, which themselves may be determined by phenomena that operate at different hierarchical levels, such as that of the whole organism (reduction in body size by paedomorphosis), at the molecular and cellular levels (by increases in genome and cell size), which lead to a reduction in the number of limb bud cells, or to a failure of blastemal masses to segment, bifurcate, or both.

VERTEBRAL JOINTS

The intervertebral joint varies in structure among salamanders in ways that to this time have defied attempts at functional interpretation. In the majority of plethodontids, the intervertebral joints are not well differentiated. The notochord, which in adults contains cartilage and is an important skeletal element, persists throughout life. Between the husklike, ossified centra lies a spindle-shaped intervertebral cartilage, which is interrupted by invasive fibrocartilage so that a kind of ball and socket joint (functional opisthocoely) results.[75] In distant outgroups, such as members of the family Salamandridae, as vertebral growth proceeds, the lateral walls of the developing centra move around longitudinally aligned blood vessels, and as the blood vessels finally reach the cartilage inside the centrum, the intervertebral cartilage ossifies basally, producing a true bony condyle, capped with cartilage only in the joint region.[66] The notochord disappears, or is reduced to a crushed remnant. An identical phenomenon has evolved in the large plethodontid salamander *Phaeognathus*, a strong burrower, and it may well be that the strong intervertebral articulation of this form has adaptive significance. But bony condyles also are encountered as rare variants in species of moderate size, such as in a specimen of *Batrachoseps wrighti* in which blood vessels have invaded the lateral surfaces of a few anterior vertebrae (Figure 4), with the posterior vertebrae retaining the apparently ancestral notochordal vertebrae.

A superficially similar but developmentally entirely different arrangement has evolved in several diminutive bolitoglossine genera.[66] In *Thorius* the nervous system is extraordinarily large compared to the skeletal system.[19,48,50,66] In hatchlings the neural canal is relatively enormous, while the centrum is what one would expect in a salamander of its size. During later ontogeny the neural canal is eroded from inside, and bone accretes to the outside, thereby accommodating the relative rapid growth of the spinal cord. As a result, the dorsal surface of the centrum, which is also the ventral floor of the vertebral canal, erodes away, and the cartilage is exposed to the blood vessels lying inside the canal. Ossification of the intervertebral cartilage ensues, and a miniature duplicate of the condyle of large species such as *Phaeognathus* occurs, by cellular processes that I suspect are identical (inferred from work of Shapiro et al.[54]), but as a result of very different morphogenetic

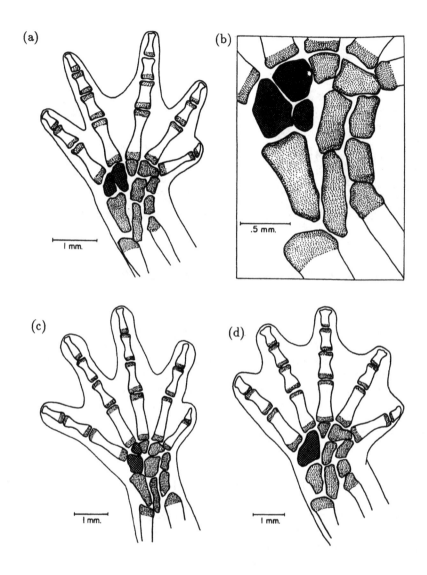

FIGURE 3 Mesopodial elements in some plethodontid salamanders. Bone is outlined, and cartilage stippled. The deeply stippled cartilaginous tarsals show different arrangements.(a) Typical ancestral pattern, as seen in *Pseudoeurycea anitae*; distal tarsal 4 (dt 4) is large, and 5 is small. (b) Unusual pattern encountered in an adult *P. anitae* (bilaterally symmetrical); dt 4 and dt 5 are of equal size, and an atavistic additional element ("m") is present. (c) Typical pattern in *Chiropterotriton* (also seen in *Aneides*); dt 5 is large, and 4 is small. (c) Typical pattern in *Bolitoglossa* (also found in several other tropical genera); dt 4, 5, and possibly element "m" are combined in a single unit.

(a)   (b)

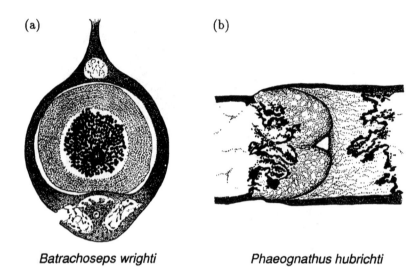

Batrachoseps wrighti   *Phaeognathus hubrichti*

FIGURE 4   Unusual intervertebral joints in plethodontid salamanders.(a) Cross-section of a trunk vertebra showing the large spinal cord filling the vertebral canal, and the small centrum underlying it. The centrum, normally filled with cartilage in plethodontid salamanders, has been invaded by lateral blood vessels and largely converted to bone, but a small notochordal remnant (circle) remains. (b) Longitudinal section of a trunk vertebra through two centra in the region of an intervertebral joint in a stout-bodied, burrowing species. The condyle (to the left) and the cotyle (to the right) of adjacent centra are cartinaginous, but the cartilage rests on an ossified cap (dark black) so that an opisthocoelous condition is attained; anterior is to the right.

pathways. Thus complex homoplasy results, with miniatures displaying parallel evolution that is simply an outcome of miniaturization, other species displaying convergent evolution (different morphogenetic pathways) that is itself subject to parallelism, and the family as a whole illustrating homoplasy in the form of reversal in comparison with distant outgroups (Figure 5).

   This example shows that even in functionally highly significant structures such as joints, specific morphology can arise for entirely incidental reasons. A complex developmental trajectory may be abbreviated or truncated (as in the ancestral plethodontids compared with salamandrid outgroups, possibly by paedomorphic truncation), but this trajectory also can be "entered" at different points in ontogeny by different morphogenetic routes. Once again the need for a multidimensional analytical approach is evident.

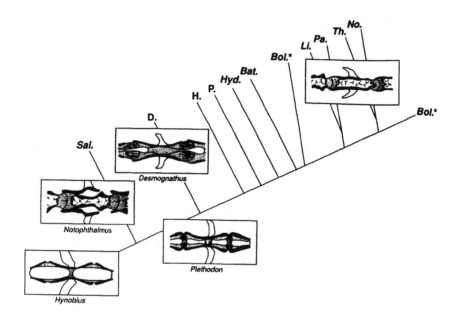

FIGURE 5   Scenario for the evolution of vertebral articulations in salamanders, based
on phylogenetic hypotheses of Duellman and Trueb[11] and Lombard and Wake.[36] The
boxes show longitudinal sections through one complete centrum and two intervertebral
joints. Dark black is bone, and stipple is cartilage. Anterior to right. The situation
in *Hynobius* and *Plethodon* is ancestral, and shows non-ossified joint regions.
Ossification has occurred in *Notophthalmus* by one mechanism, and in *Thorius* and
species of three other genera by another. A third condition is seen in *Desmognathus*,
in which the condyle is mineralized cartilage; the sister taxon of *Desmognathus*,
Phaeognathus (Figure 4), has carried this one step further to ossification, in parallel
with the situation in *Notophthalmus*. The condition in *Thorius* is convergent. Sal. =
family Salamandridae; D. = subfamily Desmognathinae of the Plethodontidae; H. = tribe
Hemidactyliini of the plethodontid subfamily Plethodontinae; P. = tribe plethodontini
of the subfamily Plethodontinae; Hyd. = *Hydromantes*; Bat. = *Batrachoseps*; Bol.*
members of the supergenus Bolitoglossini other than the following bolitoglossines (this
is shown as a paraphyletic assemblage): Li. = *Lineatriton*, Pa. = *Parvimolge*, Th. =
Thorius, No. = *Nototriton*.

ECOMORPHOLOGY AND HOMOPLASY
ELONGATION AND FOSSORIALITY

Fossorial specialization in salamanders includes elongation, attenuation, and limb reduction. Most genera of tropical salamanders display no variability in numbers of trunk vertebrae, but in out-groups even intrapopulational variability is common. This fixed number of vertebrae appears to act as a constraint on elongation, precluding selection for increased vertebral numbers. In the Mexican genus *Lineatriton*, however, a unique developmental pattern has evolved in which the individual vertebrae have elongated to produce an extremely attenuate body form.[58,76] This is a "giraffe-neck" solution to the problem of how to become elongated. The only genus of tropical salamanders which shows vertebral variability is another fossorial genus, *Oedipina*, which occurs south and east of the Isthmus of Tehuantepec. Here a common homoplasy is encountered, one seen not only in bolitoglossines (e.g., *Batrachoseps*), but also in other plethodontid taxa (e.g., *Plethodon*, *Phaeognathus*) and in other salamander families (e.g., Amphiumidae, Sirenidae), in which an evolutionary alteration has affected segmentation during development, so that elongation is accomplished by adding trunk vertebrae. At one time all tropical fossorial species (members of the currently recognized genera *Oedipina* and *Lineatriton*) were included in a single genus.[60] This was bad taxonomy (corrected by Tanner[58]), for here is an example of a true convergence, which arises from different morphogenetic mechanisms. I envision only the two morphogenetic options that have been exercised to produce elongation in salamanders—one is common and expected; the other is unique and represents a novel response to selection. All close relatives of *Lineatriton* have a fixed number of trunk vertebrae; the only variation on which selection could work was in vertebral length, not vertebral number.

In this example, homoplasy is a phenomenon of the whole organism. Related lineages independently have adapted to similar microhabitats by assuming essentially identical ecomorphologies based on fundamentally distinct morphogenetic mechanisms (change in shape versus change in number). While convergence in this instance might be interpreted as evidence of natural selection (although not within the more rigidly defined criteria of Endler[13]), there also is evidence of design limitations (the oddly elongated vertebrae are unique to *Lineatriton*) that make certain responses (increases in numbers of segments) more likely than others. In short, there is a bias to the direction of evolution.

TAIL AUTOTOMY AND DEFENSE AGAINST PREDATION

All tropical salamanders are capable of autotomizing the entire tail as a defense against predators. Under attack, a salamander disengages the tail, which then whips back and forth in a violent and dramatic manner while the animal lies quietly. At least some predators are attracted by the active tail, and when they eat it they receive a distasteful if not poisonous dose of secretion from the abundant "poison" glands.[7,27]

The functional morphology of tail autotomy is relatively well understood.[71] Typically there is a constriction at the base of the tail where one finds one or two shortened vertebrae, a shortened muscular segment, and weakened connective tissue. In contrast to the well-known convergent situation in lizards, in which an autotomy plane exists within one or more vertebrae, the vertebrae of salamanders disengage without separating into two halves. A precondition for tail autotomy restricted to a single spot is a wound-healing specialization that is widespread in plethodontids. When the tail is forcibly broken in species with this specialization, the skin breaks a full segment behind the muscle, so a sleeve of skin is left which collapses over the wound, staunching the flow of blood and facilitating blastema formation and regeneration; a perfect tail, lacking only the notochord, is regenerated. In order for this mechanism to function, a precise coordination of parts is required.

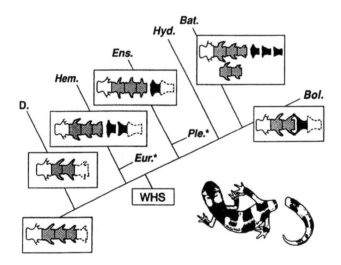

FIGURE 6 Scenario illustrating the evolution of tail autotomy in plethodontid salamanders. Collapsed cladogram (based on Lombard and Wake[36]) showing distribution of specialized zones of separation of vertebrae in the tails of plethodontid salamanders. Anterior to the left. Dorsal view. Vertebra outlined in black = sacrum. Stipple = caudosacral vertebrae. Black = vertebrae in front of which autotomy occurs. Broken line = nonautotomic caudal vertebrae. Eur* = all members of the Hemidactyliini except *Hemidactylium*. Ple* = all members of the Plethodontini except *Ensatina*. Ens. = *Ensatina*. WHS = Wound healing specialization. Other abbreviations as in Figure 1.

Tail autotomy has evolved to a high degree of specialization in plethodontids (but in no other salamanders) at least three times (*Hemidactylium, Ensatina,* supergenus *Bolitoglossa*), each event featuring a unique combination of morphological specializations associated with basal constriction (Wake and Dresner[71]; Figure 6). A different kind of specialization has evolved in *Batrachoseps*; no basal constriction is found, but specialized points of separation occur at each segmental boundary in the tail. Furthermore, in *Lineatriton* and *Oedipina* (two unrelated bolitoglossines) secondary convergence has twice given rise to the condition seen in *Batrachoseps*.

In the ancestral condition there are three caudosacral vertebrae. Two exist in desmognathines and in the supergenus *Bolitoglossa*, and two or three are found in *Batrachoseps*. The independent points of acquisition of tail autotomy are indicated in Figure 6. What appears to be a "key innovation" (Larson et al.[33]) is the evolution of a wound-healing specialization in the lineage leading to the tribes Plethodontini and Bolitoglossini. A different and less effective kind of wound-healing specialization has evolved (perhaps more than once) independently in the tribe Hemidactyliini.[71] Because tail autotomy has evolved only in species that inherited some kind of wound-healing specialization, I conclude that some such specialization in morphology is a necessary, but not sufficient, condition for the evolution of true tail autotomy.

This example shows that discrete variations on a theme occur once a particular adaptation has become fixed. These variations are limited by very specific functional and design constraints. A precise coordination of structural modifications at very specific points in the segmented body is necessary, and in species in which variation in vertebral number occurs, maladaptive organization can result (e.g., in *Ensatina*; Frolich[14]).

FOOT WEBBING, LOCOMOTION, DEVELOPMENT, AND ECOLOGY

Virtually all lowland species in the very large tropical genus *Bolitoglossa* (which includes about 20% of the species of living salamanders) are arboreal to some degree and they all have webbed hands and feet. Some of these species, usually the larger ones, have extensive interdigital webbing and are capable of generating suction.[1] However, other species, usually miniaturized species which are paedomorphic in much of their morphology, have apparently webbed feet that are in reality rather undifferentiated pads that cannot generate suction.[1,3] The limb buds of plethodontid embryos first produce a pad-like structure, and as development proceeds the digits grow out of these pads. The paedomorphic species truncate development at the pad stage, and thus as adults they have feet that are superficially similar to the highly specialized webbed feet of congeners (Figure 7), which arise from secondary growth of skin between the nearly normally developed digits. The extremely complex phylogenetic pattern is slowly emerging as monophyletic groups are being identified.[12,30,76] Often in lowland areas (as in Nuclear Central America), two

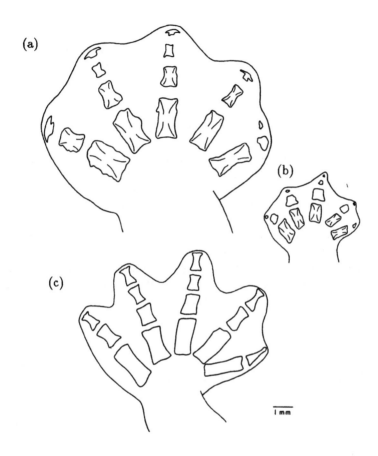

FIGURE 7 Fully webbed feet of three convergently derived taxa of tropical bolitoglossine salamanders. Only outlines of the entire foot and of the ossified parts of the digits are indicated. (a) *Bolitoglossa salvinii*, a Guatemalan species with enlarged feet in which cutaneous webbing has grown between the digits. (b) *Bolitoglossa rufescens*, Middle American species with small, pad-like feet that superficially appear to be webbed but are developmentally immature, paedomorphic structures. (c) *Chiropterotriton magnipes*, a cave-dwelling species from northern Mexico with enlarged feet in which cutaneous webbing has grown between the digits.

kinds of *Bolitoglossa* co-occur, one large with webbed feet and the other small with padded feet, and these may be close or distant relatives. When one examines another region (e.g, Talamancan Central America), one again finds this pattern, but the lineages represented are apparently independent.[76]

In the caves of northern Mexico, one encounters a very strange salamander, *Chiropterotriton magnipes*, unrelated to *Bolitoglossa*[9,42] but resembling many species of that genus in having extensive interdigital webbing in its prominent hands and feet (Figure 7). Here webbing has evolved in complete independence to that in *Bolitoglossa*, apparently as an adaptation for clinging to wet, smooth walls and ceilings of caves.

The point of this example is that both direct adaptation (an increase in webbing associated with efficient production of suction that is used to maintain perch, or with increasing surface tension to cling to smooth, wet surfaces) and indirect effects of miniaturization achieved through paedomorphosis can produce superficially similar outcomes. Both of these outcomes can facilitate ecological specialization for arboreality: large webbed feet produce suction, while miniaturized salamanders can move effectively along arboreal surfaces and hide in leaf axils. But neither is specific for arboreality—some paedomorphs are upland terrestrial forms, and some webbed forms live in caves at middle elevations, not in lowland arboreal habitats. Foot shape in paedomorphs appears to be largely unrelated to specific function, even in arboreal microhabitats, and parallelism in paedomorphs arises from design limitations (the nature of early development). In contrast, species with increased webbing are responding in parallel ways to generally similar selection for clinging ability, but in very different ecological contexts.

HIERARCHICAL FACTORS AND HOMOPLASY
TONGUE EVOLUTION

Terrestrial salamanders feed exclusively by apprehending their prey, typically small arthropods for plethodontids, with the tongue. Plethodontid salamanders have the longest, fastest, and most accurate tongues among salamanders and can fire the tongue a distance equivalent to one-third the length of the body in 7.7 msec.[32] Different taxa display differences, often subtle, in the way in which tongues are projected. Three functional classes are recognized: attached protrusible (the ancestral condition), attached projectile, and free projectile.[35] Whereas earlier workers thought that free projectile tongues had evolved only once,[59,61] each of the derived functional classes is now thought to have evolved three times within the Plethodontidae.[35,36] These arguments are based on biomechanical (e.g., the recognition that there are two folding options for skeletal components during tongue projection, both of which have been used) and phylogenetic considerations (the hypothesis of single origin of free projectile tongues is considerably less parsimonious than that of alternatives).

Elsewhere[47,68,74] it has been argued that the evolution of this complex, integrated system can best be understood by using both functionalist and structuralist perspectives. A series of necessary, but not sufficient, conditions establishes a framework within which homoplasy becomes increasingly probable (Figure 8). Loss of lungs, a synapomorphy for the Plethodontidae, frees the hyobranchial skeleton from the functional constraint of filling the lungs during respiration and makes extreme specialization possible. Loss of larvae, which has evolved independently at least three times in the Plethodontidae, frees the hyobranchial skeleton from the functional constraint of feeding in the larval stage and from the associated strongly cephalized development that is characteristic of species with larvae. As a result, the probability of heterochrony is increased. Ontogenetic repatterning (Roth and Wake[47]; Wake and Roth[77]) during direct development becomes increasingly likely as well, and there are many manifestations of this phenomenon for extreme specialization in biomechanical (e.g., the appearance of a new option for folding tongue cartilages during protraction), neurophysiological (early and extensive development of ipsilateral as well as the expected contralateral retinotectal projections,[43,46]) and behavioral (the uncoupling of forward lunging from tongue projection and the ability of salamanders to fire their tongues while maintaining a stationary body) aspects of tongue projection. Once specialization is initiated, it proceeds to a stopping point, often determined by biomechanical considerations, which can be overcome by a subsequent, usually novel, event. One such event is the disappearance of part

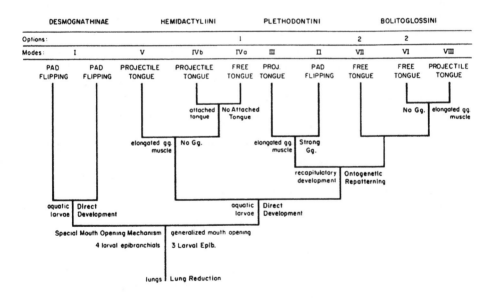

FIGURE 8 Flow diagram of tongue evolution in plethodontid salamanders. Based on phylogenetic hypothesis favored by Lombard and Wake[36] and arguments of Lombard and Wake,[35] Wake,[68] and Roth and Wake.[47] Options 1 and 2 (the latter has evolved twice) are the two biomechanical options for free tongues identified by Lombard and Wake.[35] The modes are the different kinds of functional tongues identified by Lombard and Wake.[35] Bold face indicates synapomorphies. Gg = genioglossus muscles.

of a skeletal element and the associated disarticulation of the tongue skeleton in the diminutive Mexican genus *Thorius*.[35] This produces a biomechanically efficient variation on the theme of folding option 2 that is a direct outcome of miniaturization combined with large cell volume (the result of relatively large genome size).

The point of this example is that both functionalist and structuralist perspectives are necessary for a deep understanding of how complex systems evolve. Some events are best understood through the use of functionalist explanations (e.g., evolution of direct development, miniaturization), but these in turn lead indirectly (as in the evolution of a second folding option, which does not inevitably evolve in association with larval loss) or directly (as in the case of miniaturization in *Thorius*) to morphological changes that have functional implications but are independent of the original adaptation. It is at this point that structuralist interpretations enrich our understanding of historical events.

MINIATURIZATION

Salamanders range in size from more than 1.5 m to species that become sexually mature at sizes as small as about 15 mm. Both extremes have been subject to homoplasy. Gigantism is associated with parallel neotenic trends in the families Cryptobranchidae, Amphiumidae, and Sirenidae. Miniaturization is less easily categorized. Within the family Plethodontidae, miniaturization has evolved repeatedly, in all major lineages, with concordant extensive homoplasy. Let us assume that miniaturization always is strictly adaptive, related to early sexual maturation and reproduction, use of specialized microhabitats, escape from predation, or other ecologically relevant phenomena. We can then examine the effects of organismal-wide miniaturization on parts of the organism.

Often miniatures resemble each other in unexpected ways, for when miniaturization occurs in species with large cells (as a consequence of having relatively large genomes), structural and design limitations are approached. Specific arrangements can be understood best within the analytical framework of structuralism. The phenomenon of miniaturization in plethodontids is receiving much recent attention by my colleagues James Hanken and Gerhard Roth[19,20,21,48,50]; here I present only a brief summary.

Within the bolitoglossine salamanders of the New World tropics, miniaturization has evolved independently many times; all members of the genera *Dendrotriton*, *Nototriton*, *Parvimolge*, and *Thorius* are miniaturized (i.e., sexual maturity of at least one sex is attained at a body size less than 30 mm), and miniaturized species have evolved within the genera *Bolitoglossa* and *Chiropterotriton*.[74] In *Thorius* some species become sexually mature at less than 15 mm, and one undescribed species does not exceed 20 mm.[23]

The bolitoglossines all have large to very large genomes,[52,53] so one has an *a priori* expectation that design limitations might be reached in miniaturized species. For example, in *Thorius* the organization of the head as a whole is affected.[19,20] Most species of *Thorius* inhabit small spaces under bark or surface cover, and their eyes do not protrude more than slightly beyond the limits of the head. But, because the cells are large (because the genome is large), the eye must be relatively large (in relation to the head) in order to achieve sufficient optic resolution for feeding and other functions. Therefore, the eyes must impinge on the space for the brain, which lies largely between the eyes, and the brain accordingly is deformed and displaced posteriorly.[19,46]

The neurons, as well as the other cells of *Thorius*, also are large, and there are space constraints within the confined cranial vault. There are only about 25,000 photoreceptors in the eye of *Thorius* (there are about 450,000 in *Rana*,[57] but there are 26,000 retinal ganglion cells; thus, the whole retina is a functional fovea, with maximal visual acuity.[34,48] The optic tectum, the main integration center in amphibian brains, which contains the cells that are directly related to visual function, contains only about 30,000 cells[50] in these extraordinarily tiny brains (there are about 800,000 in *Rana*[57]). Yet, although in relation to other salamanders, the eye is small, visual acuity and distance perception remain at an effective level. Vision is important in *Thorius*, which uses the most extremely specialized (in a biomechanical sense) tongue projection system known in salamanders.[35] The size of the brain areas containing visual and visuomotor centers is relatively greatly increased at the expense of other areas, especially the olfactory centers of the forebrain. Cell packing in miniaturized species with large genomes and large cells is increased at the expense of basal dendrites and glial cells, and is especially great in hatchlings.[50] The close packing of cells in the brains of *Thorius* and other miniaturized bolitoglossine salamanders has many implications for such phenomena as cell migration (which might be strongly impeded as a result), and may well contribute, together with developmental phenomena,[19] to the general impression of paedomorphosis in these species.

The nervous system is only one of the systems affected by miniaturization. The entire skeletal system is affected as well, but in different ways in different taxa. In *Thorius* the skull is retarded in development, and appears to be shrunken around the components associated with the olfactory, visual, and auditory systems (Figure 9). The upper jaws in most species are toothless, and many bones simply "float" on top of cartilage or in fibrous sheets, with no articulations.[20,64] There is an enormous cranial fontanelle, and the brain is largely not covered by bone.

In other miniaturized genera, design limitations are accommodated in different ways. In the genus *Parvimolge*, with a smaller genome than *Thorius*, the skull is not so deformed (Figure 9) and appears to be a smaller version of the skull of such out-group genera as *Chiropterotriton* and *Pseudoeurycea*. Cell density in *Parvimolge* is much lower than in *Thorius*.[50] In the western North American *Batrachoseps* (a member of the tribe Bolitoglossini), the cranial fontanelle is even larger than in *Thorius*, and the brain shares many features, but the jaws are relatively strong and well supplied with teeth.

Some miniaturized plethodontids have small genomes (*Desmognathus wrighti* and *Desmoganthus aeneus*). These species do not appear to have reached the limits that have been discussed, and they do not manifest the compromises that must be met in *Thorius*.[50]

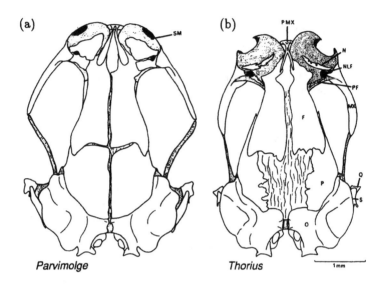

FIGURE 9 Dorsal views of the skulls of miniaturized species of tropical bolitoglossine salamanders. (a) An adult *Parvimolge townsendi* from Veracruz, Mexico, and (b) An adult of *Thorius* (undescribed species) from Oaxaca, Mexico. Cartilage is stippled. Fibrous tissue covering the brain is indicated by wavy lines. SM = septomaxillary bone, PMX = premaxillary, N = nasal, PF = prefrontal, F = frontal, MX = maxillary, P = parietal, Q = quadrate, S = squamosal, O = otic-occipital, NLF = nasolacrimal foramen.

This example shows that while miniaturization has occurred repeatedly, only when the organism as a whole is at its design limits must some structural-functional "compromise" occur. Genome and cell size are factors that mediate or exacerbate the consequences of changes in absolute body size. When compromise occurs, it often involves those features that are of apparently lesser adaptive significance—the forebrain and the teeth of *Thorius*, for example. But miniaturization also can affect structures that have no other developmental options—the loss of the fifth toe in some miniaturized taxa, for example.

HOMOPLASY—THE DILEMMA

Is the extensive homoplasy in plethodontids typical or unique? For many years my attempts to hypothesize a robust phylogeny for the plethodontids, particularly the tropical bolitoglossines, have been frustrated by homoplasy. The problem appears to be general; homoplasy is so common in salamanders that despite many efforts there is no generally accepted phylogenetic hypothesis for the Order Caudata. Each hypothesis requires extensive convergence and reversal.[11,24] New data from aligned sequences of 18s and 28s ribosomal RNA, based on statistical analysis tests using parsimony, are in deep conflict with all phylogenetic hypotheses based on morphological data (confirmed by the independent analysis of the data of Larson and Wilson[32] by Hillis[26]; see also Larson[31]), especially with regard to the phylogenetic position of the Plethodontidae. The sequence data have turned the accepted phylogenetic tree upside down by locating the plethodontids near the base of the tree and the cryptobranchoids rather far up in the tree, whereas morphologists always have concluded that plethodontids are a deeply nested taxon and the cryptobranchoids basal or nearly so. Larson[31] has reanalyzed the morphological data, and has shown that only slightly more non-molecular homoplasy is required by his most parsimonious tree than is required by other trees.

The combined morphological and molecular data reveal an extraordinary amount of homoplasy. It may be that salamanders are special in this respect, but I doubt it. We simply do not as yet have adequate molecular information for such groups as neobatrachian frogs, teleost fishes, and others in order to have accurate estimates of the relative amount of homoplasy in different taxa. If salamanders are unique, we should investigate the basis for that uniqueness, but until more groups have been investigated in comparable detail, we should not assume that salamanders are different from other vertebrate taxa.

The analysis by Sanderson and Donoghue[51] of 60 recent cladistic analyses shows that for both molecular and morphological data sets, levels of homoplasy (as measured by a consistency index) increase with increasing numbers of taxa. One obvious answer to the difficulty in finding a robust phylogenetic hypothesis for bolitoglossine salamanders and neobatrachian frogs is that there are too many species, for the probability of character-state change increases with the total number of branches in a tree. However, the other side of the issue, as these authors make clear, is that the probability of homoplasy increases if the number of possible states of characters are limited. That is the major point of the present paper (and an important point in Rieppel[45]). When these two factors combine, as in the bolitoglossines, levels of homoplasy become so high as to frustrate analysis. Perhaps in the case of the bolitoglossines, DNA sequence data will provide a new perspective on the problem, but the risks are great that the results will be no more clarifying than in the case of the salamander families, where all previous hypotheses were brought into question.

The relative balance of external (related to specific function and arising from interaction with environmental factors by means of natural selection) and internal (arising from such structuralist principles as wholeness, self-regulation, and transformation, and manifest through form generation during ontogeny) factors in the determination of biological form has yet to be determined for any group of organisms. I advocate a research program in which both avenues of explanation are pursued simultaneously, without subjugation. The goal is an eventual synthesis, such as I have attempted, in which both functionalist and structuralist perspectives are presented to explain how the selective assembly of unit structures into more complex organismal-level structures has evolved in a hierarchical manner, within a phylogenetic framework. Only when this is accomplished can the heuristic value of an analytical study of homoplasy be appreciated.

Homoplasy complicates phylogenetic analysis enormously, but at the same time it enriches our appreciation for the diversity of evolutionary processes.

ACKNOWLEDGMENTS

I am grateful to H. Greene, J. Hanken, K. Nishikawa, G. Roth, N. Shubin, and, especially, A. Larson and M. Wake, for comments on the manuscript, and to A. Larson and D. Hillis for permission to cite papers in press. Karen Klitz prepared most of the illustrations. I gratefully acknowledge the National Science Foundation for supporting my research program for many years.

REFERENCES

1. Alberch, P. "Convergence and Parallelism in Foot Morphology in the Neotropical Salamander Genus *Bolitoglossa*. I. Function." *Evolution* **35** (1981): 84–100.
2. Alberch, P. "The Logic of Monsters: Evidence for Internal Constraint in Development and Evolution." *Geobios, mem. spec.* **12** (1989): 21–57.
3. Alberch, P., and J. Alberch. "Heterochronic Mechanisms of Morphological Diversification and Evolutionary Change in the Neotropical Salamander *Bolitoglossa occidentalis* (Amphibia: Plethodontidae)." *J. Morphol.* **167** (1981): 249–264.
4. Alberch, P., and E. A. Gale. "A Developmental Analysis of an Evolutionary Trend: Digital Reduction in Amphibians." *Evolution* **39** (1985): 8–23.
5. Alberch, P., S. J. Gould, G. F. Oster, and D. B. Wake. "Size and Shape in Ontogeny and Phylogeny." *Paleobiol.* **5** (1979): 296–317
6. Brame, A. H. "Distribution of the Oregon Slender Salamander, *Batrachoseps wrighti* (Bishop)." *Bull. So. Calif. Acad. Sci.* **63** (1964): 165–170.
7. Brodie, E. D., Jr. "Antipredator Adaptations of Salamanders: Evolution and Convergence Among Terrestrial Species." In *Adaptations to Terrestrial Environments*, edited by N. S. Margaris, M. Arianoutsou-Faraggitaki, and R. J. Reiter, 109–133. New York and London: Plenum Press, 1983.
8. Bryant, S. V., D. M. Gardiner, and K. Muneoka. "Limb Development and Regeneration." *Am. Zool.* **27** (1987): 675–696.
9. Darda, D. M. "Morphological and Biochemical Evolution Within the Plethodontid Salamander Genus *Chiropterotriton*." Ph. D. dissertation in Zoology, University of California, Berkeley, 1988.

10. de Queiroz, K. "The Ontogenetic Method for Determining Character Polarity and Its Relevance to Phylogenetic Systematics." *Syst. Zool.* **34** (1985): 280–299.
11. Duellman, W. E., and L. Trueb. *Biology of Amphibians.* New York: McGraw-Hill, 1986.
12. Elias, P. "Salamanders of the Northwestern Highlands of Guatemala." *Contrib. Sci. Nat. Hist. Mus. Los Angeles Co.* **347** (1984.): 1–20.
13. Endler, J. *Natural Selection in the Wild.* Princeton, NJ: Princeton University Press, 1986.
14. Frolich, L. "Osteological Conservatism and Developmental Constraint in the Polymorphic 'Ring Species' *Ensatina eschscholtzii* (Amphibia: Plethodontidae)." *Biol. J. Linn. Soc.*, in press.
15. Goodwin, B. C. "Changing from an Evolutionary to a Generative Paradigm in Biology." In *Evolutionary Theory: Paths into the Future*, edited by J. W. Pollard, 99–120. New York: Wiley & Sons, 1984.
16. Goodwin, B. C., and L. E. H. Trainor. "The Ontogeny and Phylogeny of the Pentadactyl Limb." In *Development and Evolution*, edited by B. C. Goodwin, N. Holder, and C. C. Wylie, 75–98. Cambridge, MA: Cambridge University Press, 1983.
17. Hanken, J. "Appendicular Skeletal Morphology in Minute Salamanders, Genus *Thorius* (Amphibia: Plethodontidae): Growth Regulation, Adult Size Determination, and Natural Variation." *J. Morphol.* **174** (1982): 57–77.
18. Hanken, J. "High Incidence of Limb Skeletal Variation in a Peripheral Population of the Red-Backed Salamander, *Plethodon cinereus* (Amphibia, Plethodontidae) from Nova Scotia." *Canad. J. Zool.* **61** (1983): 1925–1931.
19. Hanken, J. "Miniaturization and Its Effects on Cranial Morphology in Plethodontid Salamanders, Genus *Thorius* (Amphibia, Plethodontidae). II. The Fate of the Brain and Sense Organs and Their Role in Skull Morphogenesis and Evolution." *J. Morph.* **177** (1983): 255–268.
20. Hanken, J. "Miniaturization and Its Effects on Cranial Morphology in Plethodontid Salamanders, Genus *Thorius* (Amphibia: Plethodontidae). I. Osteological Variation." *Biol. J. Linn. Soc.* **23** (1984): 55–75.
21. Hanken, J. "Morphological Novelty in the Limb Skeleton Accompanies Miniaturization in Salamanders." *Science* **229** (1985): 871–874.
22. Hanken, J., and C. Dinsmore. "Limb Skeletal Variation in the Red-Backed Salamander, *Plethodon cinereus*." *J. Herpetol.* **20** (1986): 97–101.
23. Hanken, J., and D. B. Wake. Unpublished data.
24. Hecht, M. K., and J. L. Edwards. "The Methodology of Phylogenetic Inference Above the Species Level." In *Major Patterns in Vertebrate Evolution*, edited by M. Hecht, P. C. Goody, and B. M. Hecht, 3–51. New York: Plenum Press, 1977.
25. Hennig, W. *Phylogenetic Systematics.* Urbana: University of Illinois Press, 1966.

26. Hillis, D. M. "The Phylogeny of Amphibians: Current Knowledge and the Role of Cytogenetics." In *Amphibian Cytogenetics and Evolution*, edited by D. M. Green and S. K. Sessions, 7–31. New York: Academic Press, 1991.

27. Hubbard, M. "Correlated Protective Devices in Some California Salamanders." *Univ. Calif. Publ. Zool.* 1 (1903): 157–170.

28. Lande, R. "Evolutionary Mechanisms of Limb Loss in Tetrapods." *Evolution* 32 (1978): 73–92.

29. Larsen, J. H., Jr., J. T. Beneski, Jr., and D. B. Wake. "Hyolingual Feeding Systems of the Plethodontidae: Comparative Kinematics of Prey Capture by Salamanders with Free and Attached Tongues." *J. Exp. Zool.* 252 (1989): 25–33.

30. Larson, A. "A Molecular Phylogenetic Perspective on the Origins of a Lowland Tropical Salamander Fauna. I. Phylogenetic Inferences from Protein Comparisons." *Herpetologica* 39 (1983): 85–99.

31. Larson, A. "A Molecular Perspective on the Evolutionary Relationships of the Salamander Families." *Evol. Biol.* 25 (1991): 211–277.

32. Larson, A., and A. C. Wilson. "Patterns of Ribosomal RNA Evolution in Salamanders." *Mol. Biol. Evol.* 6 (1989): 131–154.

33. Larson, A., D. B. Wake, L. R. Maxson, and R. Highton. "A Molecular Phylogenetic Perspective on the Origins of Morphological Novelties in the Salamanders of the Tribe Plethodontini." *Evolution* 35 (1981): 405–422.

34. Linke, R., G. Roth, and B. Rottluff. "Comparative Studies on the Eye Morphology of Lungless Salamanders, Family Plethodontidae, and the Effect of Miniaturization." *J. Morph.* 189 (1986): 131–143.

35. Lombard, R. E., and D. B. Wake. "Tongue Evolution in the Lungless Salamanders, Family Plethodontidae. II. Function and Evolutionary Diversity." *J. Morph.* 153 (1977): 39–80.

36. Lombard, R. E., and D. B. Wake. "Tongue Evolution in the Lungless Salamanders, Family Plethodontidae. IV. Phylogeny of Plethodontid Salamanders and the Evolution of Feeding Dynamics." *Syst. Zool.* 35 ((1986): 532–551.

37. McNamara, K. J. "A Guide to Nomenclature of Heterochrony." *J. Paleont.* 60 (1986): 4–13.

38. Patterson, C. "Morphological Characters and Homology." In *Problems of Phylogenetic Reconstruction*, edited by K. A. Joysey and A. E. Friday, 21–74. London: Academic Press, 1982.

39. Patterson, C. "Homology in Classical and Molecular Biology." *Mol. Biol. Evol.* 5 (1988): 603–625.

40. Piaget, J. *Structuralism*. New York: Basic Books, 1970.

41. Oster, G. F., N. Shubin, J. D. Murray, and P. Alberch. "Evolution and Morphogenetic Rules: The Shape of the Vertebrate Limb in Ontogeny and Phylogeny." *Evolution* 42 (1988): 862–884.

42. Rabb, G. B. "A New Salamander of the Genus *Chiropterotriton* (Caudata: Plethodontidae) from Mexico." *Breviora, Mus. Comp. Zool.* 235 (1965): 1–8.

43. Rettig, G., and G. Roth. "Retinofugal Projections in Salamanders of the Family Plethodontidae." *Cell Tiss. Res.* 243 (1986): 385–396.

44. Rieppel, O. C. *Fundamentals of Comparative Biology*. Basel: Birkhauser, 1988.
45. Rieppel, O. C. "Character Incongruence: Noise or Data?" *Abh. Naturwiss. Ver. Hamburg (NF)* **28** (1989): 53–62.
46. Roth, G. *Visual Behavior in Salamanders*. Berlin: Springer-Verlag, 1987.
47. Roth, G., and D. B. Wake. "Trends in the Functional Morphology and Sensorimotor Control of Feeding Behavior in Salamanders: An Example of the Role of Internal Dynamics in Evolution." *Acta Biotheor.* **34** (1985): 175–192.
48. Roth, G., B. Rottluff, and R. Linke. "Miniaturization, Genome Size and the Origin of Functional Constraints in the Visual System of Salamanders." *Naturwissenschaften* **75** (1988): 297–304.
49. Roth, G., B. Rottluff, and R. Linke. "Miniaturization, Genome Size and the Origin of Functional Constraints in the Visual System of Salamanders." *Naturwissenschaften* **75** (1988): 297–304.
50. Roth, G., B. Rottluff, W. Grunwald, J. Hanken, and R. Linke. "Miniaturization in Plethodontid Salamanders (Caudata: Plethodontidae) and its Consequences for the Brain and Visual System." *Biol. J. Linn. Soc.* **40** (1990): 165–190.
51. Sanderson, M. J., and M. J. Donoghue. "Patterns of Variation in Levels of Homoplasy." *Evolution* **43** (1989): 1781–1795.
52. Sessions, S. K. "Cytogenetics and Evolution of Salamanders." Ph.D. dissertation in Zoology, Univ. California, Berkeley, 1984.
53. Sessions, S. K., and A. Larson. "Developmental Correlates of Genome Size in Plethodontid Salamanders and Their Implications for Genome Evolution." *Evolution* **41** (1987): 1239–1251.
54. Shapiro, I. M., E. E. Golub, B. Chance, C. Piddington, O. Oshima, O. C. Tuncay, and J. C. Haselgrove. "Linkage Between Energy Status of Perivascular Cells and Mineralization of the Chick Growth Cartilage." *Devel. Biol.* **129**: 372–379.
55. Shubin, N., and P. Alberch. "A Morphogenetic Approach to the Origin and Basic Organization of the Tetrapod Limb." *Evol. Biol.* **20** (1986): 319–387.
56. Simpson, G. G. *Principles of Animal Taxonomy*. New York: Columbia University Press, 1961.
57. Szekely, G., and G. Lazar. "Cellular and Synaptic Architecture of the Optic Tectum." In *Frog Neurobiology*, edited by R. Llinas and W. Precht, 407–434. Berlin: Springer-Verlag, 1976.
58. Tanner, W. W. "A New Genus of Plethodontid Salamander from Mexico." *Great Basin Nat.* **10** (1950): 27–44.
59. Tanner, W. W. "A Comparative Study of the Throat Musculature of the Plethodontidae of Mexico and Central America." *Univ. Kans. Sci. Bull.* **34(2)** (1952): 583–677.
60. Taylor, E. H. "The Genera of Plethodont Salamanders in Mexico, Part I." *Univ. Kans. Sci. Bull.* **30(1)**: 189–232.
61. von Wahlert, G. "Biogeographische und Oekologische Tatsachen zur Phylogenie Amerikanischer Schwanzlurche." *Zool. Jb. Syst.* **85**: 253–282.

62. Wagner, G. "The Origin of Morphological Characters and the Biological Basis of Homology." *Evolution* **43** (1989): 1157–1171.

63. Wake, D. B. "Comparative Osteology of the Plethodontid Salamander Genus *Aneides*." *J. Morphol.* **113** (1963): 77–118.

64. Wake, D. B. "Comparative Osteology and Evolution of the Lungless Salamanders, Family Plethodontidae." *Mem. So. Calif. Acad. Sci.* **4** (1966): 1-1111.

65. Wake, D. B. "The Abundance and Diversity of Tropical Salamanders." *Am. Nat.* **104** (1970): 211–213.

66. Wake, D. B. "Aspects of Vertebral Evolution in the Modern Amphibia." *Forma et Functio* **3** (1970): 33–60.

67. Wake, D. B. "Functional and Evolutionary Morphology." *Persp. Biol. Med.* **25** (1982): 603–620.

68. Wake, D. B. "Functional and Developmental Constraints and Opportunities in the Evolution of Feeding Systems in Urodeles." In *Environmental Adaptation and Evolution*, edited by D. Mossakowski and G. Roth, 51–66 Stuttgart: Fischer-Verlag, 1982.

69. Wake, D. B. "Adaptive Radiation of Salamanders in Middle American Cloud Forests." *Ann. Missouri Bot. Gard.* **74** (1987): 242–264.

70. Wake, D. B. "Phylogenetic Implications of Ontogenetic Data." *Geobios, Mem. Spec.* **12** (1989): 369–378.

71. Wake, D. B., and I. G. Dresner. "Functional Morphology and Evolution of Tail Autotomy in Salamanders." *J. Morphol.* **122** (1967): 265–306.

72. Wake, D. B., and P. Elias. "New Genera and a New Species of Central American Salamanders, with a Review of the Tropical Genera (Amphibia, Caudata, Plethodontidae)." *Contrib. Sci. Mus. Nat. Hist. Los Angeles Co.* **345** (1983): 1–19.

73. Wake, D. B., and J. D. Johnson. "A New Genus and Species of Plethodontid Salamander from Chiapas, Mexico." *Contrib. Sci. Nat. Hist. Mus. Los Angeles Co.* **411** (1989): 1–10.

74. Wake, D. B., and A. Larson. "Multi-Dimensional Analysis of an Evolving Lineage." *Science* **238** (1987): 42–48.

75. Wake, D. B., and R. Lawson. "Developmental and Adult Morphology of the Vertebral Column in the Plethodontid Salamander *Eurycea bislineata*, With Comments on Vertebral Evolution in the Amphibia." *J. Morph.* **139** (1973): 251–300.

76. Wake, D. B., and J. F. Lynch. "The Distribution, Ecology and Evolutionary History of Plethodontid Salamanders in Tropical America." *Sci. Bull. Nat. Hist. Mus. Los Angeles Co.* **25** (1976): 1–65.

77. Wake, D. B., and G. Roth. "The Linkage Between Ontogeny and Phylogeny in the Evolution of Complex Systems." In *Complex Organismal Functions: Integration and Evolution in Vertebrates*, edited by D. B. Wake and G. Roth, 361–377. Chichester: Wiley & Sons, 1989.

78. Wake, D. B., T. J. Papenfuss, and J. F. Lynch. "Distribution of Salamanders Along Elevational Transects in Mexico and Guatemala." In *MesoAmerican Biogeography*. Tulane University Press, in press.
79. Webster, G. C., and B. Goodwin. "The Origin of Species: A Structuralist Approach." *J. Soc. Biol. Struct.* 5 (1982): 15–47.
80. Wiley, E. O. *Phylogenetics*. New York: Wiley-Interscience, 1981.

David G. Stork
Ricoh California Research Center, 2882 Sand Hill Road #115, Menlo Park, CA 94025-7022

Preadaptation and Principles of Organization in Organisms

Recent simulations of the evolution of the neural circuitry subserving the tailflip escape maneuver in the crayfish are discussed in relation to fundamental principles of organization in organisms. The simulations and analyses shed light on the possible origin of a "useless" synapse in the current tailflip as a vestige from a previous evolutionary epoch in which the circuit was used for swimming instead of flipping. Such preadaptation effects may underlie a broad range of neural and other structures throughout the animal world, and illustrate fundamental principles of organization in organisms, most notably the locally greedy nature of evolutionary change, the partial separability of dynamical processes at different levels of organization, which imply that under some circumstances at least, "elegance of design counts for little."

Principles of Organization in Organisms,
SFI Studies in the Sciences of Complexity, Proc. Vol. XIII,
Eds. J. Mittenthal & A. Baskin, Addison-Wesley, 1992   **205**

INTRODUCTION

The structure and function of every organism depend crucially upon its evolutionary precursors.[1] The form of the human nervous system, for example, depends upon the evolutionary history of hominids and pre-hominids.[31] Evolutionary change is so fundamental to our understanding of biology that Dawkins[5,6] claimed that life without the notion of evolution was virtually unthinkable.

Neural systems of all animals possess structure at birth,[11] and such structure is absolutely fundamental to the performance of the organism. It determines what can and what cannot be performed by the organisms, and what can and what cannot be learned from its environment.[12,32,34] Investigations into the *sources* of biological structure will help us understand principles of organization in organisms.

Because biological structure arose through evolution in complex, dynamic environments, biological solutions need not always conform to good "engineering" design principles. The research related here is directed to understanding how "inelegant"—indeed, counter intuitive, or "non-optimal"—structures might arise through evolution, even in quite simple neural systems. We argue, moreover, that "non-optimality" should be expected to be even more prevalent in complex neural structures, for instance, the human brain.

Although its roots extend back to the time of Darwin,[3] the concept of *preadaptation* has been recently elaborated by S. J. Gould, E. Mayr and others.[14,15,27] Preadaptation is used to describe the process by which an organ, behavior, neural structure, etc., which evolved to solve one set of tasks is later utilized to solve a *different* set of tasks. It illustrates the dichotomy between designed, planned, and "optimal" forms in biology on the one hand, and "non-optimal" ones on the other.

An example of preadaptation of an organ is the bird wing. The proto-bird wing was too short to be used for flight, and hence must have been used for some other task; the Darwinian fitness at that time did not depend upon flight. Theories of the use of the proto-wing include thermoregulation (the proto-bird spreads or retracts its wings to cool or warm itself), insect catching (the proto-wings are used to knock insects out of the air for food), reorientation during jumps when hunting for insects (the proto-bird can then catch insects from a larger volume of air), and others. Regardless of these theories, it is clear that the proto-wing was indisputably not used for flight. Later in evolution, as the proto-wing became longer, a behavioral threshold was reached in which the limb *could* be used for flight. At that time, then, a different set of evolutionary pressures were placed on the wing, favoring a lighter and more aerodynamic wing.

There was no way in which the proto-bird could "anticipate" the ecological niche provided by flight, of course. Thus, the later wing had to be built upon the structures that evolved for the previous task. Thus there could be structures in the current bird wing—holdovers from the earlier evolutionary epoch—that are "non-optimal" for flight.[33] In short, locally "greedy" processes (optimizing for the current use) at one structural level might be incommensurate with optimality at another.

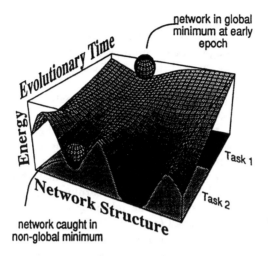

FIGURE 1   Preadaptation. Metaphorical "energy" or fitness landscape describing performance of a neural network throughout evolution. Evolutionary time runs from the back of the figure to the front; the "energy" (a measure of the contribution of the network to the fitness in the population) is vertical; some index of network structure runs left to right. At an early epoch, the network may have been optimal for solving the task at that time—Task 1—but later, the appearance of Task 2 deforms the energy landscape. The network might therefore be in a local minimum, and hence "non-optimal" for Task 2. In our typical crayfish simulations below, Task 1 is swimming and Task 2 flipping.

If such structures do not present an excessive biological "cost" (say, in metabolic resources, space, weight,...), then that structure may remain in the later system. Even if the structure does pose some cost to the organism, that structure might nevertheless remain in the later organisms, since intermediate states in its elimination may prove very detrimental to the organism. In such a case, the structure is "frozen into" the organism, a relic of the earlier evolutionary epoch, much as the appendix is "frozen into" humans.

Figure 1 illustrates metaphorically the process of preadaptation, and can be discussed in terms of neural circuitry (our primary system of interest). At an early epoch, the network solved Task 1, and might even have been optimal for it. (Optimal is, of course, dependent upon one's measure. We need not be specific here, but state roughly that a circuit which uses the minimum number of components, metabolites and structure to solve the problem without compromising the organisms ability to solve other problems can be regarded as more nearly optimal than a circuit that doesn't.) At a later evolutionary epoch, a *different* task becomes more relevant. This switch in task might be due to a changing environment, or to the network evolving such that new niches become available (as in the bird wing), and so on. The network is then under *different* evolutionary pressures, and the "energy landscape" is deformed. The network, however, must build upon structures selected based on Task 1—structures that might not be appropriate for the second task. The result is that the network may be "non-optimal" for Task 2—caught in a local minimum of the energy function.

Simulation studies of preadaptation elucidate the nature of evolutionary change and the function of biological networks (especially since such information cannot be preserved in the fossil record). Preadaptation sheds light on the principles of organization in organisms, as well as the study of artificial neural networks in at least two ways: it can help guide the "reverse engineering" of biological systems, showing which structures might or might not be relevant to the cognitive task at hand; it can suggest general hybrid evolution-learning neural networks based on biological processes.[25,36,37] It stresses the interplay of phenomena at very large (evolutionary) scales and very small (physiological) scales.

The crayfish tailflip neural circuit provides fertile material for simulation studies. First, this circuit has been extensively mapped by neurophysiologists.[42] Second, the circuit is small enough that realistic simulations—both of short-time behavioral processes, and long-term evolutionary processes—can be made. (It would be absurd to try to simulate the evolution of visual cortex, in contrast.) Third, an apparently "non-optimal" structure is evident in the circuit. Such non-optimal structures, while present, might be difficult to discover in more complicated structures, such as the brain. Fourth, the circuit is responsible for a behavior that is of the utmost survival value for the crayfish (flipping away from danger), and thus Darwinian selection pressures on the circuit are great. Fifth, a highly plausible evolutionary scenario can be made for the circuits, as our simulations will confirm. Finally, the crayfish has a phylogenetically close relative, *Anaspides tasmaniae*, which can serve as a "control" organism, since its homologous circuits lack the "non-optimal" structures found in the crayfish—a result that can easily linked to its different behavior.

CRAYFISH TAILFLIP CIRCUIT

The crayfish tail (abdomen) consists of six segments, each with its own small neural circuit linking pressure-sensitive cells ultimately to flexor muscles governing tail segment flexion (bending). The tailflip escape maneuver we shall consider (the LG-mediated flip) is effected by *flexion* of the *anterior* segments (segments 1-3) with *no* flexion in the *posterior* segments (segments 4-6). Figure 2 shows the basic structure of the actual crayfish circuits responsible for this behavior.

Segment 2, an anterior segment, consists of a simple linear sequence of neurons with excitatory synapses. An aquatic pressure pulse (e.g., from a predator) leads to excitation in the sensory interneurons, and by the circuit shown, ultimately excites the flexor muscles in segment 2. The flexion necessary for the tailflip is thereby achieved.

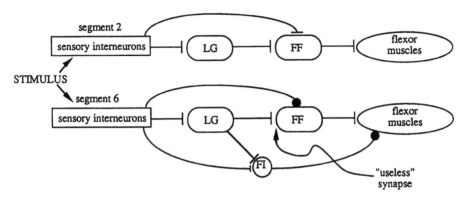

FIGURE 2   The neural circuitry subserving the tailflip in crayfish.[8] Excitatory synapses are represented by a "T" and inhibitory synapses by a •. In the event of a rapid rise in ambient water pressure (from a predator), pressure transducers yield excitatory activation in the sensory interneurons. To effect the tailflip maneuver, each anterior segment (e.g., segment 2) must flex (i.e., the flexor muscles must be excited) and each posterior segment (e.g., segment 6) must *not* flex (i.e., the flexor muscle must be inhibited). Note especially that one of the excitatory synapses in segment 6 is "useless": any time an excitatory volley passes from neuron LG to FF, the FF neuron is inhibited (via connections from the sensory interneurons), thereby rendering the excitation ineffective. Furthermore, the only projection of the FF (which is to the flexor muscles) is also overridden by inhibition from the FI neuron.

The circuit in segment 6, a posterior segment, is different. Here the aquatic pressure pulse leads to inhibition of the flexor muscles. The lack of flexion in segment 6 together with the vigorous flexion in segment 2 gives the tailflip response. A neural volley passing from the LG to the FF neuron would lead to excitation of the FF. However, this excitation is counteracted by the direct *inhibitory* connection from the sensory interneuron to the FF itself. There is, moreover, inhibition of the flexor muscle via the FI neuron.[41,42] The synapse between the LG and FF is thereby overridden; it seems to have no functional purpose. So far as is known, then, the LG⇒FF synapse is "useless"—the circuit is "non-optimal."

The question naturally arises: Why does the crayfish have this apparently "useless" synapse? What can account for such "non-optimality" in design?

PREADAPTATION HYPOTHESIS

Dumont and Robertson[8] hypothesized that the excitatory LG⇒FF synapse is a vestige from an earlier evolutionary epoch, one in which the proto-crayfish did *not* flip, but instead merely *swam*. (Simultaneous and iterated flexion in all segments would lead to swimming, as in the *Anaspides tasmaniae*, which has in each of its six tail segments a circuit homologous to those in the anterior segments of the crayfish.) The hypothesis is that the circuits in the posterior segments originally had the form at the top of Figure 2 (appropriate for swimming), but under a change or inclusion of a new task, the circuit evolved by building upon the previous ones. The LG⇒FF synapse was useful for swimming, but not for LG-mediated flipping, and the circuit evolved other connections to override that synapse. Because that synapse is no longer expressed behaviorally, it is "frozen into" the circuit—a vestige of the earlier epoch, and non-optimal in the context of the circuit's current use.

We discuss computer simulations presented more fully elsewhere[37] and further analysis in support of this hypothesis, and their relationship to general principles of organization in organisms.

SIMULATION APPROACH

The overall approach follows a classical Darwinian evolution scenario, shown in Figure 3. Each network has a haploid genome, which is expressed to yield the full network, including connectivities and neural response characteristics. Networks then respond to the environment—a simulated pressure pulse from a predator—and are selected based on their response. The selection will be based either on swimming or on flipping. The selected networks then reproduce to give the genes of the next generation, and the cycle continues.

GENOTYPE

The genetic representation and development used in the computer model are meant to capture the most biologically relevant aspects of actual crayfish. The most important question centers on that of genetic representation of neural connection strengths: is this representation *localized* (each initial connection strength determined by one or a small number of genes) or is it *distributed* (the many connection strengths determined by several genes)?

There is abundant evidence for pleiotropy and a *distributed* genetic representation in genetic neurobiology.[6,17,19,40] Most notably, there does not seem to be much evidence for "one gene-one synapse." Instead, genetic representations can act in several ways: setting *affinities* for connections, development rates, etc.[30] Furthermore, there are many cases in which mutations in a single or a small number of genes can have distributed consequences, as in many systemic neural disorders such as multiple sclerosis. On the computational and systems levels, a distributed representation has several useful properties. Perhaps most importantly, it permits mutations to make large changes in network structure, thereby leaving small refinements to be accomplished through learning.[25,28,29]

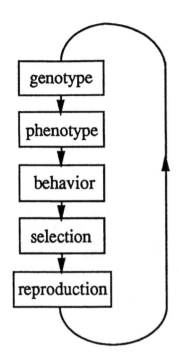

FIGURE 3 Evolutionary processes. The genes lead via development to a structured network, including interconnections (excitatory and inhibitory), neural channel properties, etc. The network then responds to the environment and is selected based on the resulting fitness score. Fitness depends upon the posited task, here either swimming or flipping. The most-fit individuals then reproduce to yield the genotypes in the next generation, and the evolutionary processes continue.

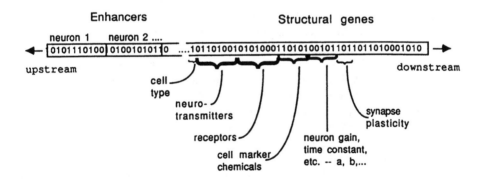

FIGURE 4 Haploid genome used in simulations. Structural genes (shown downstream, grouped for convenience) govern the phenotypic structures in the network. Enhancers (upstream, grouped by neuron for convenience) govern the expression of the structural genes.

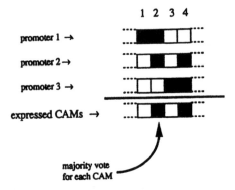

1 2 3 4

promoter 1 →

promoter 2 →

promoter 3 →

expressed CAMs →

majority vote
for each CAM

FIGURE 5 Model for the expression of cell adhesion molecules in a neuron. Suppose that for a given neuron three promoter genes are activated. In the example shown here, the first leads to activiation of the structural genes 1 and 2 (black), which would lead to CAM1 and CAM2; promoter 2 would likewise lead to CAM2 and CAM4, etc. (This relationship between promoters and these structural genes is stored in a lookup table in the simulations.) The final CAMs expressed in the neuron are the result of a majority vote for each CAM—in the case shown, CAM2 and CAM4 are expressed. (Tie votes are decided by an unbiased random decision.)

The simulations employ a distributed representation, based on properties of *control genes* and *structural genes*.[21] The structural genes code for fundamental aspects of the phenotype, here the cell type, neurotransmitters, type of synaptic receptors, etc.; the control genes guide the expression of the structural genes (Figure 4). Thus, for instance, if a particular enhancer from the control genes is activated, it will activate expression of a set of structural genes, distributed along the entire genome. This approach captures the fact that certain phenotypic features are expressed in concert. For example, a human photoreceptor contains both photopigment and disks, as well as other structures unique to photoreceptors[7]; these are all expressed together. (One typically does not find cells with photopigment but no disks, for instance.) In our model, then, several of these features are represented by a *single* structural gene; if that gene is activated, *all* of the component phenotypic features are candidates for expression.

Consider just one of the phenotypic traits: cell marker chemicals, or cell adhesion molecules (CAMs), implicated in developmental programs for connectivity.[9,10] In the model, there are four types of CAMs; during development the initial connectivity between two neurons is specified by the similarity in their CAMs, just as many biological CAMs, large cell surface glycoproteins, are homophilic. Suppose that activation of promoter 1 (also sometimes called an enhancer) would lead to the expression of CAM1 and CAM2 (Figure 5). If no other promoters are activated, the final neuron would have those two CAMs expressed. But suppose, moreover, that activation of promoter 2 would lead to CAM2 and CAM4, but *not* CAM1 and

CAM3 (and analogously for promoter 3, as shown in the figure). If all three promoters are activated, each would express its corresponding set of CAMs, but prevent other CAMs from being expressed. The final distribution of CAMs expressed in a neuron is then the result of a majority vote for each CAM, as if the promoters competed among themselves to express the individual CAMs. Similarity in the CAMs expressed in any two neurons determines the initial interconnectivity—the greater the similarity, the stronger the initial synaptic connection, in accord with homophilic properties of CAMs.[9]

A similar computation occurs for the neurotransmitter to be produced in a neuron; twelve candidate neurotransmitters are used since twelve neurotransmitters (e.g., GABA, acetylcholine,...) account for the vast majority of transmitters appearing in the animal kingdom. Only *one* transmitter is expressed (as described by Dale's Law, which is not universally obeyed). Genes coding for acetylcholine and cholineacetyltransferase have been found on two separate chomosome segments in *Drosophila melanogaster*,[16,19] and this suggests that a similar arrangement could exist in the crayfish. Grouped phenotypic features that lead to a neuron being either a sensory, or an inter-, or a motor neuron are expressed by an analogous mechanism, though with only three (exclusive) attributes rather than twelve.[11]

Neural channel properties are computed as the *average* of those from each structural gene activated. Thus, if one structural gene would lead to a large number of Na channels, while another would lead to a small number, then if both are activated, that actual number expressed will be intermediate.[20] Such features of the model are motivated by recent results on mutations in three different alleles in the Shaker locus, which led to post-synaptic potentials in muscles longer and larger than in the wild type,[24] implying a genetic representation of potassium channels.[39]

What is important here is that the relationship between genetic representation and ultimate phenotype is distributed and indirect.

PHENOTYPE

Each neuron is thus described by its global type (sensory, inter-, or motor neuron), its decay rate constant (a in equation 1, below), neural channel concentrations, which determine the excitatory and inhibitory saturation levels (b, c, d, e, below), its neurotransmitter type, its synaptic receptor type, and its complement of cell adhesion molecules.

The network as a whole is specified by the neural interconnectivities, determined by the similarities of the CAMs (computed as a Hamming distance) on each candidate pair of neurons. There is also a distance-dependent term, making neurons that are physically farther separated have lower connectivity for any given CAM similarity. Expressed networks have the form shown in Figure 9, below.

BEHAVIOR

The behavior of each neuron in the network is governed by Hodgkin-Huxley equations of the following form:[18,22]

$$\frac{dx_i}{dt} = -az_i + (b - cx_i)\Big\{ \sum_{j \in G_{ex}} z_{ij} f(x_j) + I_i \Big\} + (d - ex_i)\Big\{ \sum_{j \in G_{in}} z_{ij} f(x_j) \Big\} \quad (1)$$

where:

x_i = activity in neuron (depolarization)

$f(x_j)$ = output spike rate, a compressively nonlinear transfer function of the activity

a, b, c, d, e = constants describing ion concentrations, channel densities, etc. In particular, a describes the time constant for neural recovery, b and c together with a specify the excitatory saturation level, and likewise d, e and a specify the inhibitory saturation level

z_{ij} = strength of synapse between neurons i and j

I_i = external input for neuron i (not due to other neurons)

G_{ex} = the set of neurons connected to neuron i by synapses leading to excitation

G_{in} = the set of neurons connected to neuron i by synapses leading to inhibition

The right-hand side of the equation consists of three terms. The first denotes a relaxation decay, the second an excitation term (involving the sum over all the inputs that lead to excitation), and the third term analogously for inhibition. For our task, the input I_i is non-zero only for the sensory neuron, and in that case consists solely of a brief delta-function impulse at $t = 0$, corresponding to the pressure pulse from the predator.

SELECTION

The selection procedures are based on fitness-proportional reproduction[13]; the fitness score depends upon the task. For *swimming*, this score is equal to the maximum instantaneous excitation in the network's motor neuron (normalized over the population), corresponding (roughly) to the strength of flexion in the posterior tail segments. For *flipping*, the score is the maximum magnitude of *inhibition* in the motor neuron, corresponding (roughly) to the lack of such flexion. Although other measures of fitness are possible (motor neuron activity integrated over time, maximum value of the derivative of the activation, etc.), the one used captures the

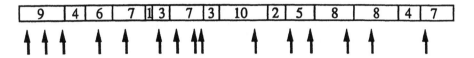

FIGURE 6 Selection for fitness-proportional reproduction. Each network is represented by a rectangle having a width equal to its fitness score. Selection is achieved by randomly chosing points along the entire population (arrows); if a randomly placed arrow points to a network's rectangle, then that network will survive and reproduce for the next generation. Thus the probability a network survives is proportional to its fitness score. It is possible—though somewhat rare—for a network with very low fitness score to be selected over a network with high fitness score. (Here the scores have been arbitrarily normalized to maximum = 10.)

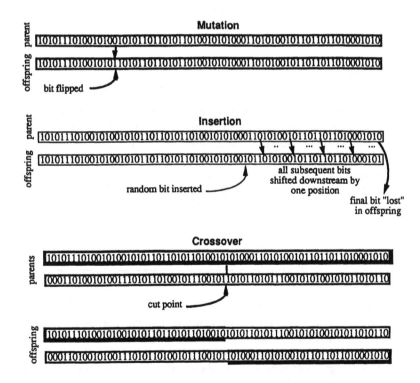

FIGURE 7 The processes of random mutation, bit insertion, and cross-over (shown) as well as replication (i.e., duplication without mutation, not shown) are used between generations.

behaviorally relevant features of flexion. This fitness function is biologically plausible, since the crayfish tailflip maneuver is fundamental to its survival. Of course, other traits confer fitness: the discussion here concentrates solely on one of the most important.

The algorithm for selection can be visualized as taking the fitness scores of each of the networks in a population and lining them up in a bar whose length is proportional to the individual scores. Then, points are chosen randomly and independently along the entire length (Figure 6). The probability of reproduction is thus proportional to the fitness—fitness-proportional reproduction. (The number of points chosen is equal to the number of individuals—the size of the population does not change from generation to generation.) The networks selected in this way are then reproduced (see subsection on reproduction, below).

Such fitness-proportional reproduction is biologically motivated and generally preferable to schemes in which merely the *most*-fit individuals are selected by truncation selection. In general, fitness-proportional reproduction helps to preserve diversity in the genome by permitting some low-fitness networks to pass on their genes.

REPRODUCTION

Those networks selected in the manner just described are reproduced using the familiar processes of replication, mutation ($p_{bit\ flip} = 10^{-2}$/bit/generation), bit insertion ($p_{bit\ insert} = 10^{-3}$/network/generation), and single-position crossover (75% of pairs), as put forth by Holland[23] and illustrated in Figure 7. The genetic algorithm parameters—in particular, the somewhat high mutation rate—were chosen in order to probe the phenomena as thoroughly as possible using our computer. The fundamental findings did not depend significantly on the choice of parameters, initial random number seeds, etc., over a wide range.

RESULTS AND ANALYSIS

All simulations were done on Connection Machine CM-2s, either at RIACS (Moffett Field, CA) or Thinking Machines Corporation (Cambridge, MA). The program consisted of roughly 12,000 lines of Cde*, the parallelized version of C; typical simulations required two to three hours.

On SIMD (single instruction multiple data) computers, there is always the question of the level at which parallelization of the problem should be made. The Connection Machine operating system and C* language permit construction of **domains**, which are processed in parallel. Candidate domains for the crayfish simulations were:

■ individual networks,

- individual neurons, and
- individual synapses.

(The temporal dynamics of the neurons are inherently serial—the integration of Eq. 1—and thus could not be parallelized. Indeed, this serial integration alone accounted for over 1/4 of the total processing time.)

For instance, if the code were parallelized at the level of individual networks, then the neurons and synapses would be serially processed. If, on the other hand, individual *neurons* were parallelized, then just the synapses and any finer grain structures would be processed serially, and so on. While parallelizing to the finest grain (here synapses) would lead to most rapid calculations, the overhead in interprocessor communication would thereby increase, since each neuron interacts with several other neurons. For the small number of neurons (7), parallelizing at the level of individuals was most efficient. Only if the number of neurons per circuit were larger (roughly 20–30) would the speedup in computation by parallelizing at the neuron level outweigh the drawbacks in communication overhead.

The parallel aspect of the the computer program is that all members of the population are calculated simultaneously in parallel on this SIMD machine. Individual neurons and synapses within a network are computed in series. The parallel data structure "*domain* individual," a C* domain allocated one processor (each with 8 kbytes of memory) per crayfish circuit. All the code was on the host VAX, while the data (synaptic strengths, neural activities, etc.) were stored on each physical processor. Whereas the statistics for larger numbers of individuals is only slightly more reliable than that presented for 200 individuals (see Figure 8), analyzing individual networks for "non-optimal" structures—which had to be done laboriously, by hand—became prohibitively time consuming. Thus, only simulations with 200 individuals are summarized here.

PREADAPTATION

Figure 8, from a typical simulation, illustrates the basic phenomenon of preadaptation.[35,36] The graph on the left shows the population average fitness as a function of generation. From generation 0 to generation 75, the task was assigned to be *swimming*; after that, the task was assigned to be *flipping*. The population's average fitness drops precipitously as the circuits previously selected for swimming are then tested and selected for flipping. Later the fitness levels off (by generation 150) to a mean score of 0.13 (in arbitrary units). The right-hand graph shows evolution in the case of rewarding flipping alone—*no* preadaptation. After 75 generations the mean score, 0.29 (in the same arbitrary units), is significantly above that of the preadapted networks in the left figure, given the same number of generations rewarding flipping. In short, evolving flipping networks from those previously selected for swimming leads to poorer performance than evolving them from the random networks present at the beginning of each of our simulations. Although, of

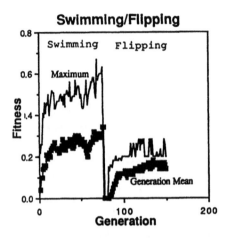
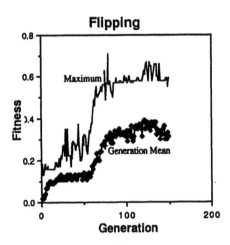

FIGURE 8 (left) The maximum individual fitness and the generation mean fitness for a population selected first for swimming and then (after generation 75) for flipping. (right) Population selected solely for flipping. Thus the final fitness for flipping is lower if the population was first selected for swimming. The minimum fitnesses were zero at virtually every generation, and hence have not been plotted. The same normalization convention was used for the graphs.

course, there is a small chance the preadapted networks (Figure 8, left graph) could spontaneously increase in fitness through a fortuitous combination of mutations or crossovers, the networks seem caught in a local minimum (cf. Figure 1).

The structure of preadapted networks differed from those not preadapted (Figure 9). In particular (based on an analysis of several dozen networks), roughly three times as many "non-optimal" structures were found in preadapted circuits as in non-preadapted circuits (other variables held constant). The structures we termed "non-optimal" included neurons unconnected to the rest of the network and synapses whose polarities (e.g., excitatory) were counterbalanced by another projection of the opposite polarity (i.e., inhibitory).

Because non-optimal structures arose more frequently in preadapted circuits in these simulations, and because several simulated circuits had non-optimal forms very closely homologous to those in the biological crayfish (compare Figures 2 and 9), our simulations provide support for an understanding of the LG⇒FF synapse in the crayfish in terms of preadaptation.

A possible objection arises: how can we be sure that the LG⇒FF synapse is indeed never used by the crayfish for some other purpose?—perhaps we simply have not been clever enough to guess a use. By analogy, very recent work on potassium channels in *Aplysia* on first analysis seemed to show that certain channels were non-functional, and hence perhaps non-optimal.[38] It was only after the ambient water temperature was raised from the (natural) 10°C to the warmer 15-20°C that these channels became active. (This suggested that the channels might help prevent convulsions in the *Aplysia*.) As F. H. C. Crick has remarked, evolution can be more creative than humans!

To such objections we respond that the manifest simplicity of the crayfish network and the restricted behavioral repertoire exhibited by the crayfish (at least evident in laboratory studies) seems to limit such hypothetical uses. Of course, a use might be found in the future. It might be possible that the "non-optimal" synapse and attendant projections give an architectural constraint of some sort, and cannot be removed without great behavioral and fitness cost. (One hypothetical "use" for the "non-optimal" circuit is for the inhibitory sensory-FF projection to limit the duration of an excitatory volley—perhaps to make a short "burst" in activity in the motor neuron. Alas, this does not appear to be the case in either the crayfish or our model networks: the inhibition of the FF neuron invariably preceeds the excitatory volley through the "useless" synapse.)

Given the simplicity and plausibility of the preadaptation scenario provided by Dumont and Robertson and by our simulations, our explanation seems far more acceptable than any current alternative.

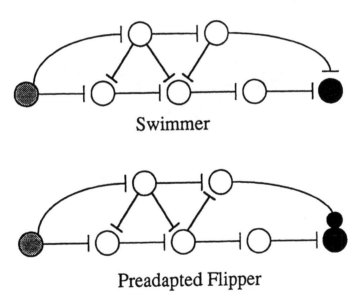

FIGURE 9   (top) Network resulting from evolution by selection for swimming alone. (bottom) Network after preadaptation scenario. Note in particular the non-optimal connections in the lower circuit. In both circuits, the sensory neuron is shown at the left and the motor neuron at the right. (As in Figure 2, a "T" represents an excitatory connection and a • an inhibitory one.)

EVOLUTIONARY MEMORY

How can we understand in a deeper way the preservation of genetic information coding for functionally useless structures? Perhaps we can consider genetic information to be "junk." But note: junk is fundamentally different from trash. The junk around our house was at one time useful, and is often stored in an attic in the possibility of being used later. Trash, however, might never have been useful, and is not useful at present. We discard trash; we save junk, even if there is but a small chance that it might be used again. Perhaps the distributed genetic information responsible for the "useless" synapse is "junk" in just this way.

In order to explore this possibility, different simulations were made. Neurons were selected first for swimming, and then for flipping (as before), thereby creating a population of networks which possessed a significant fraction of structures "non-optimal" for the flipping. The task was then changed back *again* to swimming, in order to see how rapidly and how well the population then evolved for swimming.

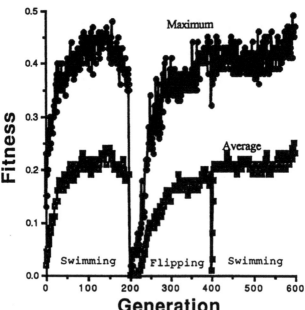

FIGURE 10   Evolutionary memory. The population was first selected for 200 generations for swimming, and then for another 200 generations flipping. At generation 400, the task was changed back to swimming. Note especially that the recovery of fitness is extremely rapid in this last epoch (i.e., after generation 400).

Figure 10 shows typical results. After selection for swimming and then flipping, the population fitness rose very *rapidly* for the subsequent swimming task. The population did this more rapidly than when it had evolved under the first swimming epoch, presumably in part because the later evolution could appropriate structures remaining from the first swimming epoch. The "junk" in the genome permits the crayfish to *rapidly* relearn how to swim, should the environment require it.

Keeping genes that were useful at previous epochs may help explain how evolution can be faster at later epochs, since the structures need only be recalled or reselected, not rebuilt *ex nihilo*.[4,40] Thus, just as the genes are not trash, but instead "junk" (possibly useful in the future), so the LG$\Rightarrow$FF synapse can be considered junk.

CONCLUSIONS AND FUTURE WORK

These simulations support an explanation that an apparently "useless" feature of the contemporary crayfish tailflip circuit arose from preadaptation, specifically, that the crayfish circuit was historically selected based on the circuit's ability to have the crayfish *swim*; later selection was based on the crayfish's ability to *flip*. As such, there are features "left over" from the earlier (*swimming*) epoch and, hence, perhaps "non-optimal" in the current (flipping) circuit. Nevertheless, genes that code for structures that are at one epoch "useless" may be expressed under different environmental circumstances, and thus permit the system to respond rapidly to changing environments.

These results, and the theories underlying them, have great import for theories of biological systems. As Dumont and Robertson write of the evolution of biological networks[8]: "As long as both the end result and all the intervening stages work, elegance of design counts for little." The same phenomena are even more likely to occur in *complex* neural systems (which have more degrees of freedom) because there are more intervening stages between the genes and the behavior they influence. Hence non-optimality may permeate neural systems in the animal world. We thus provide an alternate—but not necessarily competing—explanation to that of Edelman[9] for the large number of silent and perhaps unused synapses throughout the mammalian brain.

It has been argued pursuasively that human language has a strong innate, and hence genetic, component.[2,32] However, speech seems to have arisen fairly late in hominid evolution, roughly 100,000 years ago.[26] This epoch is very brief (on an evolutionary time scale), and surely too brief for complex language circuits to arise *ex nihilo*. Thus it appears likely that our current language circuits appropriated and built upon structures selected for tasks other than language. Perhaps the most plausible use for the circuits before language was orofacial motor control.[26,34] Generalizing and extrapolating from our crayfish analysis, we can perhaps understand why language may not be "optimal," i.e., why grammar contains quirky forms and rules, due to preadaptation.

ACKNOWLEDGMENTS

This paper is a revision and extension of that in Stork.[37] Useful discussions with several participants of the SFI Workshop in Principles of Organization in Organisms are gratefully acknowledged, most notably those with Stuart Kauffman, Jay Mittenthal, and Dave Wake.

REFERENCES

1. Bonner, J. T. *The Evolution of Complexity*. Princeton: Princeton University Press, 1988.
2. Chomsky, N. *Syntactic Structures*. Mouton: The Hague, 1957.
3. Darwin, C. R. *The Origin of Species*. 1st edition. Harmondsworth, Middlesex: Penguin, 1866. Reprinted 1968.
4. Dawkins, R. "The Evolution of Evolvability." In *Artificial Life*, edited by C. Langton, 201–220. Santa Fe Institute Studies in the Sciences of Complexity, Proc. Vol. VI. Redwood City, CA: Addison-Wesley, 1988.
5. Dawkins, R. *The Extended Phenotype*. Oxford: Oxford University Press, 1989.
6. Dawkins, R. *The Selfish Gene*. Oxford University Press, 1976.
7. Dowling, J. E. *The Retina*. Harvard: Belknap, 1987.
8. Dumont, J. P. C., and R. M. Robertson. "Neuronal Circuits: An Evolutionary Perspective." *Science* **233** (1986): 849–853.
9. Edelman, G. *Neural Darwinism*. New York: Basic Books, 1988.
10. Edelman, G. *Topobiology*. New York: Basic Books, 1988.
11. Edwards, J. S. "Pathways and Changing Connections in the Developing Insect Nervous System." In *Developmental Neuropsychobiology*, edited by W. T. Greenough and J. M. Juraska, 74–93. New York: Academic Press, 1986.
12. Gallistel, C. R. *The Organization of Learning*. Cambridge: MIT Press, 1990.
13. Goldberg, D. *Genetic Algorithms in Search, Optimization and Machine Learning*. Reading, MA: Addison-Wesley, 1989.
14. Gould, S. J. "Darwinism and the Expansion of Evolutionary Theory." *Science* **216** (1982): 380–387.
15. Gould, S. J., and E. S. Vrba. "Exaptation—A Missing Term in the Science of Form." *Paleobiology* **8** (1982): 4–15.

16. Greenspan, R. J. "Mutations of Choline Acetyltransferase and Associated Neural Defects in *Drosophila Melanogaster.*" *J. Comp. Physiology* **137** (1980): 83–92.
17. Griffiths, A. J. F., and J. McPherson. *100+ Principles of Genetics.* New York: Freeman Press, 1989.
18. Grossberg, S. *Studies in Mind and Brain.* Boston: Reidel, 1982.
19. Hall, J. C., R. J. Greenspan, and W. A. Harris. *Genetic Neurobiology.* Cambridge: MIT Press, 1982.
20. Hall, J. C., and D. R. Kankel. "Genetics of Acetylcholinesterase in *Drosophila Melanogaster.*" *Genetics* **83** (1976): 517–535.
21. Hawkins, J. D. *Gene Structure and Expression.* Cambridge: Cambridge University Press, 1986.
22. Hodgkin, A. L. *The Conduction of the Nervous Impulse.* Springfield: C. C. Thomas, 1964.
23. Holland, J. *Adaptation in Natural and Artificial Systems.* Ann Arbor: University of Michigan Press, 1975.
24. Jan, Y. N., J. Y. Jan, and M. J. Dennis. "Two Mutations of Synaptic Transmission in *Drosophila.*" *Proc. Royal Soc. B* **198** (1977): 87–108 .
25. Keesing, R., and D. G. Stork. "Evolution and Learning in Neural Networks: The Number and Distribution of Learning Trials Affect the Rate of Evolution." In *Advances in Neural Information Processing 3 (NIPS-3)*, edited by R. P. Lippmann, J. E. Moody, and D. S. Touretzky, 804–810. Palo Alto, CA: Morgan Kaufmann, 1991.
26. Lieberman, P. *The Biology and Evolution of Language.* Cambridge: Harvard University Press, 1984.
27. Mayr, E. *Evolution and the Diversity of Life.* Cambridge, MA: Belknap Press, Harvard University Press, 1976.
28. Miller, G., P. Todd, and S. Hegde. "Designing Neural Networks Using Genetic Algorithms." *Proceedings Third International Conference on Genetic Algorithms.* Palo Alto, CA: Morgan-Kaufmann, 1989.
29. Plotkin, H. C. "Learning and Evolution." In *The Role of Behavior in Evolution*, edited by H. C. Plotkin, 133–164. Cambridge: MIT Press, 1988.
30. Purves, D., and J. W. Lichtman. *Principles of Neural Development.* Sunderland, MA: Sinauer, 1985.
31. Spinelli, D. N. "A Trace of Memory: An Evolutionary Perspective on the Visual System." In *Vision, Brain and Cooperative Computation*, edited by M. A. Arbib and A. R. Hanson. Cambridge: MIT Press, 1987.
32. Stork, D. G. *Review of Parallel Distributed Processing: Explorations in the Microstructure of Cognition, Vols. 1 and 2*, edited by D. E. Rumelhart and J. L. McClelland and the PDP Research Group. *Bull. of Math. Biology* **50** (1988): 202–207.
33. Stork, D. G. "Preadaptation and Evolutionary Considerations in Neurobiolgy." In *Learning and Recognition—A Modern Approach*, edited by K. H. Zhao, C. F. Zhang, and Z. X. Zhu, 51–58. Singapore: World Scientific, 1989.

34. Stork, D. G. "Sources of Structure in Neural Networks for Speech and Language." *International Journal of Neural Systems,* in press.

35. Stork, D. G., S. Walker, M. Burns and B. Jackson. "Preadaptation in Neural Circuits." *Proceedings of the International Joint Conference on Neural Networks-90,* Washington, D.C. Vol. I, 202–205, 1990.

36. Stork, D. G., and R. Keesing. "Evolution and Learning in Neural Networks: The Number and Distribution of Learning Trials Affect the Rate of Evolution." Submitted.

37. Stork, D. G., S. Walker, M. Burns and B. Jackson. "'Non-Optimality' Via Pre-Adaptation in Simple Neural Systems." In *Artificial Life II,* edited by C. Langton, 409–429. Santa Fe Institute Studies in the Sciences of Complexity, Vol. X. Redwood City, CA: Addison-Wesley, 1991.

38. Triestman, F. M., and A. J. Grant. "Increase in Cell Size Underlies Cell-Specific Temperature Acclimation of Early Potassium Currents in *Aplysia.*" Submitted.

39. Wilkins, A. S. *Genetic Analysis of Animal Development.* New York: Wiley, 1988.

40. Wills, C. *The Wisdom of the Genes: New Pathways in Evolution.* New York: Basic Books, 1989.

41. Wine, J. J. "Escape Reflex Circuit in Crayfish: Interganglionic Interneurons Activated by the Giant Command Neurons." *Biological Bulletin* **141** (1971): 408.

42. Wine, J. J., and F. B. Krasne. "The Cellular Organization of Crayfish Escape Behavior." In *The Biology of Crusacea 4 in Neural Integration and Behavior,* edited by D. C. Sanderman and H. L. Atwood, 241–292. Academic Press, 1982.

Arnold J. Mandell and Karen A. Selz
Laboratory of Experimental and Constructive Mathematics, Departments of Mathematics and Psychology, Florida Atlantic University, Boca Raton, FL 33431

Allosteric Brain Enzyme Kinetics as Phenotypic Microevolutionary Process

INTRODUCTION

Ligand-induced, nonlinear changes in macromolecular behavior are well known in systems ranging from active site mechanisms, membrane receptors, and ion channels, to activators or repressors of molecular biological processes, and metabolically distant substances may influence both the rates and the kinetic conformations ("shapes") of the substrate-velocity curves of, for example, rate-limiting enzymes. The property of allostery in enzyme and membrane receptor proteins can be defined, generally, as the *nonlinear* sensitivity of rate-regulatory kinetic mechanisms to smaller molecules that may or may not also be acted on by binding and/or catalysis during the generation of the time-dependent observable, such as product formed or ligand bound per macromolecular equivalent.

The small molecule co-reactant may regulate *its own* enzymatic or binding-rate function by both serving as substrate supply *and* ligand influences on conformation.

Principles of Organization in Organisms,
SFI Studies in the Sciences of Complexity, Proc. Vol. XIII,
Eds. J. Mittenthal & A. Baskin, Addison-Wesley, 1992

These are two interacting, concentration-dependent effects. This self-organized, "autonomous" behavior confounds the linear laws of mass action with nonlinear regulatory influences. By definition, these effects are best observed away from equilibrium, and are studied as enzyme catalytic rates over increasing substrate concentrations, or as membrane protein-binding kinetics over increasing ligand concentrations.

In our studies of brain enzyme kinetics using physiologically relevant, nonequilibrium reactant concentrations, we have found that these allosteric Hill-plot characterizations (named for the English physiologist who used Eq. (1) to describe oxygen saturation curves imputed to the allosteric protein, hemoglobin) may be *only piecewise correct.*[2,5,7,11] We found, as did Teipel and Koshland,[18] that these nonlinear, allosteric functions *were near periodically cyclic over increasing substrate concentration.* New allosteric substrate saturation cycles emerged at apparent critical points after apparent "saturation plateaus"; that is, after $V_{gt'} = 0$, over some range of ligand or substrate concentration. It appeared that new functional potential emerged in support of another saturation segment after that which was available had become saturated.

In the language of second-order phase transition theory,[16] in the phenomenon of recurrent saturation plateaus, the system reaches a flat coexistence line between two (conformational) phases, "hesitates" while small portions of the precritical enzyme conformation are transformed successively from one phase to another, and when all the machinery is in place, the system moves into its next, "scaled-up" state of conformational catalytic capacity. Allosteric systems operating in the neighborhood of these $V_{gt'} = 0$ "critical points" can be described by a "scaling exponent," we will call it β, which dilates or contracts the system while maintaining the ratio relations

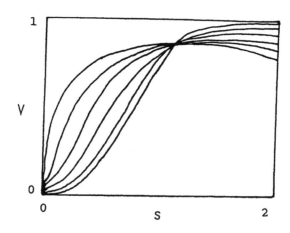

FIGURE 1   Increasing α from 0.5 to 3.0, for a fixed value of I, leads to a family of substrate-velocity functions, which converge at the $s = 1, V' = 0$ "critical point," and then diverge.

between its segments (components). This contrasts with the local nonlinearity parameter, the "allosteric" exponent, α, the *influence of which propagates across scale* (see Figure 1).

A dynamical approach to molecular biology that could account for these recurrent kinetic conformational rearrangements and that would allow increased or more efficient function to emerge "as needed" in the behavior of a hydrated, protein polymer with a genetically fixed sequence of amino acid monomers, may be relevant to the micromechanics of evolutionary theory. We remind ourselves that the machinery of the genetic process itself is composed of nucleotide polymers of fixed sequence, in water.

Is it possible that a phenotypic, recurrent, nonlinear, evolutionary scenario, with critical-point transitions, has analogical relevance to genotypic adaptation? That is, could allosteric changes in the "substrate-velocity" function constitute a new pseudo-genotypic manifold upon which the "phenotypic" time dynamics of the biochemical actions have "evolved?"

The issues in evolutionary biology that may be analogous to the behavior of these brain enzymes include some current problems for neo-Darwinian theory, such as the eonic-recursive sequence found in some phylogenetic trees of *morphological stasis, punctuated equilibria, speciation events, and return to morphological stasis.* We will explore some of the parallels that can be drawn between the phenotypic adaptation of a brain enzyme and the macromolecular, polymeric dynamics that must constitute the machinery of genotypic evolution, to see what, if anything, can be learned.

Specifically, by examining the adaptively "critical," conformational transition behavior of a class of brain macromolecules, might we be able to better imagine a "use inheritance" mechanism for fast genetic adaptation? Exploiting clonal selection and reverse transcriptase ideas, we suggest that the results of nucleotide, macromolecular "phase transitions," as seen in brain allosteric enzyme activities, would propagate via somatic-germ gene relations (see Steele,[17] for a formal development of this position using the lymphocyte-generated circulating and tissue-immunological system as a model).

MACROMOLECULAR SCALING KINETICS AS CRITICAL BEHAVIOR

A characteristic expression for the *velocity* of the product formed or ligand bound, V, as a function of macromolecular metric, g, and time, t, over increasing substrate or ligand concentration, s, in these allosteric systems is Eq. (1),

$$V(s)_{gt} = \frac{s^{\alpha}}{1 + s^{\alpha}} \tag{1}$$

with α representing the "order" of the reaction, often called its monomeric molecularity or cooperativity (nonlinearity), and l is an inertial constant. The subscript gt ($g \equiv$ metric on catalytic velocity and $t \equiv$ time) refers to conditions of the measurements (see original articles). The first derivative, describing the rate of change in velocity, V, of the reaction over increasing s,

$$V(s)'_{gt} = \frac{\alpha(s^{\alpha-1}l + s^{\alpha-1}s^{\alpha} - s^{2\alpha-1})}{(l+s^{\alpha})^3}, \tag{2}$$

is like a second derivative describing the curvature of the substrate-velocity function at $V = 0$ and the level of s at the inflection point of the curve where $V'' = 0$ of the characteristically "sigmoid" kinetic function.

Characteristic behavior of $V(s)$ across s, as in Eq. (1), is portrayed in Figure 1 at α values ranging from 0.5 to 3.0 in steps of 0.5 at 1 value of 0.2. This plot demonstrates the overlap of the six allosteric kinetic functions with increasing nonlinear molecularity exponents α, at what we call the critical point with respect to the implications it has for both the first, $V(s)'_{gt}$, and second, $V(s)'_{gt}$, derivative. Fluctuations in effective molecularity and/or nonlinearity converge here and then diverge with increasing s, in V-inverting order.

Figure 2 portrays $V(s)'_{gt}, (\partial V/\partial s)(\partial s/\partial t)$, as is implicit in Eq. (2), across 10 values of α and across fine-grained steps in substrate concentration, $s \in (0,1)$. It demonstrates the dynamical characteristic of a critical point in the normalized plot at $\alpha = 0.5$, with an apparent second-order "scaling relation" in $V(s)'_{gt}$ with respect to α, is analogous to that observed for the first-order one in the normalized plot with respect to s where $s = 1$ in Figure 1. Generally, one might anticipate a second-order phase transition in such a neighborhood.

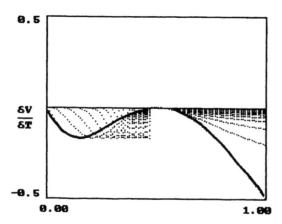

FIGURE 2 The time derivative of the velocity across $s(0,1)$ at increasing values of α, demonstrate the $V' = 0$ "critical point" as inferred from Figure 1.

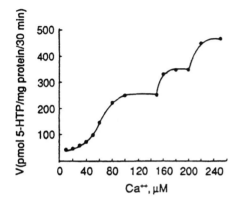

FIGURE 3 Increasing concentrations of Ca^{++} ion in rat brain stem tryptophan hydroxylase preparations demonstrate "iterative saturation plateaus," suggestive of ligand-induced phase transitions. See text.

Figure 3 demonstrates real data from a brain allosteric enzyme in a study of rat midbrain, tryptophan-hydroxylase catalytic velocity over increasing micromolar concentrations of the ion-ligand calcium, Ca^{++} (heterotropic), similar to the functions observed with increasing tryptophan (homotropic) which appear to be evidence for the interpretation of critical point, phase-transition behavior.[2]

A simple, analytic approximation of these kinetic patterns involving a combination of linear mass action, $V(s)_{gt} = ks$, allosteric nonlinearity, Eq. (1), and saturation cycles, $\cos(s)$, can be made as Eq. (3),

$$V(s)_{gt} = \frac{s^{\alpha}}{l + s^{\alpha}} + ks + \cos(s) , \qquad (3)$$

and the derivative of the velocity with respect to substrate concentrations, s, can be approximated by Eq. (4).

$$V(s)'_{gt} = \frac{s^{\alpha}\alpha}{s(l + s^{\alpha})} - \frac{(s^{\alpha})^2\alpha}{(l + s^{\alpha})^2 s} + k - \sin(s) . \qquad (4)$$

Figure 4 is a plot of Eq. (3) with $l = 0.3, \alpha = 3$, and the range of s and $V(s)_{gt}$ as indicated.

During the last few years, experiments and computer simulations of spatially extended, driven, non-equilibrium systems (here think of hydrated proteins in autonomous motion forced by increasing substrate concentration) have shown that they evolve into the most stable of the available "unstable" states (here think of dropping grains of sand on piles of near-avalanche, critical heights representing increasing substrate concentrations). When slightly more perturbed, the resulting relaxation processes to the next most stable of a set of unstable states are *scale invariant*. Their "infinite" correlation length indicates motions and "sizes" at all scales

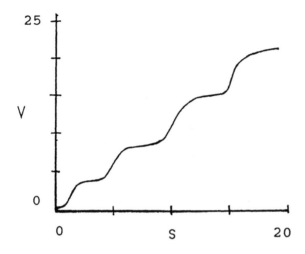

FIGURE 4   A substrate-velocity function generated by the kinetic model, Eq. (3), demonstrating the recurrent "saturation" behavior seen in the kinetics of brain stem tryptophan hydroxylase, as in Figure 3.

up to the finite limit of the system.[1] These critical states recur spontaneously with continued perturbation, and lead to the finding that the most robust characterization of the "self-organized" critical system is an (invariant) scaling law, rather than some metric quantity. An example is Eq. (5), which is the multiphasic, allosteric power law, from Eqs. (1) and (3),

$$V(s)_{gt} = \left\{ \frac{s^\alpha}{1 + s^\alpha} + \frac{ks}{\beta} + \cos\left(\frac{s}{\beta}\right) \right\}^{1+\beta} \tag{5}$$

with $1 < \beta < 2$. Higher "asymptotic plateaus" are associated with longer substrate cycles as β increases. Whereas $s = \mu$mols of amino acid tyrosine or tryptophan substrate in the brain enzyme experiments shown above with tyrosine and tryptophan hydroxylase enzyme activities, respectively, in these experiments, β (suitably normalized) represents the concentration of tetrahydrobiopterin "cofactor" in μmolar concentrations.

Figure 5(a) demonstrates the hierarchical multiplicities of the substrate curves of the rat brain stem substantia nigra tyrosine hydroxylase activities with increasing tetrahydrobiopterin cofactor concentrations, normalized between one and two as β, with each point in the plot representing the median value of five duplicate assays. The substrate-velocity curve of Figure 5(b) is a fine-grained substrate curve, in 0.2 μmolar increments, in triplicate, with the median value computed and normalized around a zero mean as determined across a range of substrate demonstrating patterns found in several neighborhoods of critical substrate concentrations. It demonstrates the characteristic nested hierarchy of amplitudes and frequencies implied by power-law scaling functions.

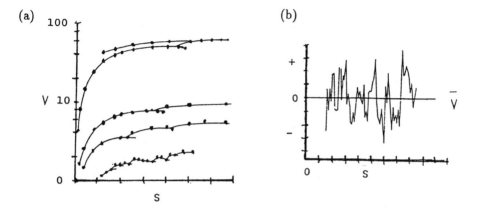

FIGURE 5 (a) Demonstrates substrate-log (velocity) curves of rat brain stem tyrosine hydroxylase activity at five increasing values of tetrahydrobiopterin concentration (parameter β), suggesting hierarchical "scaling" behavior in the kinetic segments. (b) When s was fine grained and plotted as deviations from zero mean velocity, fluctuations on at least three scales across substrate could be observed.

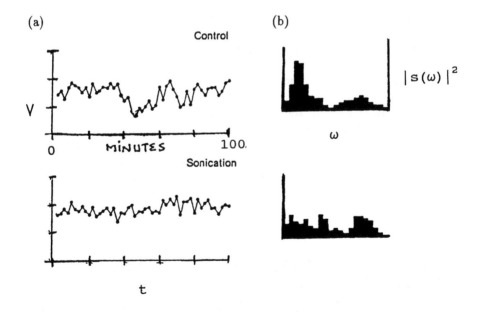

FIGURE 6 Sonication decreased α in the allosteric substrate-velocity curve (not shown) and was associated with a change in the time-dependent fluctuations in (a) discretely sampled brain stem tyrosine hydroxylase (b) and their power spectral transformations.

TIME DYNAMICS OF SCALING ALLOSTERIC KINETIC FUNCTIONS

As one might anticipate, the time-dependent dynamics of allosteric proteins such as rat brain stem tyrosine hydroxylase are subject to the nonlinear changes in kinetic conformation induced by various ligands and reaction conditions.[3] For instance, perturbation by sound waves, called sonication, increased characteristic values for α in *single-cycle* substrate-velocity curves, computed as the steepest linear ascent in a log-log plot of $V(s)_{gt}$ versus s functions. Figure 6 shows that sonication also changed the time course of the fluctuations in catalytic velocity, as indicated by both the time series and the Fourier Transform of their autocorrelation functions, in a comparison of control and post-experimental tyrosine hydroxylase fluctuations.[12]

One approach to the dynamical modeling of Eq. (5) is via the iteration of a discrete system which recursively represents the time dependence of the parameter and substrate-dependent nonlinear system where s_t represents $V(s)_{gt}$ and $s_{t+1} = f(s_t)$, as in Eq. (6).

$$s = \left\{ \frac{s_t^{\alpha}}{l + s_t^{\alpha}} + \frac{k s_t}{\beta} + \cos\left(\frac{s_t}{\beta}\right) \right\}^{1+\beta} \tag{6}$$

Making comparisons between the substrate-velocity functions at three values of β, as in Eq. (5), we studied the comparable dynamical model system in Eq. (6) heuristically, with $\alpha = 4$ (the relationship between monomeric composition in relationship to its experimental α is nonlinear; for example, the generic allosteric protein, hemoglobin, has four monomers but an α of 2.7), $l = 5$, $k = 1$, s varying from 0.1 to 20, and $\beta = 1.25$, 1.50, and 1.75. Figure 7 demonstrates the substrate saturation curves of Eq. (5), the s_t versus s_{t+1} return maps of Eq. (6), and the Fourier transformations of the autocorrelation sequence of the time series of Eq. (6) at the three indicated values of β. It is clear and not surprising that the dynamical patterns of allosteric protein activity can be "tuned" by both the local allosteric parameter α as in Eq. (1) and Figure 2, and the global scaling exponent β, as in Eqs. (5) and (6) and in Figure 7.

SYMBOLIC DYNAMICS AND RUN GENERATION BY ALLOSTERIC DYNAMICS

We have shown previously that polypeptide sequences in hormones and proteins can be encoded, using a hydrophobic sequence code, into four families of five amino acids each (and almost equally well into a binary partition with ten amino acids each). This symbolic dynamic coding led to the successful prediction of function for even apparently unrelated congeners of growth-hormone-releasing factors, corticotropin-releasing factors, calcitonins, membrane receptors, and allosteric proteins.[4,6,8,21]

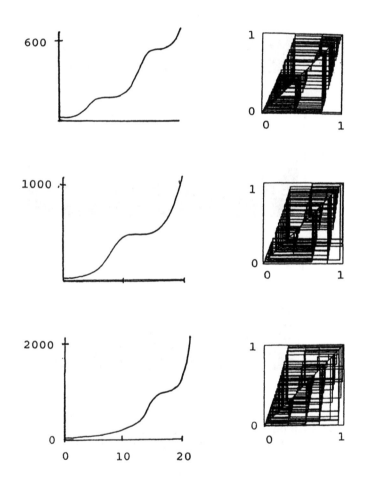

FIGURE 7 (a) The relationship between substrate velocity fluctuations generated by Eq. (5) at β = 1.25, 1.50, and 1.75 (with l = 5 and α = 3) from top to bottom Left, and the return maps of the time dynamics of Eq. (6) for these values on the Right. (continued)

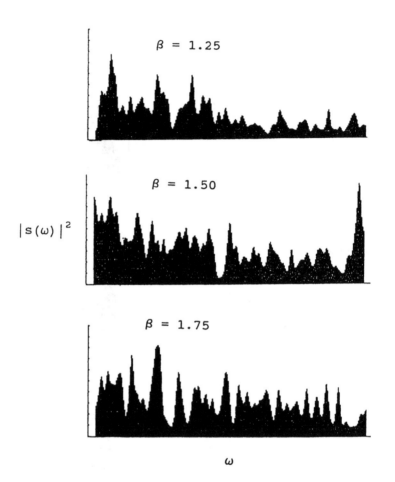

FIGURE 7 (continued) (b) The power spectral transformations of the time series of Figure 7(a), at the indicated values for β (see text), demonstrate how allosteric ligands influencing the "shape" of substrate-velocity functions can alter the frequency structure of the catalytic fluctuations, as in the real system of Figures 5 and 6.

Our polypeptide-protein pattern-matching category system is based on the Fourier Transformation of the autocorrelation transformation of the sequence of hydrophobic values of the series of amino acids in the peptide, in which "runs" of various lengths of either hydrophobic or hydrophilic residues are represented as Fourier modes in the power spectrum. We have shown this Fourier mode approach to binary run distributions in amino acid sequences to successfully group polypeptides with common actions, and categorize many proteins by symbolically implied and actual function. For example, the amino acid elements of β-strands oscillate in hydrophobicity with an average "period" of about 2.2 residues; the α-helix in 3.6 residues; the growth hormone releasing factor-glucagon family, 4.0, etc. Proteins can also be characterized by their "longest period" $\cong$ "longest run." For example, transmembrane regions in proteins involved with second messenger phosphorylation functions may have "runs" of as many as 15–18 hydrophobic residues; concavalin A manifests runs of 11 residues, thermolysin has 8, whereas hemoglobin is dominated by runs of 3 to 4 residues (probably α-helixes). *Generally, proteins with more pronounced dynamical requirements have "run distributions" of shorter average length whereas structural proteins manifest longer runs.*

That this encoding scheme has functional significance is the frequent finding, from matching hydrophobic amino acid residue, power spectral Fourier modes among polypeptide ligands and receptors that "bind" to each other. Figure 8 is an example. A monomer from the cholinergic receptor of the "electrical eel," *Electrophorus*, demonstrates α and β modes of 2.0 and 3.64 residues representing two statistically dominant "run lengths." The snake venoms that bind and inactivate this cholinergic receptor, α-cobratoxin and erabutoxin, demonstrate similar power spectral transformations of their hydrophobic sequence. Similar examples can be found among growth factors and their receptors, insulin and its receptor, and other brain polypeptides and their respective receptors.

The heterogeneity among cholinergic receptor monomers is well known, and Figure 9 reflects this in a comparison of the power spectral transformations of the hydrophobic sequences in an *electrophorus* monomer with monomers from *Torpedo* and *Calf* cholinergic receptors. They share the 3.64 and 2.0 modes with additional longer wave "run length" values.

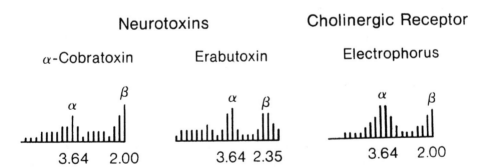

FIGURE 8   Power spectral transformations of amino acid sequences as "runs" of similar hydrophobic values in the acetylcholinergic receptor have "modes" very similar to the peptide snake venoms that bind them.

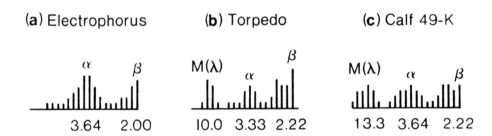

(a) Electrophorus   (b) Torpedo   (c) Calf 49-K

FIGURE 9 The acetylcholine receptor monomers from *calf* and *Torpedo* marine animal differ with respect to their largest mode from the *Electrophorus*.

TABLE 1

name	% runs	mean	σ^2	σ^3	σ^4
concav-a	0.345	1.854	2.256	1.811	6.616
thermoly	0.247	1.936	1.996	0.939	3.156

Using the presence of 5 and 10 residue run length modes in fibroblast growth factor, FGF, and a similar mode structure in the enzyme ribonuclease, we successfully predicted that FGF would significantly decrease the turnover time of template RNA in neuroendocrine cell preparations.[21] Like allosteric regulation, this suggests how macromolecular dynamics generating polypeptide messengers can influence other macromolecular mechanisms *simultaneously and at a distance*.

Functionally relevant categorization using a binary coding of polypeptide sequences opened up the possibility that generic dynamical systems, which were low dimensional, "universal" representatives of higher-dimensional dynamical systems could be deterministic generators of molecular biological sequences in the same way that chemical-parameter-dependent, symbolic dynamical sequences have successfully represented complex multicomponent chemical reactions[14] up to a monotonic, parameter-dependent sequence of ever lengthening, perfectly predictable words composed of two letters.

The relative usefulness of metaphorically generic nonlinear models, compared with specific differential equations, in representing complex systems, is an issue of great current interest in theoretical physics. Hydrodynamic turbulence approached directly by the Navier-Stokes equations has proven to be a formidable problem, even in these days of numerical analyses on Cray computers, whereas remarkably

TABLE 2

α	β	% runs	mean run
3.1	1.25	0.439	1.274
3.2	1.25	0.428	1.333
3.3	1.25	0.409	1.43
3.4	1.27	0.371	1.505
3.5	1.27	0.345	1.891

"representative" lower-dimensional systems appear to be quite good at depicting and predicting reality.[10] It is in that spirit that we study the allosteric macromolecular dynamics in time of Eq. (6) to demonstrate that with two parametric controls, α and β ("co-dimension two"); run length distributions, resembling those seen in macromolecular sequences, can be specified.[13]

The dynamical system requirements for this monotonic, parameter-dependent "U-sequence" (universal sequence) is a one-dimensional, single-maximum map with certain differentiability conditions on its critical point at the top. Our allosteric dynamical system is not such a system, but rather it is one which expands *nonuniformly* on the torus via *intermittent dynamics*, not unlike those characterized by Manneville and Pomeau,[9] in the context of vortex generation in relatively low Reynold's number, hydrodynamic turbulence.

Table 1 demonstrates the amino acid hydrophobic-hydrophilic run distribution metric from real examples. We compare them in the protein with the lowest (concavalin A) and highest (thermolysin) x-ray "fractal dimension" indicating structural complexity of a "self-affine" sort (i.e., 1.22 versus 1.78).[20] As we noted above, the longest run of thermolysin is 8 whereas that of concavalin A is 11.

Partitioning the unit interval of map, Eq. (6), at 0.5, followed by binary encoding, (0,1), it is not surprising that run length can be under *monotonic parametric control in co-dimension two* by changing the α and β parameters as desired (see Table 2).

MACROMOLECULAR CODING VIA ALLOSTERY AND THE "BALDWIN EFFECT"

The "Baldwin effect," thought to be trivial with respect to evolutionary influence, describes the *simultaneous adaptation of a somatic cell factor with that occurring in germ-cell mechanism which is then selected by natural factors*.[15,19] The obvious problems include those of communication between somatic and germ cells and/or the issue of simultaneity of change among spatially disparate but functionally related chemical regulatory or morphogenic mechanisms. Underlying both objections is the additional issue of changes in macromolecular functions with no mutational changes in the sequence of residues (amino acids or nucleotides) in the polymer. The strong hint of a Lamarckian mechanism and the violation of the thesis of resistance of gonadal mechanisms to "normal" fluctuations in the environment were enough to earn rejection of the Baldwin effect by neo-Darwinian theorists.

The modern era of omni-present polypeptide ligands, with their influence on macromolecular mechanisms at long distances, notably the neurohormones of brain found throughout the body (as represented by the life's work of Roger Guillemin and others), have partially solved the problem of simultaneous, long-range influences. Allosteric regulation, of both the local and global sorts, help explain how it is that the same residue sequences in proteins and nucleotides may generate different behavior and products dependent upon their dynamics. The recent powerful work demonstrating universal low-dimensional dynamics that can generate the same behaviors as very high-dimensional complex systems, and that dynamics in time can be represented by one-dimensional dynamics in space via symbolic dynamics, lead us to an abstract model for how "Baldwin effects" might work.

We have used real data from brain allosteric enzyme systems, polypeptide sequence coding, and substrate-velocity, and a time-dependent model of macromolecular function generating "runs" as sequence codes to very roughly sketch how such a Baldwin system might work. Contemporaneous biochemical adaptation might influence somatic and genetic systems simultaneously and propagate through generations, thus legitimizing the Lamarckian "Baldwin effect."

Of course the possible and the probable are two very different issues. It may require a change in the current molecular biological, neo-Darwinian political climate in order to discriminate what can happen from what does happen in the context of suitably designed experiments.

ACKNOWLEDGMENTS

This work is supported by the Office of Naval Research, Systems' Biophysics and Computational Neuroscience. The authors also wish to express their gratitude to J. Mittenthal for his patient editorial efforts.

REFERENCES

1. Bak, P., C. Tang, and K. Wiesenfeld. "Self-Organized Criticality: An Explanation of $1/f$ Noise." *Phys. Rev. Lett.* **59** (1987): 381–384.
2. Knapp, S., and A. J. Mandell. "Scattering Kinetics in a Complex Tryptophan Hydroxylase Preparation from Rat Brain Stem Raphe Nuclei: Statistical Evidence that the Lithium-Induced Sigmoid Velocity Function Reflects Two States of of Available Catalytic Potential." *J. Neural. Trans.* **58** (1983): 169–182.
3. Kuczenski, R., and A. J. Mandell. "Regulatory Properties of Soluable and Particulate Rat Brain Tyrosine Hydroxlase." *J. Biol. Chem.* **247** (1972): 3114–3122.
4. Mandell, A. J. "From Chemical Homology to Topological Temperature: A Notion Relating the Structure and Function of Brain Polypeptides." In *Synergetics of the Brain*, edited by E. Basar, H. Flohr, H. Haken, and A. J. Mandell. Berlin: Springer-Verlag, 1983.
5. Mandell, A. J. "Non-Equilibrium Behavior of Some Brain Enzyme and Receptor Systems." *Ann. Rev. Pharm.* **24** (1984): 237–274.
6. Mandell, A. J. "The Source and Characteristics of Normal Modes in Molecular Biology." In *Chaos and Order in Natural Sciences*, edited by M. Velarde. Berlin: Springer-Verlag, 1988.
7. Mandell, A. J., and P. V. Russo. "Striatal Tyrosine Hydroxolase Activity: Multiple Conformational Kinetic Oscillations and Product Concentration Frequencies." *J. Neurosci.* **1** (1984): 380–389.
8. Mandell, A. J., P. Russo, and B. W. Blomgren. "Geometric Universality in Brain Allosteric Protein Dynamics: Complex Hydrophobic Transformation Predicts Mutual Recognition by Polypeptides and Proteins." *Ann. N.Y.A.S.* **504** (1987): 88–117.
9. Manneville, P., and Y. Pomeau. "Different Ways to Turbulence in Dissipative Dynamical Systems." *Physica* **1D** (1980): 219–226.
10. Ruelle, D. *Elements of Differentiable Dynamicsand Bifurcation Theory.* Boston: Academic Press, 1989.
11. Russo, P. V., and A. J. Mandell. "Metrics from Nonlinear Dynamics Adapted for Characterizing the Behavior of Nonequilibrium Enzymatic Rate Functions." *Annal. Biochemistry* **139** (1984): 91–99.
12. Russo, P. V., and A. J. Mandell. "Sonication, Calcium, and Peptides Have Systematic Effects on the Nonlinear Dynamics of Tyrosive Hydroxylase Activity." *Nemochem. Internat.* **9** (1986): 171–176.
13. Selz, K. A., and A. J. Mandell. "Bernoulli Partition-Equivalence of Intermittent Neuronal Discharge Patterns." *Int. J. Bifurcation Chaos* **1(3)** (1991): in press.
14. Simoyi, R. H., A. Wolf, and H. L. Swinney. "One-Dimensional Dynamics in a Multicomponent Chemical Reaction." *Phys. Rev. Lett.* **49** (1982): 245–248.
15. Simpson, G. G. "The Baldwin Effect." *Evolution* **7** (1953): 110–117.

16. Stanley, E. H. *Introduction to Phase Transitions and Critical Phenomena.* Oxford: Oxford University Press, 1971.
17. Steele, E. J. *Somatic Selection and Adaptive Evolution.* Chicago: University of Chicago Press, 1981.
18. Teipel, J., and D. E. Koshland. "The Significance of Intermediary Plateau Regions in Enzyme Saturation Curves." *Biochemistry* **8** (1969): 4656–4663.
19. Waddington, C. H. "The Baldwin Effect, Genetic Assimilation, and Homeostasis." *Evolution* **7** (1953): 386–387.
20. Wagner, G. C., J. T. Calvin, J. P. Allen, and H. J. Stapleton. "Fractal Models of Protein Structure, Dynamics, and Magnetic Relaxation." *J. Am. Chem. Soc.* **107** (1985): 5589–5594.
21. Zeytin, F. N., S. F. Rusk, V. Raymond, and A. J. Mandell. "Fibroblastic Growth Factor Stabilizes Riboneucleic Acid and Regulates Differentiated Functions in a Multipeptide-Secreting Neuroendocrine Cell Line." *Endocrinology* **122** (1988): 1121–1128.

S. A. Newman
Department of Cell Biology and Anatomy, New York Medical College, Valhalla, New York 10595

Generic Physical Mechanisms of Morphogenesis and Pattern Formation as Determinants in the Evolution of Multicellular Organization

Development starts from a more or less spherical egg, and from this there develops an animal which is anything but spherical.... One cannot account for this by any theory which confines itself to chemical statements, such as that genes control the synthesis of particular proteins. Somehow or other we must find how to bring into the story the physical forces which are necessary to push the material about into the appropriate places and mould it into the correct shapes.

> — C.H. Waddington
> *The Nature of Life*[80]

INTRODUCTION

Discussions of biological development and evolution often assume that the genome of each multicellular organism contains a set of instructions for the generation of the organism's overall form and the pattern of arrangement of its tissues. The origin of novel, heritable morphological phenotypes is, in this view, considered to be

Principles of Organization in Organisms,
SFI Studies in the Sciences of Complexity, Proc. Vol. XIII,
Eds. J. Mittenthal & A. Baskin, Addison-Wesley, 1992   **241**

due to alterations in the hypothesized "genetic program" brought about by random mutations in the germ-line DNA. Morphological evolution, i.e., the establishment of new morphological phenotypes at the population level, is attributed solely to differential reproductive success of the randomly produced variants, with the implication that the mechanisms of selection are entirely disconnected from those that generate variation. Thus, extant morphological phenotypes, while acknowledged to be subject to the laws of physics and chemistry, are held to owe their structural organization to the immediate physical environment to no greater degree than do machines, such as clocks or automobiles.

There are few *a priori* constraints on the kinds of morphological phenotypes that are possible when development is considered to be dictated by a genetic program. Much as a sophisticated computer graphics program can represent any conceivable structure on a video screen, a genetic program would seem to be capable of generating any arbitrary arrangement of proteins, polysaccharides, nucleic acids, lipids, minerals, and even metals, all of which may participate in biochemical pathways. Given enough time, anything is possible. If there are any reasons, apart from time limitations, for the absence of reproducing chain saws or compact disc players among the Earth's organisms, they are not contained in the genetic program concept.

More recently this extreme viewpoint has been tempered by the recognition that random genetic changes do not lead to equally random phenotypic changes. Because the array of viable forms that can be generated by embryological mechanisms is limited, phenotypic variation is said to be subject to "developmental constraints."[28] But developmental mechanisms themselves must have come into existence sometime after the evolution of cellular life. Did they arise purely by chance, or was their emergence in some sense "inevitable?" One possible answer to this question is the suggestion that basic mechanisms of morphogenesis and pattern formation inescapably arise from properties of cell aggregates considered as physical matter. The implication is that biological forms and patterns, particularly in the early embryo, are, to an important extent, predictable from the interactions of such matter with the immediate physical environment. This is a concept that has some precedents in the scientific literature, in the work of D'Arcy W. Thompson,[75] for example. But because of the non-obvious connection of this idea to genetic analyses of both development and evolution, it has been relegated to the margins of biology during most of this century. This is unfortunate, because, as will be discussed below, the recognition that certain morphogenetic and pattern-forming mechanisms are physically inevitable is entirely consistent with our understanding of gene-dependent biological processes. Moreover, unlike the standard notion of a randomly arrived at genetic program for the construction of organisms, this idea can provide insight into the question of why organisms are organized in the fashion they are.

DEFICIENCIES OF THE "GENETIC PROGRAM" METAPHOR

The genetic program concept of development is widely held and taught (e.g., "We know that the instructions for how the egg develops are written in the linear sequence of bases along the DNA of the germ cells"[83]), although it has also been criticized[38,60,62,64,74,84,87] and it is doubtful whether it has ever played a significant role in elucidating any developmental mechanism. Each embryonic cell that will participate in forming the *somatic* or bodily structures of an organism contains a virtually identical set of DNA sequences. Some of these sequences correspond to, and provide templates for, the sequences of RNA molecules that perform various structural and catalytic functions in the cell, and help specify the primary structure of all of the organism's proteins. Since DNA segments are selectively expressed in each cell type, the differential regulation of gene activity through time and space is a central problem of developmental biology, and the one which the genetic program metaphor was invented to explain.

It is clear that during development the set of biochemical and physical conditions that prevails in the embryo at any moment can lead to changes in composition and distribution of cellular components. These, in turn, bring about a new set of biochemical and physical conditions, leading to further changes. Genes, as the embryo's record of the primary sequences of the RNAs and proteins it can produce, are essential participants in this cascade of reactions, but they in no way embody a program for them. The deficiency of this metaphor extends well beyond the inability of DNA *per se* to act as the program's storage medium. The concept of a program, understood as an intricately organized set of instructions acting on a collection of variables to bring about a well-defined outcome, would not be particularly helpful in characterizing development even if the "software" were considered to include cellular components such as ions, vitamins, and water, and not only the genes and what they specify. Just because a process (such as embryogenesis) has a reliably repeated outcome does not require that its behavior be specified by a set of instructions. The return of Halley's comet to our solar system every 76 years depends on precise physical interactions, but on no instructions or program.

In addition to being a misleading *post hoc* description of the results of developmental investigations, the genetic program idea may encourage misconceived research strategies. For example, the notion that form and pattern are coherently represented like a map in each of the embryo's cells, and biologically realized as a point-to-point "interpretation" by the genome of a simple, chemically defined coordinate system,[86] raises the hope that a small set of genes, such as those that specify proteins containing the "homeobox" DNA-binding motif, might provide a "Rosetta Stone" for pattern formation.[70] Indeed, one frequently expressed rationale for the Human Genome Initiative is the expectation that a comprehensive knowledge of the relative locations of DNA sequences will reveal the instructions for embryonic development.[10]

GENERIC PHYSICAL PROCESSES AS ORGANIZING
PRINCIPLES OF BIOLOGICAL FORM AND PATTERN

In his book *On Growth and Form*, D'Arcy Thompson considered the role of physical forces in the generation of biological patterns and structures: "Cell and tissue, shell and bone, leaf and flower, are so many portions of matter, and it is in obedience to the laws of physics that their particles have been moved, molded and conformed."[75] His attempts to explain structure-function relationships and evolutionary modification of biological form in terms of physical and mathematical laws led to a number of striking insights. For example, features of the vertebrate skeletal system, ranging in scale from the gross anatomy of the spinal column to the microscopic configuration of trabeculae in spongy bone, were seen to resemble the arrangement of load-bearing elements in a cantilever bridge. In addition, differences in body shapes of related fish could be derived from one another by simple coordinate transformations that were plausibly tied to different strategies of accommodation to external forces, such as friction.

Unfortunately, D'Arcy Thompson largely neglected mechanisms of development and inheritance, both of which were subjects of vigorous research during the course of his work. This certainly contributed to the failure of his important ideas to enter the mainstream of biology. Even if a structure can be rationalized on the basis of forces acting on adult forms, this provides little insight into the generation of such a structure. Why, for example, should skeletal elements become arranged during development *in utero* in a manner appropriate to bearing weight, when the fluid environment ensures that the developing bones are never exposed to such stresses? Baldwin[5] proposed a process of "organic selection" by virtue of which fortuitously appearing gene combinations which produced a structure identical to one that would be produced by mechanical factors would be advantageous, and spread through the population. In a similar fashion Waddington[79] demonstrated how a purely "physiological" response to environmental stimuli might be incorporated into an organism's developmental repertoire under certain circumstances, a mechanism he referred to as "genetic assimilation." Simpson considered this class of phenomena (which he referred to as the "Baldwin effect") to be a "relatively minor outcome of the theory [of evolution by natural selection]."[69] In contrast, I will suggest that a generalization of this effect, and the recognition that its influence on organisms differed at different stages of their evolutionary history, necessitates a thoroughgoing revision of the neo-Darwinian paradigm.

Leaving aside, for the moment, the puzzle of the initial evolution of embryonic development of traits whose only rationale is in the biology of the adult form, we can consider correspondences in the embryo itself between organismal structure and "external" influences. Unlike adult organisms, the forms and functions of which are of a scale, composition, and level of integration that make them resistant to gross transformation by simple physical effects, embryos are at least potentially subject to such forces. This was recognized by Balfour as early as 1875, when he observed that "...there is no question that during their embryonic existence animals are more susceptible to external forces than after they have become fully grown,"[4] and following the first edition of *On Growth and Form* in 1917, Needham called for an extension of D'Arcy Thompson's analysis to the realm of embryology.[54]

Studies using this approach have subsequently appeared sporadically in the scientific literature. Steinberg and his coworkers have shown, for example, that isolated embryonic tissue fragments round up, take on "equilibrium" shapes, and spread upon and engulf one another, as if they were immiscible elasticoviscous liquid droplets with characteristic *surface tensions*.[72,73] Morphogenetic phenomena such as *gastrulation*, the formation of a new layer of embryonic tissue by the intrusion of the edge of a cell sheet between two pre-existing layers, can thus be viewed as occurring for reasons similar to the "wetting"[9,17] by one fluid of the interface between a second fluid and its substratum (Figure 1; see also Phillips and Davis[65]).

Gravity is another physical force that has been proposed to play a role in early development. For example, Ancel and Vintemberger suggested that the ooplasmic rearrangement that occurs following fertilization in the eggs of anuran amphibians might be caused by density differences[2] (Figure 2). *Diffusion* is usually thought of as a process by which uniform distributions of chemical substances are

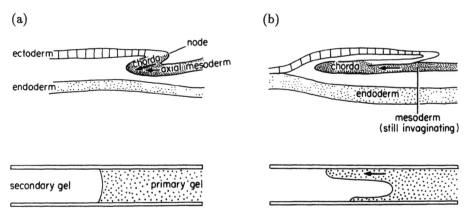

FIGURE 1 (a) Stages in mammalian gastrulation viewed in median section. (Left) Axial mesoderm has begun to enter the primitive pit, forming the notochordal process, or chorda. (Right) Chorda advances beneath the ectoderm followed by additional mesoderm, which then spreads laterally (out of the plane of section). Redrawn from Deuchar.[12] (b) Spreading and rearrangement by "matrix-driven translocation" of two partially assembled type I collagen gels, between two parallel polystyrene plates.[17,58] Primary gel contains heparin-coated cell-sized polystyrene latex beads. Secondary gel contains the heparin-binding adhesive glycoprotein fibronectin. Relative movement of the two matrix regions is dependent on "phase separation" of the two gels and differential adhesion.

brought about, but Turing demonstrated that the kinetic coupling of diffusion with chemical reaction or biosynthesis can lead to nonuniform concentration distributions of reactants.[76] One possible outcome is *striping:* the establishment of parallel bands of a chemical substance separated by bands lacking that substance (Figure 3). A demonstration of the Turing phenomenon in nonliving polymer gels has recently been described,[8] and this pattern-forming mechanism has been shown to be biochemically plausible in several animal and plant systems.[21,30,43,44,49,56,59]

We have called phenomena such as surface tension, phase separation, gravity, and reaction-diffusion coupling, "generic" physical mechanisms of morphogenesis and pattern formation, to emphasize their broad applicability to living and nonliving systems.[61] Generic mechanisms are contrasted to "genetic" processes, a term we have reserved for a distinct phenomenon in biological systems: assemblies of intricately organized macromolecules that have coevolved to carry out highly specific functions, e.g., cytoplasmic "motors,"[77] regulated membrane channels,[33] or gene promoter elements and their associated *trans*-acting regulatory factors.[25,71] Because generic physical mechanisms are plausible determinants of many morphological features of early embryos, it is reasonable to hypothesize that they play important developmental roles. But when this notion has been put to the experimental test, the story turns out to be less straightforward.

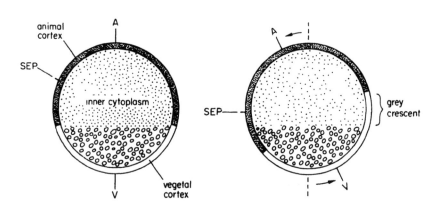

FIGURE 2   The cortical/cytoplasmic rotation in anuran amphibian embryos. A 30° rotation of the cortex relative to the inner cytoplasm is required for normal dorsoventral polarity to be established. Diagrammatic sections are shown before (left) and after (right) rotation. The cortex rotates so that the sperm entry point (SEP) moves vegetally. The grey crescent visible in *Rana pipiens* embryos is formed by the overlapping of pigmented animal hemisphere cytoplasm by nonpigmented vegetal hemisphere cytoplasm. Redrawn from Elinson and Rowning.[14]

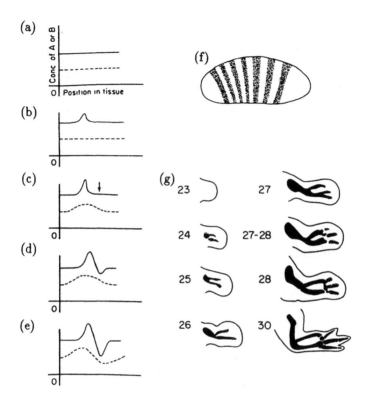

FIGURE 3 Chemical wave generation by a reaction-diffusion mechanism, and examples of stripe patterns during development. (a-e) Graphical representation of chemical wave formation, based on Maynard Smith.[48] It is assumed that two substances, a and b, which influence one another's synthesis, are produced throughout a row of cells, and that there is a balance in the rates of synthesis and utilization of a and b (i.e., they are at a *steady state*) The steady state shown in Panel a is *spatially uniform:* the concentrations of both a and b are unvarying along the row of cells. Under certain conditions a *spatially nonuniform* stationary state (panel e) can be achieved by the growth and stabilization of a fluctuation (panels b-d).[76] The following conditions are sufficient to bring about this phenomenon: substance a has a positive effect on the synthesis of both itself and substance b; substance b has an inhibitory effect on the synthesis of A; the diffusion rate of b is greater than that of a. Arrow in panel c indicates the point at which a reduction of the concentration of a to below its uniform steady state level will be initiated on the basis of the assumptions above. The number of peaks and valleys of a and b that will be in place when the system finally reaches the new steady state will depend on reaction and diffusion rates, the size and shape of the spatial domain in which these events are occurring, and the modes of

FIGURE 3 (cont'd.) utilization of a and b at the boundaries of the domain. See Turing,[76] Maynard Smith,[48] and Newman and Comper[61] for further details.
(f) *Drosophila* blastoderm-stage embryo showing early *even-skipped* protein pattern (stippled stripes). Based on Frasch et al.[18] (g) Progress of chondrogenesis in the chick wing bud between 4 and 7 days of development. Solid black regions represent definitive cartilage; stippled areas represent early cartilage. Stages are those of Hamburger and Hamilton.[31] Panel g from Newman and Frisch,[56] copyright © 1979 by the AAAS.

MORPHOGENESIS AND PATTERN FORMATION ARE NOT SIMPLY GENERIC

In several cases that have been studied in detail, it has been found that patterns and forms that could potentially be determined by generic physical effects are driven (in addition? exclusively?) by intricate, highly evolved molecular machinery. After fertilization in the amphibian egg, for example, there is a 30° rotation of the dense cortical cytoplasm relative to the less dense, deeper cytoplasm. In some species this rotation has the effect of revealing a lightly pigmented region of deep cytoplasm on one side of the egg, the "grey crescent." This region, or its equivalent, marks the future dorsal area of the embryo (Figure 2). As noted above, Ancel and Vintemberger suggested that the relatively rigid egg cortex, upon mechanical release from the underlying cytoplasmic core midway during the first cell cycle, slips to one side under the influence of gravity in normally oriented eggs.[2] The ability of unit gravity to effect a proper overlap between cortical and deep cytoplasm in eggs whose normal rotation is inhibited by UV light[55] is consistent with a role for gravity in normal axis specification. However, a mechanism of axis specification driven exclusively by gravity is contradicted by results of Vincent et al.[78] who embedded *Xenopus* eggs in gelatin so that the cortex could not move, and found that the cytoplasmic core, including the denser vegetal regions, rotated up one side of the cortex by 30°, working against gravity. Clearly the egg must also have a means other than gravity to drive the rotation. The presence of an oriented array of microtubules in the shear plane between the cortex and subcortical cytoplasm[14] suggests that the force-generating mechanism might be similar to the energy-consuming microtubule-dependent organelle transport systems found in other cell types.[77]

Another phenomenon that has been hypothesized to be due, at least in part, to a generic physical mechanism, is the formation of striped patterns of expression of the primary "pair-rule" genes in *Drosophila* embryos. The pattern of expression depicted for the *even-skipped* product shown in Figure 3(b) would seem to be a likely candidate for explanation by a Turing-type mechanism, and indeed it has been the subject of several such models.[45,49] Nonetheless, experiments in which pair-rule genes with mutated promoter sequences were introduced as "P-elements" into *Drosophila* embryos and subsequently expressed, have demonstrated convincingly that individual stripes, rather than arising collectively from a generic dynamical process, are each specified by a unique complex of DNA-binding proteins acting in *trans* to activate stripe-specific promoters.[25,71] These regulatory proteins, which include the "gap" gene products, are not themselves distributed in a striped pattern, but as simpler gradients. In contrast to the self-organizing stripe-forming processes represented by Turing-type mechanisms, the recent genetic results point to the embryo's use of Rube Goldbergian molecular machinery for "making stripes inelegantly."[1]

What are the implications of the recognition that whereas organisms *could* produce their major morphological and patterning features by generic mechanisms, contemporary organisms rarely exhibit such simple developmental processes? It is likely that the complexity of modern ontogeny is centrally tied to the fact that organisms are the results of evolutionary processes.[6] Indeed the evolution of organisms virtually assures that their structural and functional organization will often be "overdetermined," the outcome of several relatively independent mechanisms. This "suspenders and belt" phenomenon is familiar to all students of experimental embryology, and is highly relevant to the question of whether generic physical mechanisms contribute importantly to determining why organisms are organized in characteristic ways.

WHY DO ORGANISMS LOOK THE WAY THEY DO?

The evolution of a novel form or pattern must occur in the context of all previously existing properties of an organism. That the eggs of most marine organisms are spherical, for example, must have originally had something to do with the fact that the equilibrium configuration of a droplet of cytoplasm surrounded by a lipid membrane, immersed in an aqueous medium, is a sphere. The evolution of cytoskeletal proteins and the elaboration of cytoarchitecture resistant to mechanical deformation would certainly be of selective advantage to phylogenetic lineages in which they occurred. It makes sense to think of this evolution of genes and the associated genetically based structures as having been guided by generic forces: eggs that were both generically *and* genetically determined to take on a spherical shape would be much more likely to maintain this shape in the face of environmental change than those formed *only* by generic forces. Lineages represented by such organisms may indeed evolve to a point where the cytoarchitecture becomes more important *functionally* in determining the egg's shape than do the egg's equilibrium properties as a droplet of fluid. However, if we want to understand why the egg looks like it does, we must understand the origin of the shape, and the nature of the generic "template" that presumably guided the evolution of the cytoarchitectural support system.

Conservation and reinforcement of "successful" forms by gene evolution would contribute to morphological stasis in phylogenetic lineages, but radical changes in body plan could also be precipitated by small genetic changes that happened to bring new external forces into play. During early evolution of multicellular forms these external forces would often be generic physical processes. It is plausible, for example, that the biochemical evolution of yolk phosphoproteins was driven mainly by selection for efficiency of nutrient storage. However, the resulting yolk platelets would sediment within the egg by virtue of their incidental property of having greater density than other cytoplasmic components, and for the phylogenetic lineages that evolved them, serve as a basis for spatial differentiation within the oocyte and any multicellular form to which it gave rise.

In larger tissue masses an analogous harnessing of generic physical effects could also have taken place. For instance, Holtfreter demonstrated that amphibian endodermal cells containing an endogenous surface coat will spread over uncoated endoderm, and conversely, an aggregate of uncoated endoderm will sink into a layer of coated endoderm.[35] Such effects may serve to initiate gastrulation in the hollow sphere that constitutes the early amphibian embryo.[36,37] Steinberg and his coworkers interpreted such engulfment phenomena as arising from adhesivity-cohesivity differentials, and suggested that even *quantitative* differences in intercellular adhesive strengths are enough to bring these effects into play.[72,73] Recent experiments have verified this conclusion.[20] Whatever stabilizing machinery may have subsequently evolved to ensure the reliability of gastrulation under a wide variety of environmental conditions, it is plausible that the generic mechanism of differential adhesion was the origin of cellular ingress and tissue invagination in multicellular aggregates.

Another illuminating example involves the recognition that whereas the products of as many as 25 different genes appear to be involved in the segmentation of the *Drosophila* body plan, it is inconceivable that this network could have arisen *de novo* during evolution. In fact, this system is redundant and overdetermined. This can be seen, for instance, in the interaction of *nanos* and *hunchback*, two of the genes in the segmentation regulatory network. The gene product of *nanos* apparently serves only to repress the translation of mRNA specified by the maternal *hunchback* gene.[40] Embryos that lack both maternal *hunchback* gene product and a functional *nanos* gene develop in an apparently normal fashion, although the loss of *nanos* alone is lethal.[39]

The formation of chemical stripes in the early embryos of *Drosophila*'s ancestors can have been due originally to a reaction-diffusion mechanism if (i) there were an appropriate ratio between the spatial scale of the tissue domain to be so organized and the diffusivities of candidate morphogens, and (ii) at least some of the putative morphogens positively enhanced their own synthesis. Each of these conditions apparently holds for some systems in contemporary *Drosophila* embryos.[18,34] Cellular release (and thus the potential for diffusion) of growth and differentiation factors, and auto-stimulation of the production of such factors, may have evolved separately as means for large-scale coordination of tissue function and for amplification of signals. When appropriately tuned, biochemical circuits with *both* these features can spontaneously form stripes. Such stripes may have been fortuitously advantageous in promoting, say, redundancy of structure in the insect's wormlike ancestor. The problem with reaction-diffusion coupling as a mechanism for such patterning is that it is temperature dependent (as are all diffusion-based processes), often sensitive to domain size and shape, and thus generally unreliable. The superimposition by genetic evolution of reinforcing circuitry for the generically selected pattern would undoubtedly have favored certain phylogenetic sublineages, although such evolution would not, in general, be reflected in the morphological fossil record.

What is clear is that virtually all morphogenetic and patterning effects seen during early development (see Gilbert[22] for a review) can, *in principle*, result from the action of generic processes on embryonic cells or tissues. *Ooplasmic rearrangement* could have its origin in density differences of physically distinct intracellular determinants. Tissue immiscibility and the engulfment and spreading effects resulting from differential adhesion could have been the origin of a variety of morphogenetic movements, including *epiboly*, the concerted movement of epithelial sheets to enclose deeper layers of the embryo; *invagination*, the infolding of a region of tissue, like the indenting of a rubber ball; *involution*, the inturning of an expanding outer layer so that it spreads over the internal surface of the remaining external cells; *ingression*, the migration of individual cells from the surface layers into the interior of the embryo; and *delamination*, the splitting of one cellular sheet into two or more parallel sheets. Convective effects, particulary in conjunction with gravity or surface tension, can give rise to *microfingering*, the interpenetration of parallel protrusions of distinct cytoplasmic materials or tissues.[59] And reaction-diffusion mechanisms could have provided the original basis for *striping*, as described above.

While it is all but impossible to reconstruct the phylogeny of present-day developmental mechanisms, it may be feasible to experimentally "deconstruct" modern

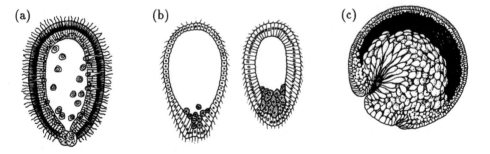

FIGURE 4 Different modes of gastrulation: (A) Multipolar introgression, in the sponge *Leucosolenia blanca* (from Minchin[52]); (B) two stages of unipolar introgression, in the hydroid *Aqueorea forskalea* (from Mergner[50]); (C) emboly, in the amphibian *Amblystoma punctatum* (from Holtfreter[36]).

organisms to determine the minimal requirements for generating specific forms or patterns. Putative secondarily acquired or co-opted genetic reinforcements for morphogenetic processes could be removed experimentally by mutation (as in the studies of the *nanos/hunchback* system),[39,41] or by specific drugs directed, for example, against cytoplasmic motors.[77] In many cases the organism may have evolved to a point at which its early embryo is no longer suceptible to the forces that originally engendered a given pattern or form. But in other cases the relevant morphogenetic effect might still occur, albeit roughly or unreliably.

IMPLICATIONS OF GENERIC-GENETIC COUPLING FOR THE EVOLUTION OF EARLY DEVELOPMENT

The previous discussion suggests that at some fundamental level the embryonic organization of organisms is an almost predictable function of the physical world of which they are a part. Morphological evolution, at least during the early radiation of multicellular forms, can be viewed in terms of the reiterative application of a small set of generic physical processes. Prior to the evolution of a high degree of genetic reinforcement of a particular subset of forms, a vast array of morphologies must have been possible. But all such forms would have had the stamp of the generic processes and forces to which semisolid, chemically reactive matter is subject: gravity, adhesion, surface tension, convection, and the interaction between reaction and diffusion.

The view outlined here has a number of implications, both specific and general:

i. *Some form of gastrulation will be among the earliest evolutionary manifestations of metazoan development.* Multicellularity presupposes that a cell surface system mediating intercellular adhesion already exists. Indeed, such a system must be the defining characteristic of the original metazoa.[6] To ensure that a cell cluster forms an undistorted sphere, strict regulation of the quantitative strength of this adhesion would be required. But in the absence of such tight controls, random attenuation of adhesive strength would cause individual cells to slough off, or alternatively, burrow within the aggregate, as in *multipolar ingression.* If a *patch* of cells in the aggregate were all to exhibit reduced adhesive strength (perhaps by virtue of having incorporated some inhibitory substance that was nonuniformly distributed in the cytoplasm of the aggregate's founder cell), *unipolar ingression* of individual cells would be likely. If the cells of the patch were more strongly adhesive to each other than were the surrounding cells, *emboly,* or inpocketing, would occur (Figure 4). It should be noted that the capacity to undergo one or another of these types of gastrulation would be *inherited* not as a genetic program, but rather as the outcome of generic physicochemical effects acting on multicellular aggregates.

In his discussion of the possible origin of gastrulation, Buss[7] proposes that the protistan world was the arena of a conflict between cell lineages "in their quest for increased replication" in the face of a constraint against simultaneous ciliation and cell division. Gastrulae, or multilayered organisms, are held to be the solution to this "problem." In contrast to this notion, I suggest that gastrulation is the inevitable consequence of nonuniform adhesive interactions in solid or hollow cell aggregates. The significance of the resulting forms for the origin of biological individuality, as persuasively laid out by Buss, would certainly pertain. Gastrulae would arise, however, not as the solution to a pre-existing problem, but inescapably, as loci of new biological possibilities.

ii. *Once organisms with a multiple germ-layer organization had evolved, a wide spectrum of body types would have become possible.* Molecular mechanisms for the establishment of tissue segregation or immiscibility by differential adhesion, a generic requirement for gastrulation according to the hypothesis above, could be used reiteratively to create additional tissue boundaries and independent cell lineages. For this to occur in a reproducible fashion, factors that influenced the quantitative expression of cellular adhesion molecules would need to become nonuniformly distributed. This might happen as the result of sedimentation of cytoplasmic determinants in the colony's founder cell, for example, or by appropriately tuned reaction-diffusion coupling between cell products that are positively autoregulating and ones that are secreted and diffusible. The profusion of metazoan body forms found in the Precambrian Ediacara formation[16,23] and in the Burgess Shale formation from the Middle Cambrian period[29,85] can plausibly be seen as manifestations of an exploration of the particular universe of morphological phenotypes generated in primitive embryos by the iteration and permutation of a small number of generic physical processes.

iii. *The evolutionary persistence of a subset of morphological phenotypes will depend not only on the relative adaptation of body forms to ecological conditions, but also on the facility by which genetic mechanisms can be recruited to stabilize and reinforce such forms.* It is expected that morphogenetic mechanisms based on generic forces such as gravity, surface tension, and reaction-diffusion effects would produce outcomes relatively sensitive to externalities like orientation and temperature. We can postulate a kind of "covert evolution" by which molecular processes of the organism that predated the origin of a generically conditioned morphology might become linked to the production of that same morphology by random mutation of DNA and protein structure. This could lead to integration of various subsystems with little morphological consequence. In this model, random genetic change does not produce a biological form by increments. The form, presumed to have been brought about by generic forces, and to have established itself in the ecosystem, already exists. Natural selection in this covertly evolving population would not be based on ecological discrimination among minor morphological differences, but on the ability to maintain successful morphological phenotypes in changing environments (i.e., Schmalhausen's "stabilizing selection"[68]). But the more that genetic mechanisms are mobilized for the preservation of a particular morphology, the less the developmental pathway in question will be divertible by environmental perturbation. This phenomenon has been termed "canalization" by Waddington,[79] and "autoregulation" by Schmalhausen.[68] Put another way, after the initial profusion of body forms morphological innovation will be more and more the exception; selection will typically be for morphological stasis.

The paleontological data on the profusion of body forms during early metazoan evolution are consistent with this view. The amount of time separating the earliest known unicellular eukaryotes (approximately 1.4 billion years ago) from the earliest recognizable members of extant metazoan phyla (from the Precambrian Ediacara formation) was about 700 million years. By 100 million years later, all extant phyla were apparently established. Yet in the 600 million years that separated the Cambrian from the Recent, no new body plans arose.[16,23,29,85]

iv. *Certain body plans arising during the original phase of "morphological profusion" would be intrinsically more susceptible to reinforcement and stabilization by genetic mechanisms than others, leading to a culling of the original array of morphologies in changing environments.* Although this is a speculation that needs to be tested by appropriate modeling, the following example can serve to illustrate what is meant. Reaction-diffusion processes can readily give rise to stripes, spots, or spirals[29,30] of a diffusible molecule. Were such a molecule to induce the enhanced expression of a cell adhesion molecule, a patterned set of "compartments," with no cell mixing across boundaries, would form. Thus any of these pattern types could define an embryonic *bauplan*. Let us assume that some of these *bauplans* suited organisms for reproductive success, or for the occupation of a particular ecological niche. A body plan based on a striping mechanism could be readily reinforced and stabilized by multiple promoter elements in *cis* to the gene specifying the protein in question,[25,71] which could readily arise by random mutational events. For instance, if other factors were present that were, for incidental reasons (gravity, diffusion), distributed in simple gradients in the direction orthogonal to the stripes, promoters that were activated by a combination of such factors so as to reinforce a given stripe would be retained by stabilizing selection, whereas promoters that disrupted the generically templated striping pattern would be selected against. The evolution of analogous molecular mechanisms to reinforce and stabilize a particular set of spots or spirals would be much more formidable. A segmental body plan, which could originally have been set on its evolutionary path by a stripe-forming process, might thus be more susceptible to genetic stabilization, and thus evolutionarily more persistent, than a "checkerboard" or a "pinwheel" plan, which may have originally been established by spot-forming or spiral-forming processes.

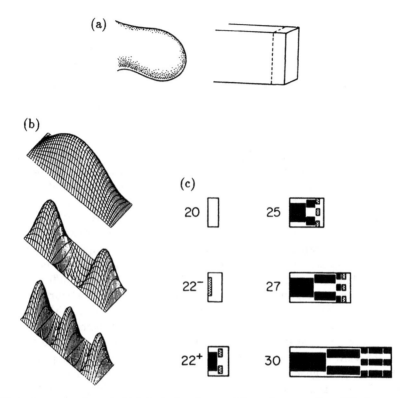

FIGURE 5 Interpretation of chick limb development based on reaction-diffusion mechanism. (A) (Left) Drawing of a 5-day wing bud. (Right) Schematic representation of 5-day wing bud with as yet unpatterned distal mesenchyme demarcated by dashed line. (B) Top to bottom: graphs representing predicted distribution of TGF-β and fibronectin in the prechondrogenic distal mesenchyme at three successive stages of development. (C) Predicted cartilage pattern based on schematic model. Figure from Newman and Comper,[61] based on Newman and Frisch.[56]

It is significant that many of the metazoan body plans of the Ediacaran fauna (some indeed reminiscent of checkerboards and pinwheels)[16,23] and a large proportion of those from the later Burgess shale fossil bed,[29,85] defy placement into any of the 30 or so currently extant phyla. Taken together with the failure of new *bauplans* to emerge since that period, it appears that a large proportion of viable body plans fell by the wayside after their emergence during the "Cambrian explosion." This culling may have been purely serendipitous, as has been suggested.[29] Alternatively, it could have resulted from an intrinsic bias in the capacity of random gene mutation to evolve stabilizing mechanisms for certain generically produced morphologies.

v. *Phylogenetic lineages unable to make use of a particular generic mechanism of development would have a profoundly different evolutionary history from lineages for which all such mechanisms were available.* The acquisition of rigid cell walls by some taxa would make the generic processes based on tissue fluidity and differential adhesion less important in setting trends for morphogenetic change. The early separation of plant and animal kingdoms once multicellular forms became established may reflect a division based primarily on the foreclosure of this important generic mechanism of development in the plant lineages.

vi. *At any particular stage of generic/genetic coevolution, more than one developmental pathway, leading to quite distinct body types, may be available to the developing organism.* This is exemplified by the phenomenon of the *phenocopy* in *Drosophila*, in which a wild-type fly that experiences a temperature or chemical perturbation as an embryo, without undergoing genetic alteration, develops to resemble a mutant fly.[24] An even more striking set of examples is provided by organisms, ranging from insects, to tunicates, to amphibians, that undergo metamorphosis. This phenomenon demonstrates that different body forms are compatible with a single genotype.[47] (These can be radically different, as seen in tunicate metamorphosis, where a chordate-like larva is transformed into a sessile adult with no obvious axial organization). Moreover, as the phenomenon of "direct development" in sea urchins[66] or frogs[13] demonstrates, the larval forms are not obligatory precursors to the adult forms, as would be suggested on the view that development occurs according to a "genetic program." With the concept of generic/genetic coupling, the existence of divergent pathways of morphogenesis and pattern formation in genetically identical individuals is to be expected, as is the possibility of changes in the order of expression of such pathways ("heterochrony").[26] Direct development, instead of being considered the result of the evolutionary "loss" of the set of steps that leads to the larval stages,[66] might be more productively analyzed in terms of the action of generically based processes of early development in a different physical context (e.g., a much larger egg[13,66]). Of course, at present this viewpoint represents indications for further study rather than a set of conclusions.

GENERIC PHYSICAL EFFECTS IN LATER ONTOGENY AND PHYLOGENY

Once the major features of metazoan body organization became established and genetically reinforced, the locus of further morphological evolution would, if the view described above is correct, shift to small tissue primordia which were physically susceptible to generic physical effects similar to those that previously guided the development of the embryo as a whole. This corresponds to the phase of *organogenesis* that follows the establishment of the body plan in the more elaborate phyla. If we are concerned merely with form and pattern, the generic mechanisms may indeed be identical to some of those seen to be virtually inevitable in the developing egg: differential adhesion and reaction-diffusion coupling. (Gravity would probably have a negligible role at these later stages).

To take one example of organ formation, the developing vertebrate limb, the acquisition of differential adhesivity by subpopulations of cells along the body wall has plausibly been proposed as the mechanism by which the limb buds individuate as blobs of tissue immiscible with the surrounding flank.[32] Moreover, a reaction-diffusion mechanism, based on the autostimulatory activity of a TGF-β-like morphogen, coupled with the ability of this factor to stimulate production of the adhesion protein fibronectin, has been proposed to account for the pattern of skeletal elements[46,56,59] (Figure 5). It is significant that this mechanism of limb skeletal pattern formation implies that the number of parallel elements that develop at any proximodistal level in the limb can vary by the simple agency of increase or decrease in the anteroposterior dimension of the limb bud.[56] If this is true, then abrupt changes in limb morphology during evolution need not be the result of gradual, incremental change. "Jumps" between varieties with different digit numbers, for example, can occur in a single step, by virtue of tiny changes in the growth control of the limb buds.[57]

Like the inheritance of the body plan as a whole, the inheritance of the capacity to reproducibly form organs such as the limbs, heart, or kidneys, requires not a genomic representation of these structures, but rather the production of an appropriate set of components (tissue masses, for example) subject to relevant generic effects.

A feature, which evidently took on increasing importance as organs became ever more elaborate, is the process of *differentiation*. This was encountered earlier in this discussion as a matter of *quantitative* regulation of certain genes by nonuniformly distributed factors, by which intercellular adhesion, for example, was modulated in strength. With the occasional duplication and random drift of the "structural" or protein-specifying portion of a regulated gene, gene products specific to particular pattern elements could arise ("segment identity" proteins in *Drosophila*, cartilage-type collagen, cardiac myosin) and could be brought into play in a modular fashion,[53] without disrupting the regulation of the ancestral gene.

The hypothesis that gene duplication was important in the evolution of differentiation is, of course, not new.[63] In relation to the general framework presented here, however, some new implications emerge. The regulation of differentiation involves the *conditional* expression of genes. For example, in the scenario described above for the evolution of gastrulation, intercellular adhesivity was proposed to be modulated by spatially nonuniform intraembryonic factors that were put in place originally by generic effects such as sedimentation or reaction-diffusion coupling. The cartilage pattern in the developing limb was similarly suggested to be the result of gene expression conditional on a pattern of inducers that was originally templated by generic effects. The basic body plans and organ patterns would thus bear the stamp of generic processes pertaining to the small elasticoviscous fluid droplets that eggs and organ primordia are. This would remain so even if rein-forcing and stabilizing genetic processes were eventually to "take over," and by the processes of *heterostasy* (action under changed conditions) and heterochrony, induce the generically templated pattern in the absence of the original determi-nant. In large adult animals, cell responses such as differentiation can also occur in response to "generic" external effects appropriate to the scale of these larger objects. Thus, *weight bearing,* by means of physiological adaptation mechanisms, induces reorganization of long bone trabeculae, and *friction* induces thickening of the soles of the feet. Since these conditional responses to new generic effects may be reinforced by covert "stabilizing" or "canalizing" evolution, they can come to depend as much on reinforcing genetic mechanisms as on the original generic ones. As a result of heterostasy and heterochrony, features like bone trabeculae suitable for weight bearing,[75] or thickened soles,[53] can be induced *in utero*, in the absence of the generic processes that originally caused them. This process, termed "genetic assimilation" by Waddington,[79] is thus a special case of generic/genetic coupling, applicable to later stages of evolution.

HOMOPLASY AND THE GENETICS OF MORPHOLOGICAL CHANGE

The viewpoint discussed above suggests strongly that the initiation of major mor-phological change during evolution depends upon the early embryo, or upon or-gan primordia of comparable size and composition to the early embryo, becoming susceptible to fresh generic effects. This could be due either to genetic modifica-tion of an organism, or to an alteration in its environment. Morphological changes brought about by environmental effects would be heritable only if the new environ-ment were relatively permanent (and therefore itself inherited).[42,64] Morphological changes brought about by genetic mutation in an unchanged environment would, of course, also be inherited. But, if the ideas presented here are valid, such changes can only be of a relatively limited variety. Indeed, if the generic mechanisms that may be brought into play during early development include only gravity, surface tension, and related phase separation phenomena, reaction-diffusion coupling, and convection, we might expect to see additional manifestations in the new body plans of gradients and stripes, of "compartments" defined by adhesive differentials con-forming to such chemical prepatterns, and perhaps of microfingers, brought about by interfacial tension or density differentials. Beyond this, it is difficult to imag-ine what else might arise. After the capacity for cells to specialize with regard to functions other than adhesion had evolved, we would expect to see organs in which differentiated tissues (e.g., muscle, cartilage) were arranged according to patterns originally templated by gradients or stripes of chemical concentration, or by the spreading and engulfment effects of differential adhesion. In essence, the vast ma-jority of random genetic changes with overt effects on form or pattern would have consequences that, regardless of how profound they were, would also be stereotypical.

Taken together with the proposal, discussed in an earlier section, that genetic changes of a canalizing or stabilizing nature would be the ones most readily retained by natural selection, the stereotypical nature of morphological alterations that may indeed arise by mutation implies that there will be only a loose connection between gene evolution and morphological evolution. Although the presumed strength of such a connection has long been a tenet of neo-Darwinism, it continues to come up against incompatible findings. A recent study analyzed inherited differences in the morphology of the mandible in inbred strains of mice in relation to the genetic divergence between the strains. It was concluded that there was little correspondence between morphological and genetic divergence.[3] Another recent study demonstrated that extensive morphological evolution in fish had taken place in a period of about 200,000 years with only minor genetic change.[51] Such findings are entirely consistent with, and indeed expected, on the basis of the concept of generic/genetic interaction presented here.

Homoplasy is morphological similarity of a feature in divergent lineages whose common ancestor was not similar to either of the lineages in this trait. For example, a reduction in the number of hind limb digits from five to four has occurred independently in three different lineages of plethodontid salamanders.[81] Such phenomena have traditionally (within the neo-Darwinian framework) been considered to result from convergent evolution based on functional adaptation to similar environments. If homoplasy is rare, the expectation is that morphological divergence will provide an accurate assessment of genealogical relationships. But in fact, homoplasy is rampant in salamander taxa,[81,82] and recent extensive revisions, based on nucleic acid data, of the phylogenetic trees for these amphibia, embody hypotheses that require extensive convergence and supposed reversals in the evolution of morphological phenotypes.[11,82] And there is no reason to think that these taxa are exceptional with regard to their degree of homoplasy.[67]

I want to propose here that homoplasy is as pervasive as it appears to be because morphological features are generically templated. Instead of being viewed as an exception to an expected concordance of "morphological distance" with "genetic distance," homoplasy would be seen rather as the exploration by various sublineages of a delimited universe of morphological phenotypes. Such exploration would account for the profusion of body forms at the Precambrian/Cambrian transition, but also encompass later phenomena like the explosive speciation of fish in Lake Victoria[51] and the genealogically uncorrelated morphological divergence of mandible form,[3] of which homoplasy can be considered an extreme case. This interpretation generalizes the notion of "developmental constraint" in the determination of adaptively inexplicable structures,[28] by incorporating the hypothesized basis of the "developmental rules" themselves. Thus, while agreeing with the statement of Wake and Larson that convergent evolution may represent recurrent production of discrete alternative phenotypes that are intrinsic to the generative system,[81] I would go further, and say that the generative systems themselves are, in an important sense, intrinsic to the material properties of early embryos and organ primordia. On this basis we might hope to explain not only why certain morphological variations show up recurrently in parallel lineages, but why body plans, and the various organs and appendages, took the forms they did in the first place.

CONCLUSIONS

I have presented a view of the relationship between development and evolution that entails significant departures from the standard neo-Darwinian model. The new view utilizes the distinction between "generic" and "genetic" mechanisms of development[61] to synthesize disparate insights of Baldwin,[5] Waddington,[79] Schmalhausen,[68] and D'Arcy Thompson,[75] among others, with recent findings on physical and molecular mechanisms of morphogenesis and pattern formation. If this view is correct, it provides the beginnings of an account of why biological organization has the particular character it does, and helps resolve certain difficulties for the neo-Darwinian view, such as the tempo and mode of evolutionary change.[26]

I will list the main elements of the framework outlined here, emphasizing, for clarity, the points of divergence from the standard neo-Darwinian picture. It should be kept in mind, however, that nothing in the view presented here prohibits gradual morphological evolution by standard Darwinian mechanisms from taking place; such effects are just suggested to be of less importance than generally considered.

i.   Biological forms and patterns are "generically templated" by physical processes acting on multicellular aggregates. These processes include rearrangement of ooplasmic determinants due to gravity, boundary formation, spreading and engulfment effects in tissues due to differential adhesion, and the formation of biochemical stripes due to reaction-diffusion coupling.

ii. Numerous morphological phenotypes would be consistent with a given genotype, particularly during early stages of evolution. This is because the phenotype is not "programmed" by the genotype but arises from interactions between an organism containing a particular set of genes and a variable environment.

iii. If some of the numerous possible phenotypes are well suited to survival and reproductive success, and are of a type to which genetic reinforcing mechanisms can be recruited by random mutation, relatively rapid, covert genetic evolution will take place, which will have the effect of making the expression of these phenotypes more reliable, and thus more *heritable*. Such evolution "selects" pre-existing, generically templated forms, rather than incrementally moving from one adaptive peak to another through nonadaptive intermediates.

iv. Stabilizing evolution will make subsequent morphological evolution less likely. In addition, reinforcing genetic mechanisms can "take over," by being triggered under different conditions (heterostasy), and at different times during development (heterochrony), thus bringing about an outcome that was originally generically templated in the absence of the original generic stimulus.

v. Either minor genetic changes *or* environmental changes will bring about alterations in the spectrum of morphological phenotypes available to the organism if they bring fresh generic processes into play. Like the alterations caused by genetic change, those induced by environmental change will be heritable if the new stimulus persists.

vi. Initial speciation would take place with little or no genetic change. Subsequent genetic evolution would occur rapidly, but with little or no additional morphological diversification between the new taxa.

vii. Minor morphological variations within modern species, such as those studied by Mendel, utilized by animal and plant breeders, and considered to be the raw material of macroevolution by Darwin, in actuality represent the limited scope of generic/genetic interactions in highly canalized taxa. Such variations would not be expected to lead to extensive morphological diversification.

A number of specific predictions, and explanations of well-accepted phenomena that are difficult to reconcile with the neo-Darwinian perspective, flow in a straightforward fashion from the statements above:

i. Gastrulation, of one form or another, is an inevitable consequence of multicellularity and differential adhesion, and should have arisen numerous times during early metazoan evolution.

ii. Organisms with a wide spectrum of body plans (e.g., segmented, "checkerboard," annular, and "pinwheel") would all be expected to arise from the interaction of differential adhesion effects with gravity-driven and reaction-diffusion mechanisms, and to have proliferated during early metazoan evolution. But because only some of these *bauplans* would be susceptible to reinforcement by standard genetic mechanisms such as promoter selectivity, a culling from the original array would occur. Once significant stabilizing selection of the relevant body plans had taken place, no new ones would emerge.

iii. Later structural specialization, including organogenesis in animals, would evolve by the action on small tissue primordia of some of the same generic mechanisms as those which originally acted on whole embryos. Generically templated organ forms and patterns (like generically templated *bauplans*) would initially be expressed in a highly diversified, but nonetheless stereotypical, fashion. Eventually, a subset of "reinforceable" morphologies will be stabilized by covert gene evolution and may come to be expressed during development in the absence of their original generic determinants.

iv. *Homoplasy* and other discordances between genealogy and morphology would be the rule rather than the exception, particularly at the origin of new taxa. Correspondingly, phylogenetic trees constructed on the basis of morphology alone would be inescapably flawed.

v. The tempo of allelic change at the population level would vary as a function of the capacity of the external environment to elicit new morphological phenotypes. It would also be influenced by internally dictated possibilities for genetic reinforcement of some of these phenotypes. Because genetic variations would be retained or eliminated according to their ability to stabilize discrete morphological phenotypes, genetic evolution should tend to accelerate *after* major speciation events.

vi. Evolution of basic body form, and of later structural specializations, would be characterized by rapid morphological diversification (in the absence of significant genetic change) separated by long periods of morphological stasis (accompanied by a great deal of genetic change).

vii. During early periods in the evolution of metazoa, before very extensive genetic stabilizing mechanisms for generic developmental processes had been acquired, development would have been more subject to environmental perturbation than is the case at present. Therefore, global climate changes, which have occurred episodically during the history of life, would be expected to have led to mass extinctions of whole taxa, leaving others relatively untouched, by virtue of selective effects on early, vulnerable stages of development.

ACKNOWLEDGMENTS

This paper is a synthesis of material presented by the author at the workshop on "Principles of Organization of Organisms" at the Santa Fe Institute, Santa Fe, New Mexico, June 1990, and at the Mahabaleshwar Seminar on Modern Biology, on "Patterning in Biology," Pachmarhi, India, October, 1990. Discussions with John Tyler Bonner, Jay Mittenthal, Vidyanand Nanjundiah, David Wake, and other participants of these meetings are gratefully acknowledged. Wayne Comper, Gabor Forgacs, and Ruth Hubbard also provided valuable comments. This work was supported by the National Science Foundation.

REFERENCES

1. Akam, M. "Making Stripes Inelegantly." *Nature* **341** (1989): 282–283.
2. Ancel, P., and P. Vintemberger. "Recherches sur le Déterminism de la Symmétrie Bilatérale dans l'Oeuf Des Amphibiens." *Bull. Biol. France Belg.* **31** (Suppl.) (1948): 1–182.
3. Atchley, W. R., S. Newman, and D. E. Cowley. "Genetic Divergence in Mandible Form in Relation to Molecular Divergence in Inbred Mouse Strains." *Genetics* **120** (1988): 239–253.
4. Balfour, F. M. "A Comparison of the Early Stages in the Development of Vertebrates." In *The Works of Francis Maitland Balfour*, edited by M. Foster and A. Sedgwick, Vol. 1, 112–133. London: Macmillan, 1885. Originally published, 1875.
5. Baldwin, J. M. *Development and Evolution.* New York: Macmillan, 1902.
6. Bonner, J. T. *The Evolution of Complexity.* Princeton: Princeton University Press, 1988.
7. Buss, L. *The Evolution of Individuality.* Princeton: Princeton University Press, 1987.
8. Castets, V., E. Dulos, J. Boissonade, and P. DeKepper. "Experimental Evidence of a Sustained Standing Turing-type Nonequilibrium Chemical Pattern." *Phys. Rev. Lett.* **64** (1990): 2953–2956.
9. De Gennes, P. G. "Wetting: Statics and Dynamics." *Rev. Mod. Physics* **57** (1985): 827–863.
10. DeLisi, C. "The Human Genome Project." *Amer. Sci.* **76** (1988): 488–493.
11. Duellman, W. E. and L. Trueb. *Biology of Amphibians.* New York: McGraw-Hill, 1986.
12. Deuchar, E. *Cellular Interactions in Animal Development.* New York: Halstead, 1975.
13. Elinson, R. P. "Change in Developmental Patterns: Embryos of Amphibians with Large Eggs." In *Development as an Evolutionary Process*, edited by R. A. Raff and E. C. Raff, 1–21. New York: A. R. Liss, 1987.
14. Elinson, R. P., and B. Rowning. "A Transient Array of Parallel Microtubules in Frog Eggs: Potential Tracks for a Cytoplasmic Rotation that Specifies the Dorso-Ventral Axis." *Develop. Biol.* **128** (1988): 185–197.
15. Epstein, I. R. "Spiral Waves in Chemistry and Biology." *Science* **252** (1991): 67.
16. Fedonkin, M. A. "Precambrian Metazoans: The Problems of Preservation, Systematics and Evolution." *Phil. Trans. R. Soc. Lond. B* **311** (1985): 27–45.
17. Forgacs, G., N. S. Jaikaria, H. L. Frisch, and S. A. Newman. "Wetting, Percolation and Morphogenesis in a Model Tissue System." *J. Theoret. Biol.* **140** (1989): 417–430.
18. Frasch, M., T. Hoey, C. Rushlow, H. Doyle, and M. Levine. "Characterization and Localization of the *Even-Skipped* Protein of *Drosophila.*" *EMBO J.* **6** (1987): 749–759.

19. Frasch, M., and M. Levine. "Complementary Patterns of *Even-Skipped* and *fushi tarazu* Expression Involve Their Differential Regulation by a Common Set of Segmentation Genes in *Drosophila*." *Genes. Dev.* **1** (1987): 981–995.

20. Friedlander, D. R., R.-M. Mège, B. A. Cunningham, and G. M. Edelman. "Cell Sorting-Out is Modulated by Both the Specificity and Amount of Different Cell Adhesion Molecules (CAMs) Expressed on Cell Surfaces." *Proc. Nat. Acad. Sci. USA* **86** (1989): 7043–7047.

21. Gierer, A., and H. Meinhardt. "A Theory of Biological Pattern Formation." *Kybernetik* **12** (1972): 30–39.

22. Gilbert, S. F. *Developmental Biology*, 3rd edition. Sunderland, MA: Sinauer, 1991.

23. Glaessner, M. F. *The Dawn of Animal Life*. Cambridge: Cambridge University Press, 1984.

24. Goldschmidt, R. *Physiological Genetics*. New York: McGraw-Hill, 1938.

25. Goto, T., P. MacDonald, and T. Maniatis. "Early and Late Periodic Patterns of Even Skipped Expression are Controlled by Distinct Regulatory Elements that Respond to Different Spatial Cues." *Cell* **57** (1989): 413–422.

26. Gould, S. J., and N. Eldredge. "Punctuated Equilibria: The Tempo and Mode of Evolution Reconsidered." *Paleobiology* **3** (1977): 115–151.

27. Gould, S. J. *Ontogeny and Phylogeny*. Cambridge, MA: Harvard University Press, 1977.

28. Gould, S. J., and R. C. Lewontin. "The Spandrels of San Marco and the Panglossian Paradigm." *Proc. Roy. Soc. London* **205** (1979): 581–598.

29. Gould, S. J. *Wonderful Life*. New York: W.W. Norton, 1989.

30. Harrison, L. G., and N. A. Hillier. "Quantitative Control of *Acetabularia* Morphogenesis by Extracellular Calcium: A Test of Kinetic Theory." *J. Theor. Biol.* **114** (1985): 177–192.

31. Hamburger, V., and H. L. Hamilton. "A Series of Normal Stages in the Development of the Chick Embryo." *J. Morphol.* **88** (1951): 49–92.

32. Heintzelman, K. F., H. M. Phillips, and G. S. Davis. "Liquid-Tissue Behavior and Differential Cohesiveness During Chick Limb Budding." *J. Embryol. Exp. Morphol.* **47** (1978): 1–15.

33. Hille, B. *Ionic Channels of Excitable Membranes*. Sunderland, MA: Sinauer, 1984.

34. Hiromi, Y., and W. J. Gehring. "Regulation and Function of the *Drosophila* Segmentation Gene *fushi tarazu*." *Cell* **50** (1987): 963–974.

35. Holtfreter, J. "Properties and Functions of the Surface Coat in Amphibian Embryos." *J. Exp. Zool.* **93** (1943): 251–323.

36. Holtfreter, J. "A Study of the Mechanics of Gastrulation." Part I. *J. Exp. Zool.* **94** (1943): 261–318.

37. Holtfreter, J. "A Study of the Mechanics of Gastrulation." Part II. *J. Exp. Zool.* **95** (1944): 171–212.

38. Hubbard, R. "The Theory and Practice of Genetic Reductionism—From Mendel's Laws to Genetic Engineering." In *Towards a Liberatory Biology*, edited by. S. Rose, 62–78. London: Allison and Busby, 1982.

39. Hülskamp, M., C. Schröder, C. Pfeifle, H. Jäckle, and D. Tautz. "Posterior Segmentation of the *Drosophila* Embryo in the Absence of a Maternal Posterior Organizer Gene." *Nature* **338** (1989): 629–632.
40. Ingham, P. W. "The Molecular Genetics of Embryonic Pattern Formation in *Drosophila*." *Nature* **335** (1988): 25–34.
41. Irish, V., R. Lehmann, and M. Akam. "The *Drosophila* Posterior-Group Gene *Nanos* Functions by Repressing *Hunchback* Activity." *Nature* **338** (1989): 646–648.
42. Johnston, T. D., and G. Gottlieb. "Neophenogenesis: A Developmental Theory of Phenotypic Evolution." *J. Theor. Biol.* **147** (1990): 471–495.
43. Kauffman, S. A., R. M. Shymko, and K. Trabert. "Control of Sequential Compartment Formation in *Drosophila*." *Science* **199** (1978): 259–270.
44. Lacalli, T. C., and L. G. Harrison. "The Regulatory Capacity of Turing's Model for Morphogenesis, with Application to Slime Molds." *J. Theor. Biol.* **70** (1978): 273–295.
45. Lacalli, T. C., D. A. Wilkinson, and L. G. Harrison. "Theoretical Aspects of Stripe Formation in Relation to *Drosophila* Segmentation." *Development* **103** (1988): 105–113.
46. Leonard, C. M., H. M. Fuld, D. A. Frenz, S. A. Downie, J. Massagué, and S. A. Newman. "Role of Transforming Growth Factor-β in Chondrogenic Pattern Formation in the Developing Limb: Stimulation of Mesenchymal Condensation and Fibronectin Gene Expression by Exogenous TGF-β and Evidence for Endogenous TGF-β-like Activity." *Develop. Biol.* **145** (1991): 99–109.
47. Matsuda, R. *Animal Evolution in Changing Environments with Special Reference to Abnormal Metamorphosis.* New York: Wiley, 1987.
48. Maynard Smith, J. *Mathematical Ideas in Biology.* Cambridge: Cambridge University Press, 1968.
49. Meinhardt, H. "Models for Maternally Supplied Positional Information and the Activation of Segmentation Genes in *Drosophila* Embryogenesis." *Development* (Suppl.) **104** (1988): 95–110.
50. Mergner, H. "Cnidaria." In *Experimental Embryology of Marine and Fresh Water Invertebrates*, edited by G. Reverberi. New York: Elsevier, 1971.
51. Meyer, A., T. D. Kocher, P. Basasibwaki, and A. C. Wilson. "Monophyletic Origins of Lake Victoria Cichlid Fishes Suggested by Mitochondrial DNA Sequences." *Nature* **347** (1990): 550–553.
52. Minchin, E. A. "Sponges." In *A Treatise on Zoology*, edited by. E. R. Lankester. London: Adam and Charles Black, 1900.
53. Mittenthal, J. E. "Physical Aspects of the Organization of Development." In *Lectures in the Sciences of Complexity*, edited by D. Stein, 491–528. Santa Fe Institute Studies in the Sciences of Complexity, Lect. Vol. I. Reading, MA: Addison-Wesley, 1989.
54. Needham, J. *Order and Life*, 38. New Haven: Yale Univ. Press, 1936.
55. Neff, A. W., M. Wakahara, A. Jurand, and G. M. Malacinski. "Experimental Analyses of Cytoplasmic Rearrangements Which Follow Fertilization and

Accompany Symmetrization of Inverted *Xenopus* Eggs." *J. Embryol. Exp. Morph.* **80** (1984): 197–224.

56. Newman, S. A., and H. L. Frisch. "Dynamics of Skeletal Pattern Formation in Developing Chick Limb." *Science* **205** (1979): 662-668.
57. Newman, S. A. "Vertebrate Bones and Violin Tones: Music and the Making of Limbs." *The Sciences* (N.Y. Acad. of Sciences) **24** (1984): 38–43.
58. Newman, S. A., D. A. Frenz, J. J. Tomasek, and D. D. Rabuzzi. "Matrix-Driven Translocation of Cells and Nonliving Particles." *Science* **228** (1985): 885–889.
59. Newman, S. A., H. L. Frisch, and J. K. Percus. "On the Stationary State Analysis of Reaction-Diffusion Mechanisms for Biological Pattern Formation." *J. Theoret. Biol.* **134** (1988): 183–197.
60. Newman, S. A. "Idealist Biology." *Persp. Biol. Med.* **31** (1988): 353–368.
61. Newman, S. A., and W. D. Comper. "'Generic' Physical Mechanisms of Morphogenesis and Pattern Formation." *Development* **110** (1990): 1–18.
62. Nijhout, H. F. "Metaphors and the Role of Genes in Development." *BioEssays* **12** (1990): 441–446.
63. Ohno, S. *Evolution by Gene Duplication.* Heidelberg: Springer-Verlag, 1970.
64. Oyama, S. *The Ontogeny of Information.* Cambridge: Cambridge University Press, 1985.
65. Phillips, H. M., and G. S. Davis. "Liquid-Tissue Mechanics in Amphibian Gastrulation: Germ-Layer Assembly in *Rana pipiens.*" *Amer. Zool.* **18** (1978): 81–93.
66. Raff, R. A. "Constraint, Flexibility, and Phylogenetic History in the Evolution of Direct Development in Sea Urchins." *Develop. Biol.* **119** (1987): 6–19.
67. Sanderson, M. J., and M. J. Donoghue. "Patterns of Variation in Levels of Homoplasy." *Evolution* **43** (1989): 1781–1795.
68. Schmalhausen, I. I. *Factors of Evolution.* (Trans. I. Dordick) Philadelphia: Blakiston, 1949.
69. Simpson, G. G. "The Baldwin Effect." *Evolution* **7** (1953): 110–117.
70. Slack, J. "A Rosetta Stone for Pattern Formation in Animals?" *Nature* **310** (1984): 364–365.
71. Stanojević, D., T. Hoey, and M. Levine. "Sequence-Specific DNA-Binding Activities of the Gap Proteins Encoded by *Hunchback* and *Krüppel* in *Drosophila.*" *Nature* **341** (1989): 331–335.
72. Steinberg, M. S. "Specific Cell Ligands and the Differential Adhesion Hypothesis: How Do They Fit Together?" In *Specificity of Embryological Interactions,* edited by D. R. Garrod, 97–130. London: Chapman and Hall, 1978.
73. Steinberg, M. S., and T. J. Poole. "Liquid Behavior of Embryonic Tissues." In *Cell Behavior,* edited by R. Bellairs and A. S. G. Curtis, 583–607. Cambridge: Cambridge University Press, 1982.
74. Stent, G. S. "Thinking in One Dimension: The Impact of Molecular Biology on Development." *Cell* **40** (1985): 1–2.
75. Thompson, D. W. *On Growth and Form.* New York: Cambridge University Press, 1942.

76. Turing, A. "The Chemical Basis of Morphogenesis." *Phil. Trans. Roy. Soc. Lond.* **B237** (1952): 37–72.

77. Vale, R. D. "Intracellular Transport Using Microtubule-Based Motors." *Ann. Rev. Cell Biol.* **3** (1987): 347–378.

78. Vincent, J. -P., G. F. Oster, and J. C. Gerhart. "Kinematics of Gray Crescent Formation in *Xenopus* Eggs: The Displacement of Subcortical Cytoplasm Relative to the Egg Surface." *Develop. Biol.* **113** (1986): 484–500.

79. Waddington, C. H. *The Strategy of the Genes.* London: Allen and Unwin, 1957.

80. Waddington, C. H. *The Nature of Life*, 67–68. London: Allen and Unwin, 1961.

81. Wake, D. B., and A. Larson. "Multidimensional Analysis of An Evolving Lineage." *Science* **238** (1987): 42–48.

82. Wake, D. B. "Homoplasy: The Result of Natural Selection or Evidence of Design Limitations?" To be published.

83. Watson, J. D., N. H. Hopkins, J. W. Roberts, J. A. Steitz, and A. M. Weiner. *Molecular Biology of the Gene.* Fourth edition, 747. Menlo Park, CA: Benjamin/Cummings, 1987.

84. Webster, G., and B. Goodwin. "History and Structure in Biology." In *Towards a Liberatory Biology*, edited by S. Rose, 103–119. London: Allison and Busby, 1982..

85. Whittington, H. B. *The Burgess Shale.* New Haven: Yale University Press, 1985.

86. Wolpert, L. "Positional Information and the Spatial Pattern of Cellular Differentiation." *J. Theoret. Biol.* **25** (1969): 1–47.

87. Wright, B. E., and B. F. Davison. "Mechanisms of Development and Aging." *Mech. Ageing and Devel.* **12** *(1980): 213–219.*

Brian Goodwin† and Stuart A. Kauffman‡
†Developmental Dynamics Research Group, Department of Biology, The Open University, Milton Keynes, MK7 6AA and ‡Santa Fe Institute, 1660 Old Pecos Trail, Suite A, Santa Fe, NM 87501

Deletions and Mirror Symmetries in *Drosophila* Segmentation Mutants Reveal Generic Properties of Epigenetic Mappings

The rapid accumulation of detailed molecular data on the spatia-temporal patterning of gene products involved in the segmentation process in *Drosophila* presents theoreticians with a challenge and an opportunity. Does this empirical data make it largely self-evident how gene products act in specifying the segmented body pattern of this insect, or is it necessary to introduce concepts that make sense of the data within particular theoretical contexts? This paper presents a case for the latter position in relation to a striking but hitherto unexplained set of observations relating to mutations in *Drosophila* that produce both deletions and mirror symmetries in segmental patterns on four different length scales. The proposed model is based on the observed spatial periodicities of the four categories of segmentation gene product together with inferences about the way in which gene products interact in the specification of differentiated states. From this is deduced a mapping from the spatial pattens of gene products to morphogenetic patterns. This epigenetic mapping involves a topological discontinuity which provides the explanation for the observed discontinuities of mutant morphology (deletions to mirror symmetries) produced by mutant alleles of the same gene. Some general deductions are drawn from

Principles of Organization in Organisms,
SFI Studies in the Sciences of Complexity, Proc. Vol. XIII,
Eds. J. Mittenthal & A. Baskin, Addison-Wesley, 1992

this model suggesting that the widely accepted distinction between pre-pattern formation and its interpretation by genes during embryogenesis is an unnecessary dualism that can be replaced by the progressive global-to-local properties of pattern formation as a hierarchically organized dynamic process.

INTRODUCTION

The segmented, bilaterally symmetrical structure of *Drosophila* is representative of one of the basic animal body plans. The wealth of the data now available on morphological mutants and the spatial patterns of segmentation gene products in *Drosophila* embryos provides an unprecedented molecular close-up of a fundamental epigenetic process. The challenge is to make morphological sense of what has been revealed. This paper focuses on segmentation mutants and gene product distributions that suggest generic properties of spatial patterning processes and lead to a proposal for topological features of the epigenetic mapping from gene activity to morphology.

What emerges from the analysis is the necessity to understand the influence of gene products within the context of a pattern-generating process with intrinsic properties. Therefore, pattern formation is not to be understood as a combinatorial consequence of individual gene product contributions to cell differentiation in different spatial domains of the embryo, a description encouraged by the view that embryonesis is the result of a genetic program. Rather, genes function in cooperation with the dynamic and topological properties of the pattern-generating process. Discovering and describing these properties is one of the tasks of biological modeling, which can then lead to insights into the universality of the generative processes described. At the end of the paper, we generalize from the *Drosophila* analysis, speculating on some general properties of the morphogenetic process that are implied by the model.

MUTATIONS RESULTING IN MIRROR SYMMETRIES

In Figure 1 is shown the metameric pattern of a wild-type *Drosophila* larva, with the distinctive pattern of denticle bands from anterior to posterior; and beneath it is the striking mirror-symmetric pattern of a larva from a homozygous *bicaudal*-D mother.[8] Somewhat less than half of the larval form is reflected in the ventral region, the segments being identified as a rudimentary abdominal 4 in the middle, with segments 5, 6, 7, and 8 and the terminal telson arranged in mirror-symmetric array. Dorsally, segments 4 and 5 are missing, but the organization of the form from dorsal to ventral is perfectly coherent and without discontinuities. This is only one of a great variety of symmetric *bicaudal* mutant morphologies, which can have forms reflected about any A-P position between abdominal segment 5 and 8—i.e., with anywhere from two to five segments mirror-symmeterized ventrally and with corresponding reductions dorsally. Thus there is no preferred mirror reflection plane, and the patterns are always coherently arranged from dorsal to ventral. Furthermore, asymmetric mirror-duplicated morphologies are equally common with, say, four segments on one side and one on the other side of the reflection plane.[32] Weaker alleles of *bicaudal* result in a switch to a qualitatively different type of mutant morphology. Instead of mirror duplications, there are simply anterior deletions of different degrees of severity, from headless embryos with complete thoracic and abdominal segments, to embryos in which head and thoracic segments are absent.[32] In the latter category are embryos in which rudimentary posterior spiracles are found at the anterior end of a complete abdominal region, which itself terminates posteriorly with well-developed, normal spiracles. Rudimentary mouthparts, characteristic of the anterior extremity of the embryo are occasionally found next to the duplicated spiracles. The spatial juxtaposition of structures that are normally located at the opposite ends of the embryo is a very interesting observation that will be considered later. On the face of it, this seems to require a sharp discontinuity of state, since one expects the ends of the embryo to be maximally different from one another. However, it will emerge that this unexpected result is what is predicted by the explanation that will be advanced for the process that leads naturally to mirror duplication.

There are other even more striking discontinuities of pattern in mutant *bicaudal* phenotypes. It is possible for one side of an embryo to show a mirror duplication of some abdominal segments while the other is a headless phenotype with all eight abdominal segments and no mirror symmetry (see Nusslein-Volhard,[33] Figure 2(e)). This means that the "character" of a segment changes in the midline from being,

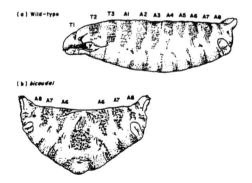

FIGURE 1   (a) Normal cuticular pattern in first instar *Drosophila* larva; (b) mirror symmetric *bicaudal* phenotype.

say, abdominal 2 on the left of the embryo to abdominal 3 on the right, with a smooth transition between them. Even more remarkable are cases such as those observed in other insects, such as *Callosobruchus*[42] in which induced bicaudal phenotypes include embryos which are fully normal except for a longitudinal stripe in which a mirror duplication has occurred. How can such striking discontinuities arise? Clearly there are no well-defined compartments or sub-divisions of the embryo that specify the positions of mirror reflection. Furthermore, there are substantial transformations of the components of overt morphology, such as the denticle band patterns. For example, in a particular mutant the bands are identified as two fused abdominal 6s in the center, with abdominals 7 and 8 in mirror-symmetric array. However, none of these segments can strictly be identified with normals; they are transformed in such a way that, if isolated, it would be difficult to name them. Mutations at a single locus can thus generate a considerable range of phenotypes, involving a broad range of transformed denticle band patterns. This defines a transformation set with respect to short-range order such as denticle band patterns, and long-range order as in mirror-symmetric phenotypes, together with all their intermediates. The dynamics of these transformations are clearly of the essence in developmental processes.

If the posterior part of a *Drosophila* embryo can be mirror reflected, it is to be expected that the same could happen to the anterior part so the embryo has two anteriors and no posterior.

The gene whose mutant phenotypes have this form is called *dicephalic*.[23] Both *bicaudal* and *dicephalic* are examples of maternal effect genes: it is the genotype of the mother that determines the phenotype of the embryo, resulting from the transfer of mRNA from the cells of the maternal ovary into the maturing oocyte. *Dicephalic* also shows the same spectrum of defects as *bicaudal*, strong alleles resulting in mirror-duplicated forms, both symmetric and asymmetric, while weaker alleles produce posterior deletions of different degrees of severity.

Strong alleles of the mutant *Krüppel* result in mirror-symmetric patterns, a typical phenotype lacking thoracic and anterior abdominal segments, replaced by a partial mirror-image duplication of the remaining posterior abdomen which can include a set of rudimentary posterior spiracles near the head.[46] Weaker alleles of *Krüppel* result in deletions of thoracic and anterior abdominal segments but no mirror symmetries. This is the same spectrum of transformations as *bicaudal*, but over a shorter spatial domain. A similar pattern to *Krüppel* is observed with the mutant *hunchback*, whose effects are centered on head and thoracic segments, these being either deleted by weaker alleles or replaced by mirror-image duplications of the anterior abdominal segments in cases of stronger alleles. *Hunchback* mutants also have a posterior domain of action, abdominal segments A7 and A8 being deleted. This is an important observation to which we shall return shortly. These two mutants belong to the category of segmentation gene called gap because of the characteristic pattern deletions, extending over four to seven segments.

The next group of segmentation genes, known as pair rules, act over a wavelength of two segments. Mutant alleles of these genes result in, for example, deletions of every other segment: in *odd-skipped* there is an absence of segments T2,

A1, A3, A5, and A7; while *even-skipped* has T1, T3, A2, A4, A6, and A8 missing. Other members of the set generate deletions on the same wavelength of two segments, but phase-shifted so that different parts of adjacent segments are missing. Once again the now-familiar pattern of deletions in weak and mirror symmetries in strong alleles shows up. For example, strong *runt* alleles result in mirror-image duplications replacing the deleted domains, with the consequence that the eleven normal denticle belts are replaced by six, five of which are mirror-imaged.[9] The exact positions of the lines of mirror reflection vary from one embryo to the next, as in the gap and *bicaudal* examples of mirror-symmetric patterns. A careful analysis of the *runt* pattern[9] showed that the mirror-image duplications are unlikely to be caused by cell death and regeneration, but reveal a characteristic aspect of pattern transformation.

The last level of the hierarchy of segmentation genes, operating over wavelengths of a single segment, is in fact characterized by the property of mirror duplication. Members of this class are called segment polarity genes, because the typical effect of mutant alleles is a replacement of the posterior part of the denticle band by a mirror-symmetric duplication.[34] The line of mirror reflection differs for each of the different genes, showing that the domains of gene influence within each segment are phase-shifted with respect to one another. There is no evidence of any primary reference point within segments, these being generated by a periodic pattern with all positions (phases) as potential mirror-reflection lines. In the case of the mutant *patch*, for example, the duplicated domain involves structures of two adjacent segments so that, despite the presence of a normal number of denticle bands, there are twice the normal number of segment boundaries since each duplicated region includes this structure.

These observations show that there is a hierarchy of genetic effects on the segmentation process in *Drosophila* that is characterized by spatial patterns with distinct domains or wavelengths of influence. These extend from single segments up to nearly half of the embryo. Furthermore, within each of the four categories of segmentation gene, each operating over characteristic embryonic domains, there are qualitatively similar patterns of mutant phenotypes, weak alleles resulting in pattern deletions and strong alleles giving mirror duplications. It is necessary now to look beyond the phenotypes to the spatial distribution of gene products to see if there is a pattern that could provide a basis for understanding these intriguing observations.

SPATIAL PERIODICITIES AND HARMONICS IN GENE PRODUCT DISTRIBUTIONS

The last two categories of segmentation genes, pair rules and segment polarity, make it obvious that the segmentation process in *Drosophila* involves the occurrence of spatial periodicities of gene influence that are simple multiples of one another, with two-segment and one-segment wavelengths. Studies of the spatial patterns of gene transcription have amply confirmed the expectation from the mutant data that gene products should be found distributed in patterns of two-segment and one-segment wavelength for pair-rule and segment polarity genes, respectively. For example, Macdonald et al.[24] showed that the pair-rule genes *even-skipped and fushi tarazu* have well-defined two-segment periodicities of gene transcription at the late syncytial blastoderm stage (early in the 14th mitotic cycle), with a relative phase shift that correlates well, though not precisely, with the deletion zones of the mutants. *Engrailed*, classified as both pair-rule and segment polarity because of the range of its mutant defects, has a one-segment pattern of transcripts at the gastrula stage.[45]

What about the other categories of segmentation gene discussed above? In view of their longer wavelength effects, one would expect these to show correspondingly extended gene product distribution patterns. However, there is nothing in the mutant phenotypes of *Krüppel*, for instance, to lead one to expect a periodic spatial pattern of gene expression, since deletions and mirror symmetries occur only in one region of the embryo, located centrally. On the other hand, *hunchback* has two domains of influence, anterior and posterior, as remarked earlier, so we expect two domains of gene transcription. This has been confined: in early cycle 14, *hunchback* transcripts show well-defined anterior and posterior transcription domains, separated by about six to eight segments.[19] But it turns out that the *Krüppel* transcription pattern is also spatially periodic: not only is there a broad central transcription domain centered on the region where mutant defects occur; there are also smaller anterior and posterior regions of transcription at early cycle 14, separated from the central region by six to eight segments. Finally, a third major gap gene, *knirps*, is of interest in relation to spatial periodicities. Mutant alleles of this gene result in a single six-segment deletion domain centered on abdominal 4. Does its gene transcript pattern correspond to its single domain of influence; or is it, like *Krüppel*, spatially periodic, one of the domains of transcription being functionally silent? From an analysis of a wide range of mutant phenotypes and transcript patterns of segmentation genes, Goodwin and Kauffman[14] predicted that *knirps* transcripts would show a periodic pattern with two bands, one centered on the posterior deletion zone, as expected, and the other located anteriorly, some six to eight segments away, where there is no evidence of a mutant influence. This has since been confirmed.[39] A full analysis of this data is presented in Goodwin and Kauffman.[12,13]

SPATIAL PERIODICITIES AND MIRROR DUPLICATIONS

Since the same phenomenon arises on different wavelengths, it does not matter which example is used to relate spatial periodities to mirror duplications. Furthermore, the model that will now be constructed to explain the characteristic spectrum of mutant phenotype, from deletions to mirror symmetries, is independent of how the patterns of segmentation gene products are generated. It requires only that the genes within any segmentation category have domains of expression that are systematically phase-shifted relative to one another, together with another property of the map from gene products to expressed patterns that will shortly be described. For purposes of clarity, consider the two-segment periodicities that are observed at cycle 14 in pair-rule gene transcripts, or the same pattern in the corresponding translation products, the pair-rule proteins. Each of the different gene products has a characteristic phase relation to the others, the maxima and minima being shifted relative to one another so that they span the two-segment cycle in a well-defined order. To demonstrate the principle involved, consider the primary pair-rule genes *hairy*, *runt*, and *even-skipped*. These have a relative phase shifts as shown in the schematic periodicities of Figure 2 which represent the concentration of protein products in relation to position in the embryo. The curves are idealized, but, as the argument that follows is essentially topological, this does not affect the deductions. The positions of the denticle bands, using Al, A2, A3 for illustration, are shown in relation to the pair-rule gene product periodicities, there being two of these per cycle.

Now plot *eve* against *hairy* over one cycle, as shown in Figure 3. The result is a closed circle in which spatial position in the embryo (the abscissa in Figure 2) is represented by position on the circle following the arrows for the direction anterior to posterior. The positions of the denticle bands, of which there are two per pair-rule gene cycle, are as shown. Since the cycle repeats along the embryo, mapping two segments per pair-rule cycle, we can assign segment identities to the denticle bands in groups as shown. T1, T3, and the even-numbered abdominal bands are mapped at one position on the repeating cycle, while T2 and the odd-numbered

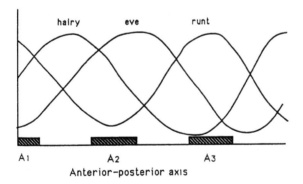

FIGURE 2 Spatial pattern of the three major pair-rule genes, *hairy*, *eve*, and *runt*, showing their approximate phase relationships.

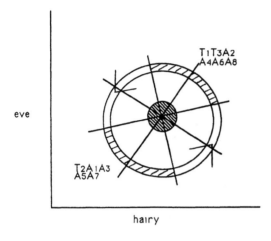

eve

hairy

FIGURE 3 Plot of *eve* against *hairy*, giving a pair-rule color wheel showing the positions of the segmental denticle bands.

abdominal segments are identified with the other. The pair-rule genes do not, of course, themselves assign identities to the segments, which result from the combined influence of all segmentation genes together with the homeotics.

READING GENE PRODUCTS AS RATIOS

It is necessary now to consider what general type of functional relationship there is between gene products and the patterns of cell differentiation for which they code. Some observations of Gergen and Wieschaus[9] and Gergen et al.[8] are particularly relevant to this question. Studies of embryos with different numbers of wild-type *runt* alleles showed that when *runt+* copy level is decreased, the odd-numbered denticle bands are deleted, giving a phenotype in the general *odd-skipped* category. However, when the copy level of *runt+* is increased, a complementary anti-runt phenotype arises in which deletions occur in the even-numbered denticle bands. So overproduction of *runt* results in a phenotype similar to that produced by loss of function at other pair-rule loci such as *even-skipped or paired*; i.e., too much *runt* is like too little *eve* or *prd*. Too little *runt*, on the other hand, results in an *odd-skipped* phenotype. Genes evidently do not affect pattern by a cumulative linear measure along a single scale of influence, such as progressive effects on the same pattern elements, but interact in some more complex manner.

Another striking observation,[5] relates to the consequence of double mutants of *even-skipped* and *odd-skipped*. From the deletion effects of these two genes taken individually, the double mutant would be expected to lack both even- and odd-numbered denticle bands. Instead, the double mutant has eight partial denticle bands, each of which is a small mirror duplication of the normal patterns of abdominal segments 1–8. So the elimination of a second gene can result in the recovery of pattern elements. These and the runt results suggest that genes influence pattern elements via the ratios of their concentrations, or some other function that is not a linear measure of product concentrations. We can now return to Figure 3 and complete the description of the relationship between the spatial distributions of pair-rule gene products and the specification of pattern. Let us make two assumptions:

1. Pathways of cell differentiation, hence the contributions of cells to patterns, are specified by ratios of gene products.
2. The ratios are measured from a point within the circle as shown in Figure 3.

It is instructive to determine the consequences of these assumptions first, and then to examine more closely exactly what they imply and how they may be modified.

A line through the common point of origin in Figure 3 is defined by a ratio $y/x = a$, where a is the slope of the line. Taking y as the appropriate concentration of *eve* product, measured relative to the point of origin within the circle, and x that for *hairy*, all points on such a line are equivalent in terms of pattern specification by these two genes. Hence, measuring gene products as ratios from a common origin within the circle has the consequence that the two-segment circle is divided into pairs of equivalent points and so maps two identical segmental patterns relative to the two genes under consideration. The point of intersection of the lines has all ratios and so cannot specify any pattern element. Since ratios are equivalent to an angle (slope) or a phase, we can call this singularity a phaseless point. The pattern specification process has a limited degree of resolution relative to ratios, so this phaseless point will effectively be a null domain of finite extent—the disc of Figure 3—within which no specific pattern elements can be discriminated. Similarly, there will be discrete sectors defined by ranges of phase (ratio) within which no discriminations are made, all cells within such a range following the same pathway of differentiation in terms of the contributions of these two genes. So the whole space, called tissue specificity space (TSS) by Winfree,[47] is quantized. In Kauffman and Goodwin[12,13] these quantized sectors were identified metaphorically with the color spectrum, and the sectored cycle was described as a color wheel. The language reflects a topologically similar analysis carried out by Winfree[48] on the periodic temporal organization of organisms, particularly biological clocks, in which the concept of the isochron was introduced to describe states that map into points of equal time in the dynamic space of biological oscillators. The description given here of the dynamics of spatial organization in organisms by periodicities in space rather than in time has deep qualitative similarities to that presented by Winfree, and some of these properties will be developed further in a later section. Clearly an objective is to join the two in a unified analysis of the space-time organization of developing and behaving organisms as dynamic systems of a particular kind, with generic properties that come from basic topological features. We are now in a position to see how these relate to deletions and mirror symmetries.

Referring to Figure 4(a), loss of function mutations in *eve* correspond to displacement of the circle towards the *hairy* axis and a flattening that describes a reduction in amplitude of the *eve* product. When this displaced curve meets the null disc, as shown by the elliptical curve, pattern deletions will occur in the even-numbered denticle bands. The odd-numbered bands will continue to be generated, since the dotted curve intersects the radial lines defining the mapping for those pattern elements outside the null disc. Thus the *even-skipped* phenotype results.

Now we carry out a similar analysis for *hairy*. The phenotype of a typical weak mutant is characterized by loss of the anterior part of each even-numbered abdominal segment and the posterior part of each odd-numbered segment, with corresponding deletions to the thoracic segments. The effect of such a mutation is a distortion of the color wheel towards the *eve* axis, resulting from decreased levels of *hairy* products. When this distorted curve intersects the null disc (see Figure 4(b)), there will be a loss of anterior parts of even-numbered segments and posterior parts of odd-numbered segments, as observed. In stronger alleles, the curve

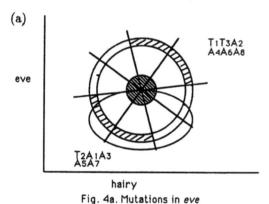

(a)

eve

T₁T₃A₂
A₄A₆A₈

T₂A₁A₃
A₅A₇

hairy

Fig. 4a. Mutations in *eve*

(b)

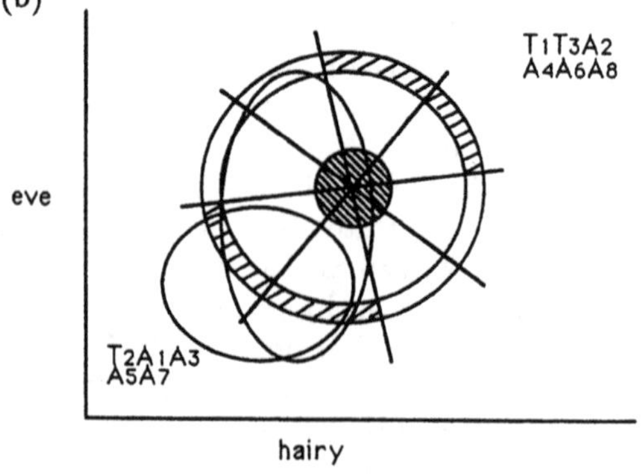

eve

T₁T₃A₂
A₄A₆A₈

T₂A₁A₃
A₅A₇

hairy

FIGURE 4 (a) Effect of a mutation in *eve* on the color wheel, showing a deletion pattern with loss of even-numbered segments. (b) Effects of mutations in *hairy* showing both deletions and a mirror-symmetric pattern.

will cross the null disc and come to lie entirely to one side of it (Figure 4(b)), so that Tl, T3, and the even-numbered denticle bands entirely disappear. Furthermore, the curve will now intersect each of the lines in two points. Following the curve around and identifying the pattern specified, it is evident that one part of the curve specifies the odd-numbered abdominal segments in the normal spatial order while the other part, intersecting the same lines in the opposite direction, specifies the same pattern elements in mirror-symmetric order (and similarly for T2). So strong *hairy* alleles should result in mirror-symmetric, odd-numbered abdominal denticle bands, which they do.[18] However, this phenotype has no naked cuticle between the mirror-symmetric denticle bands, which constitute a continuous mass of setae. So the curve is displaced into the lower left-hand quadrant, as shown. This is actually what is expected if we take account of the pattern of regulatory interactions believed to exist between the pair-rule genes: *hairy* represses *runt* which represses *eve*.[4] Therefore, decreased levels of *hairy* product will result in elevated *runt*, hence repressed *eve*. So the observed strong *hairy* phenotype is what is expected from the model. The importance of taking account of the higher-dimensional space in which gene products interact and specify patterns will become more evident as this analysis proceeds.

This procedure can now be used to explain the effects of *runt* alleles. *Runt* is plotted against *hairy* to give the relationships shown in Figure 5, with the denticle band patterns located on the circle as indicated. Too much *runt* means moving the curve towards the right and down, since *runt* represses *hairy*. When this intersects the null disc, deletions of the even-numbered denticle bands occur, giving the even-skipped phenotype. Conversely, too little *runt* corresponds to displacement of the curve up and to the left, which results in odd-numbered denticle band deletions, the complementary effect, as observed. But it is now possible to understand also the mirror-symmetric phenotype of strong *runt* alleles. These carry the curve to the left of the null disc, as shown. Following this curve around in the direction of the arrow (the A-P direction) gives a mirror-symmetric map: the rising part of the curve passes across the sectors that map the anterior part of the even-numbered denticle bands and the posterior part of the adjacent (anterior) segment, as described[9] for *runt*[YE96]. So we get very naturally the full range of mutant phenotypes described for *runt* which at first sight seemed so strange. The phenomena are generic consequences of spatial periodicities in gene products and using ratios of gene product concentrations to determine patterns of cell differentiation.

The term generic means typical or characteristic, a property that belongs to the class of structure that is under consideration, in this case the mapping between spatial patterns of gene products in embryos and the generated phenotype.

Figures 3 and 5 are really two-dimensional projections of a multi-dimensional map of all the pair-rule genes into the space of cell differentiation patterns. If only the three genes represented in Figure 2 are used to construct such a map, then it

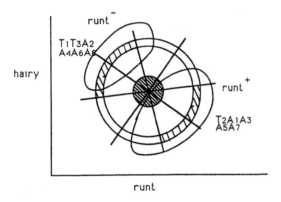

FIGURE 5 Distortions of the color wheel for too much and too little *runt* product, showing deletions and mirror symmetry.

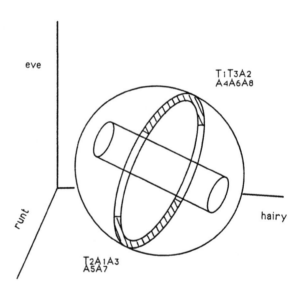

FIGURE 6 The pair-rule color wheel in three dimensions, showing the null cylinder that projects to null discs in two dimensions.

will have three dimensions. The closed curve representing the differentiated spatial pattern of cells will be on same surface, which can be described topologically as a sphere (Figure 6). The null discs then become topologically a cylinder that passes through the sphere, defining the states in which no discriminations about pattern can occur because the ratios or phases cannot be discriminated. The overall space will also be quantized, which can be described by sectors extending out from the null cylinder and defining regions like sections of an orange. In going to N dimensions, for N genes, it can be shown[47] that the dimension of the null "cylinder" is $N-2$ and the closed curve of pattern states winds its way around this $(N-2)$-dimensional cylinder with two degrees of freedom. So the overall topology of the two-dimensional projections of Figures 3 and 5 is preserved. These relations are also described in Goodwin and Kauffman[14] and Kauffman and Goodwin.[20]

The three major gap genes, *hunchback*, *Krüppel*, and *knirps*, have phase-shifted distributions of gene products during cycle 14 that cover the segmentation domain in the manner shown in Figure 7. Plotting *hb* against *Kr* and locating the segments on the closed curve that now represents part of the anteroposterior axis of the embryo gives us Figure 8(a). Mutations in *hb* result in the oval curves shown, with deletions occurring in the head and thoracic segments as well as posterior abdomen (A7, A8) while mirror-symmetric patterns involve the anterior abdominal segments, as observed. The equivalent plot of *knirps* and *Krüppel* is given in Figure 8(b), which shows weak mutants in *Kr* deleting the thoracic and anterior abdominal segments, while strong ones result only in A8 and A7 and a mirror-symmetric A6. These two-dimensional projections need to be put into a three-dimensional context, in the form shown in Figure 6, to identify the detailed characteristics of the mutant phenotypes involving interactions among all three of the primary gap genes. This should be supplemented by the further contributory effects of the other member of this group such as *giant* and *tailless*, and possibly *unpaired* and *hopscotch*.

The three primary gap genes, *hb*, *Kr*, and *kni*, are intimately involved in the process whereby the periodic pattern of pair-rule gene products is generated. This emerges from the effects on pair-rule gene activities in mutants of both the maternal

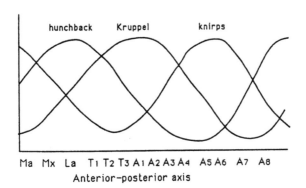

FIGURE 7 Spatial pattern of the three major gap genes, showing their phase relationships.

(a)

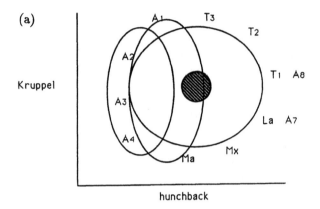

Kruppel

hunchback

(b)

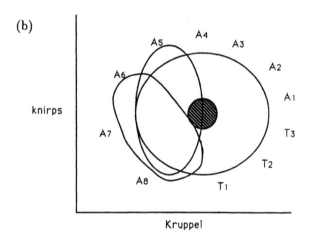

knirps

Kruppel

FIGURE 8   (a) The gap color wheel with *hunchback* mutations giving deletions and mirror symmetry. (b) The gap color wheel with *Krüppel* mutations giving deletions and mirror symmetry.

genes, which affect gap gene activity patterns, and mutations in the gap genes themselves.[3,4,7] How a periodic pattern of seven stripes arises within a domain with a distribution of primary gap genes of the type shown in Figure 7 remains a puzzle that has led Gaul and Jackle[7] to propose that "gap gene concentration levels, and ratios between concentration levels, have to be taken with account." The position adopted by Goodwin and Kauffman,[12,13] Hunding et al.,[17] and Lacalli[21] is that it is also necessary to take account of the intrinsic dynamic of the global patterning processes which spontaneously generate spatial periodicities in the field variables. As described in more detail below, it is the cooperative interaction of gene activities, read as ratios, and the global spatio-temporal dynamic of a hierarchically organized system, that results in the robust patterns of morphogenesis.

Applying these principles to maternal genes requires a modification of detail, but not of topological principle. The protein product of the *bicoid* gene is distributed in an exponential gradient with maximum at the anterior pole of the embryo and it is assumed that *bicaudal* is similar. Other maternals have similar spatial patterns though with maxima at other positions. *Oskar*, for example, is taken to have its maximum at the posterior pole. Plotting *bicaudal* against *oskar*, for instance, gives a curve of the general type shown in Figure 9. The origin of coordinates is taken to be near the mean value of the range of the two variables. Quantizing the domain to specify the ranges of resolution of the variables results in a fan rather than a color wheel. What happens now in a bicaudal deficiency mutant? The level of bic product will clearly fall, but now oskar product starts to rise at the head end, where it is normally repressed, and so the curve distorts towards the origin as shown in Figure 10. With greater reduction in X, the distortion increases and the curve passes through the null zone giving anterior deletions. Further distortion results in a loop that cuts the posterior sections twice, giving a mirror-imaged bicaudal embryo. The mirror-imaged domain can be symmetric or asymmetric, depending on the extent of distortion of the curve. The curious phenomenon of duplicated spiracles at the anterior end of a headless embryo, with mouthparts adjacent to the duplicated spiracles, also find an explanation in Figure 10. The curve that just touches the lower part of the null disc describes an embryo lacking anterior parts (say, head and thorax missing). However, the curve cuts the spiracle domain (shown as a line) and, next to that, the mouthpart domain (also shown as a line, though both are really sectors). The latter is the continuation of the normal mouthpart domain on the opposite side of the null disc. This interpretation of the pattern of such embryos is possible only if the origin of coordinates for gene ratios is in a position like that shown in Figure 10.

Turning to the other striking categories of mutant morphology mentioned earlier, such as those with mirror duplications on one side but deletions on the other, it is now evident how this comes about in terms of the model proposed. The distance between curves describing deletions and mirror symmetries (cf. Figure 8(a), 10) is small in terms of concentration differences of gene products, so neighboring regions of the embryo can undergo such transitions of morphology. Similarly, phenotypes with a longitudinal strip of mirror-symmetric patterning can be understood as regions where the curve in tissue specificity space gets distorted locally, passing across the null zone and into the mirror-symmetry region by a perturbation. So the topological properties of these mappings with singularities makes the range and mixture of phenotypes observed much easier to understand than interpretations of gene products read as concentrations.

The above discussion has shown how the general principles of the analysis extend to the longest wavelength category of segmentation genes, the maternals. Going in the direction of shorter wavelengths, from pair rules to the segment polarity genes, clearly presents no difficulties since these are treated in the same way as pair rules, but on a wavelength of one segment. Mirror symmetry is the defining character of mutations in these genes, polarity reversal of part of the segmental pattern being their distinctive phenotypic character.

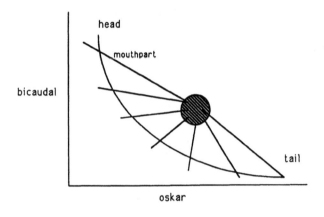

FIGURE 9 The color fan for maternal genes *bicaudal* and *oskar*.

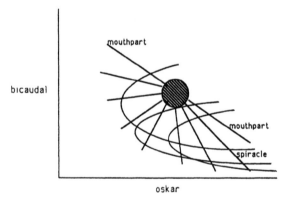

FIGURE 10 *Bicaudal* mutants showing how anterior spiracles can be produced adjacent to mouthparts in headless embryos. The most severe mutation produces a mirror-symmetric embryo.

The overall result is a set of four "color wheels" that define four multidimensional mappings of spatial distributions of gene products, each with a characteristic wavelength, onto the final spatial pattern of differential cells in the *Drosophila* larva. Each color wheel, from maternals to segment polarity genes, specifies pattern on finer and finer spatial grids, the whole hierarchy constituting a coherently organized, nested dynamical set. This coherence comes from interactions between and within members of the different groups of genes that affect their spatial and temporal patterns of activity, as previously described. So segmentation genes not only influence pathways of cell differentiation, resulting in spatial patterns, they also modulate one another's activity, resulting in coordinated spatial and temporal patterns.

The unfolding dynamics of this patterning process also involves other variables beside gene products. These are the physiological variables serving general regulatory functions, such as calcium and other ions, cAMP, DAG, IP$_3$, and the related constellation of intermediary metabolites. The evidence for this comes from characteristic morphological abnormalities induced in developing *Drosophila* embryos by stimuli that disturb these variables, such as ether,[16] ionphores and ion channel blockers, pH shock, etc.[26,25] The particular spectrum of perturbations is usually stage and stimulus dependent, some abnormalities being like genetic perturbations ("phenocopies") while others are unlike anything resulting from known mutations. This is to be expected if morphogenesis has its own intrinsic dynamic and topological properties, since, then, genetic and metabolic perturbations act upon the same morphogenetic system, resulting in an overlapping but not identical spectrum of perturbed morphologies. Models that involve metabolic variables as elements of the global spatial-patterning process, to which gene products contribute as amplifiers and sharpeners of spatial pattern, have been proposed by different investigators.[17,22,31] All of these employ reaction-diffusion processes of the Turing type (Turing,[41] also Meinhardt[28] and Murray[30]) as the generators of periodic spatial patterns in metabolites over different wavelengths corresponding to the segmentation gene patterns, providing a global context of order within which gene activities are expressed. Hunding et al.[17] use a hierarchical scheme in which gene products associated with longer wavelength patterns act as parameters of reaction-diffusion systems generating shorter wavelength patterns, so that the whole set is organized into a coherent system that is very robust in its general properties and provides a possible explanation for the transient period-doubling patterns observed in gene products, as described earlier. It predicts that there will be characteristic spatial periodicities in physiological variables associated with the development of periodicities in gene product distributions. As mentioned in Goodwin and Kauffman,[12,13] the dynamics of the pattern-generating process is not necessarily a reaction-diffusion mechanism. It could equally well be a mechanochemical process such as the calcium-cytoskeletal model described by Odell et al.,[35] and by Goodwin and Trainor,[15] with calcium and associated metabolites (cAMP, IP3 etc.) as primary pattern variables that initiate the spatial periodicities which are then amplified and stabilized by gene products.

CHANGE OF ORIGIN; SAME TOPOLOGY

The second of the assumptions—ratios are measured from a point within the circle— is a strong one, and may seem implausible. Why should gene products be measured from a point within the circle of Figure 3, which corresponds to a mean value of the spatial range shown for the variables in Figure 2? This means that values above the mean are read as positive while those below are seen as negative. The ratio of two values is the same if each of the values undergoes a change of sign from positive to negative. So every straight line through the origin in Figure 3 has a constant ratio, giving identical values to the points where they intersect the circle. What does this imply about the activity of gene products?

If we regard the two gene products as repressors of other genes, then one interpretation of the assumption follows. Where both gene products are above a critical value, they inactivate one another so that any gene they control is released from repression and so is switched on. If both are below a critical value, they fail to repress and so again genes under their control will be switched on. However, if either one is above a threshold while the other is below, repression will occur. Logically, this is the inclusive AND function in which the controlled gene is active whenever input 1 and input 2 are in the same state. So the assumption implies that this or some equivalent pattern of interactions is realized by pairs of genes. It is a somewhat restrictive, but not unreasonable, assumption. And if it is in operation, then there should be a category of mutation in which the gene products lose their property of mutual inactivation so that the controlled genes are on only when the two pair rules are below threshold. The result is a loss of pattern elements from 1/4 of the wavelength of the gene product distribution. For pair rules this would be 1/2 of every other segment. This is similar to normal pair-rule mutant phenotypes, with the difference that the extent of the deletion is only half of every other segment rather than the whole. Another feature of such pair-rule-type mutations is that they should never have mirror-symmetric mutant phenotypes.

The question now arises whether this assumption about the origin of coordinates for reading gene products can be altered without affecting the basic topological analysis of the deletion-to-mirror-symmetry spectrum of mutant phenotypes. It turns out that it can: the topological principles are not altered by a coordinate transformation. To see this, consider a map of *hairy* and *runt* in which the origin of coordinates is where one expects it to be: at zero concentration for the two gene products. Lines of constant ratio are now straight lines from the point $(0, 0)$ as shown in Figure 11(a). We now make another assumption, which corresponds to what is generally accepted about the way segmentation genes influence pattern formation. This is that only when a gene product is above a particular threshold level does it contribute to pattern specification. The map from gene product space to pattern element space is then as in Figure 11(a): only half of the color wheel is now used to specify a segment. The next segment is specified by other pair-rule combinations that give a map complementary to that of *hairy* and *runt*.

Consider the result of mutations in *hairy*. The half-circle distorts towards the *runt* axis, with *runt* increasing and *hairy* decreasing because of their mutually repressive interactions. This results in deletions in the odd-thoracic and even-abdominal denticle bands, as observed. Stronger alleles result in a further displacement of the half-circle and a rotation so that the curve cuts the lines of constant ratio in two points, giving a mirror-symmetric mapping of T2 and the odd-abdominal denticle bands, as in strong *hairy* phenotypes.[18] *Runt* mutants follow a similar pattern of distortions, but in the opposite direction (Figure 11(b)). So again we observe the same pattern of deletions and mirror symmetries with the singularity now located at the origin.

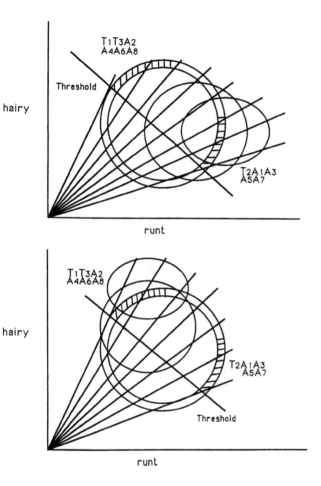

FIGURE 11 Deletions
and mirror-symmetric
patterns in color
wheels with lines of
equal ratio originating
at the origin, together
with thresholds.

This analysis can be applied to the other categories of segmentation gene, so that the sequence from deletions to mirror symmetries for mutant phenotypes again emerges as a generic property of the map describing genetic influences on pattern formation. Measuring gene products as ratios remains a basic assumption, but now a threshold value, together with a singularity at the origin, define a kind of extended singularity domain, or null region of gene product space. This allows us to recover the earlier arguments in modified form. The generalizations to higher dimensions are also valid, with the modification that thresholds are understood to define the range of activity of the different variables.

WHY RATIOS?

The topological properties of the map from gene products (and other variables) to differentiated pattern is dependent upon the use of ratios for the measure of genetic influence. It is shown in Kauffman and Goodwin[20] that direct measures of gene concentrations, combined with thresholds that result in on-off readings of effective gene influence, do not give the characteristic deletion to mirror-symmetry sequence of mutant phenotypes observed within segmentation mutants. The argument that the color wheel model captures a basic property of the mapping from genotype to phenotype space implies that there is something generic about ratios or their generalization. What could this be? Two different types of explanation will now be considered.

A primary role of gene products is the regulation of other gene activities by inductive and repressive interactions. This has been well established for the segmentation genes themselves, which interact with one another both within and between the major categories (see, e.g., Gaul and Jackle,[7] and Carroll and Vavra[4]). Since the regulative action of genes depends upon the formation of association complexes between gene products and their target sites, the influence exerted on rates of transcription of message from DNA is expressed in terms of expressions that characterize the equilibrium kinetics of binding and dissociation on polymer surfaces. If Y acts as an inducer and X a repressor, both acting on the same DNA control site with stoichiometries n and m respectively, then the transcription rate from the controlled gene can be expressed in terms of a function of the general form

$$f(X,Y) = \frac{A + K_1 Y^n}{B + K_2 X^m + K_3 Y^n} . \tag{1}$$

The exact form of the expression depends upon the details of the kinetics and the type of interaction between the inducer or the repressor and the target; however, the rate function will, in general, be a ratio of rational polynomials. The question we want to ask is: What type of curve (functional relationship between X and Y) is defined by setting expression (1) equal to a constant, which means a constant influence of the two gene products as they vary along the curve? Previously it was simply asserted that genes exert their influence in terms of ratios, X which is a special case of Eq. (1) in which $n = m = 1$, $A = B = K_3 = 0$, $K_l = K_2$.

The class of curves to be considered is of the form

$$\frac{A + K_1 Y^n}{B + K_2 X^m + K_3 Y^n} = \frac{1}{C}, \quad C \text{ a constant,}$$

which gives

$$AC - B = K_2 X^m + (K_3 - CK_1)Y^n . \tag{2}$$

We assume that $C > 1$. Furthermore, $K_1 > K_3$ since the function is sigmoid in Y. The result is $(K_3 - CK_1) < 0$. For the case in which $n = m = 2$, the curves are $a = K_2 X^2 - KY^2$ where $a = AC - B$ and $K = CK_1 - K_3$. These are hyperbole. The topological properties of the map shown in Figure 11 are not altered by this, since the family of lines is simply replaced by a family of hyperbole. Other integral values of n and m give different curves, but they all show a similar property: they are asymptotically open curves that typically either intersect a closed curve located in the positive quadrant in two places, or in none. This preserves the topological properties required for the deletion-to-mirror-symmetry mutation sequences. If, however, $K_1 = 0$ (both genes acts as repressors), then this property is lost. If both genes act as inducers, there is a similar loss. However, these are two-dimensional projections of higher-dimensional mappings. In the higher-order spaces, the required topological properties of the mapping will be preserved if, within every multidimensional color wheel, some gene products act as inducers and some as repressors of pathways of differentiation, leading to pattern elements. This is a quite general property, so the generalized form of Eq. (2) to higher dimensions may be the molecular basis of both ratio—like mappings and their morphological consequences when combined with the spatially periodic distributions of segmentation gene products.

A very different approach to understanding the significance of ratios is the following. When processes occurring in space and time, described by partial differential equations, are studied for their qualitative or generic properties, a procedure is followed that casts the equations into a non-dimensional form. This involves getting rid of specific measures of time and space, which are arbitrary units, by defining new dimensionless variables and parameters which are expressed as ratios of the original ones. For example, in reaction diffusion systems involving two substances with diffusion constants D_x and D_y, the dimensionless form of the equation will have a ratio D_y/D_x. Similarly, the dimensionless rate constants will be ratios of the original rates, and similarly for affinity coefficients describing the regulatory activities of gene products. Examples of this can be found in the reaction-diffusion studies described by Murray.[30] The new parameters thus obtained can be more complex than simple ratios of two quantities, since the original parameters themselves may be products of other independently variable quantities.

What is obtained by this procedure are equations that describe the whole class of processes under investigation, and the dimensionless parameters then assume a special significance. In the study of fluid flow, for example, one of these parameters is called the Reynolds number, a pure number that describes essential features of the behavior of fluids. As this is varied, there are critical values at which transitions from linear to periodic stream flow occur, and from periodic to turbulent, irrespective of the particular properties of the fluid (e.g., viscosity) or its physical state (temperature, pressure), which have all been normalized by casting the equations into dimensionless form. Gene products act as determinants of many of the parameters in biological processes, such as enzyme activities, activities of membrane pumps and channels, and regulators of other gene activities. In studying the generic properties of epigenetic processes, therefore, it is ratios of these quantities that are most significant. What has been shown is that the mapping from gene product space to phenotype space involves certain qualitative features, the transition from deletions to mirror-symmetric patterns, when the mapping is expressed in terms of gene product ratios. As these ratios are altered by mutation, the mapping shows characteristic transition points, as in fluid flow by altering the Reynolds number. It would be very interesting to extract more precisely the dimensionless parameters that underlie the qualitative transitions during morphogenesis, irrespective of the particular details of the species. This is what a dimensionless, qualitative analysis gives: insight into morphogenetic universals, insofar as they exist.

One final point should be stressed in this context. We have used the conventional terms genotype and phenotype to identify the spaces between which the mapping, involving ratios of gene products, has been described. In this mapping, different morphological states on the color wheels correspond to the different regions of gene product state space. This must now be recognized as a simplification. The mapping is in fact from epigenetic space, not gene product state space, to morphological space (Winfree's tissue specificity space). Epigenetic space (see Goodwin[11]) involves many variables other than gene products, including all the primary regulatory metabolites and ions that have been previously mentioned (cAMP, IP_3, DAG, Ca^{2+}, H^+, mechanical state of the cytoskeleton, and a host of other variables). Gene products constitute only one set of variables of the process, the ones that become most evident in genetic analysis, hence the stress in the present chapter, which has focussed on a generic property revealed most clearly by genetic studies. However, the existence of a set of variables complementary to gene products within the epigenetic system is evident from the perturbation studies that define classical embryology, and particularly the phenomenon of so-called phenocopies. Perturbation of development by mutation can be mimicked by physical or chemical stimuli applied to genetically normal embryos, diverting the epigenetic system into pathways other than the normal.[10,16,44] These stimuli can also produce phenotypes not yet observed in mutants, so the spectrum of possible epigenetic pathways is greater than that identified by genetic analysis, which is now reaching saturation with respect to segmentation phenotypes in *Drosophila*. The relevance of these observations to the study of generic properties of epigenetic mappings is the recognition that the object of the whole exercise is not to identify specifically how gene products exert their influence on development, but how *any* quantities acting as parameters are likely to enter into the determination of generic features of the mapping. The conclusions about gene products extends to any other variables that exert an influence on some aspect of epigenetic dynamics—e.g., change of membrane fluidity by organic solvents, change of ion flow densities by static magnetic fields, local changes of strain of the cytoskeleton by alteration of cell shape, etc. So it is necessary to keep this broader dynamic context in mind in any study of epigenetic mappings. This leads to the final lesson about development to be extracted from this analysis.

GLOBAL AND LOCAL ASPECTS OF EPIGENETIC MAPS

Developing organisms have both large-scale, spatially extended order and local expression of that order which becomes, at some stage in development, locally autonomous. This autonomy is most clearly observed by genetic analysis in the form of mutant cells or clones that express a phenotype distinct from that of their normal neighbors, but spatially coherent with it. The first geneticist to draw out the implications of such observations was Curt Stern,[37] from his studies of mutant gynanders of *Drosophila* in which clones of genetically male cells in a female fly express cuticle hair patterns distinct from, but inter-digiting and coherent with, those of their genetically female neighbors. Stern drew the conclusion that the same spatially extended, covert pre-pattern underlies both overt patterns, thus distinguishing between a global map and its local expression. Wolpert founded his theory of positional information and interpretation on this distinction. However, Stern described a considerably more sophisticated model of global and local order that came initially from Driesch (see Sander[36]): "development...(is)...a sequence of pre-patterns, each one being a realized pattern as compared to its predecessor, and a new pre-pattern as the basis of its successor. At any stage in the sequence, differential genic response to a pre-pattern will create differential pre-patterns for the next stage." Such a hierarchical process of pattern emergence in the developing embryo is what has been revealed in the hierarchy of segmentation genes, and it is reflected in the progressively shorter wavelengths of the four color wheels model. With a slight but significant change in Sander's paraphase of Driesch, we can now take the step that allows us to escape from the ghost of genetic programming that still hovers over descriptions of pattern formation. The importance of this lies not simply in resolving the sharp dualism between the epigenetic and the genetic, and so providing an integrated view of the phenomena now revealed in *Drosophila*. Of equal significance, it allows us to use the same conceptual structure to understand all developmental processes, whether in multicellular or in unicellular organisms.

The hierarchical nature of embryonic development, long established by classical embryological studies and now clearly supported by the genetic evidence in *Drosophila*, must be a dynamically robust process since it is one of the universal characteristics of epigenesis. There are obvious evolutionary advantages to this type of system, hierarchically organized in space and in time, since variations can occur in different levels of the hierarchy without disrupting the dynamic coherence of the whole (see, e.g., Maynard Smith[27] and Mittenthal[29] for clear descriptions of this principle). Robust dynamical systems are expected to have widespread occurrence in nature, so we might expect to find non-living processes that mimic the hierarchical order of epigenesis. An obvious place to look for such parallels is D'Arcy Thompson's[40] classic *On Growth and Form*, where we find examples in plenty. Perhaps the best known of those is the drop of oil falling through water, which undergoes a sequence of bifurcations to produce a form that is remarkably like that of a coelenterate medusa. Here is a process expressing characteristic properties of hydrodynamic fields and revealing the hierarchical spatial order that results from a fundamental dynamic.

It is now widely recognized that the generation of complexity in space and in time is a result of bifurcations and symmetry breakings, often in very simple dynamical systems. The march to deterministic temporal chaos is well described and understood in these terms[38] and an equivalent process occurs in spatially extended dynamical systems, such as fluids. The oil drop pattern illustrates this. Another example is the progression from smooth flow to turbulence through progressive complexity, involving a sequence of spatial periodicities of decreasing wavelength and increasing amplitude as the Reynolds number is increased.[2] This again reveals a qualitative similarity to the sequence of spatial patterns observed in the gene product distributions in *Drosophila*. So it is reasonable to propose that developing organisms express a general dynamic pattern, a march to complexity via a sequence of bifurcations that naturally results in a hierarchical emergence of spatial detail. This provides us with a principle of continuity from the non-living to the living realm, well illustrated by D'Arcy Thompson's work.

The progression from global to local order is then a generic property of this extended class of dynamical process. Within the context of epigenesis, global and local are always relative terms. As spatial order emerges progressively, gene action takes place over the wavelengths of the normal modes of the system, whatever may be their primary determinants. Gene products contribute to the formation, selection, and stabilization of the trajectories that gradually emerge as the nonlinearities of the system are expressed and basins of attraction become more clearly defined. The shorter wavelengths resulting from the bifurcations of the system partition earlier basins of attraction into a further sub-divided, quantized process with more attractors on a spatially finer grid. Within the context of the type of model we are considering, such a process is not correctly described as one in which pre-patterns are interpreted by genes. This introduces a false dualism of pre-pattern and interpretation which is foreign to a dynamic description in which there is a sequence of bifurcations in a spatially extended dynamic (a field) together with attractors.

Precisely because biological pattern formation proceeds hierarchically and takes time, unlike the printing of a die in which all the detail is generated simultaneously, it tends to be described as a process in which a pre-pattern gives rise to an overt pattern. However, the position we are adopting here is that this should not be understood to involve qualitatively different steps in the process, such as the setting up of a global spatial pattern and then the detailed local interpretation of the pattern. This would be like saying that there is something basically different between the hydrodynamic field of the oil droplet before it starts to bifurcate and after it initiates local droplet formation. Here it is clear that there is no dualism between global field properties and local expression after symmetry is broken and detailed local form is expressed. It is a continuous expression of a single field, though its boundaries become more complex as the form unfolds. Similarly in a developing organism there is no dualism of morphogenetic field and interpretation. The field is its expression, at all stages.

This type of description need not be restricted to multicellular organisms. In unicellulars such as the ciliate protozoa, there is also a hierarchy of spatial patterns from global positioning of the oral apparatus in a developing daughter cell to the local structure of kinetics.[6] These are influenced by different genes whose effects are exerted via their products in the cytoplasm. Here it is clear that the distinction between global and local is not dependent upon interpretation by genes of global morphogen prepatterns, since the genes are centrally located. Another example is the unicellular algae *Acerabularia*, which regenerates complex patterns in the total absence of a nucleus, though nuclear products are required in the cytoplasm. The hierarchical sequence of morphogenetic events resulting in regeneration of apical structures unfolds in invariant order whether the cell has a nucleus or not.[12,13]

Although the intrinsic dynamic order of morphogenesis is heavily overwritten by gene activity in *Drosophila*, we suggest that the process is essentially an expression of a bifurcating dynamic, resulting in a hierarchical unfolding of spatial order. Our nested, quantized color wheels describe the hierarchy of attractors and some of their generic properties. But it does not define the fundamental dynamic. Elsewhere[17] a suggestion is made about the origins of such a dynamic in terms of a nested set of fields of Turing type that spontaneously generate a hierarchy of spatial patterns of the kind observed in *Drosophila* via a sequence of bifurcations. A similar model is described by Lacalli.[21] These cannot be other than tentative guesses in the direction of a dynamic theory of epigenesis. However, comparative studies on spatial patterns of gene activities, made possible by an impressive array of probes that are beginning to show how conservative is the molecular substatum of epigenesis across a diversity of species, promise to reveal much more about the generic properties of developmental dynamics.

REFERENCES

1. Akam, M. "The Molecular Basis for Metameric Pattern in the *Drosophila* Embryo." *Development* **101** (1987): 1–22.
2. Batchelor, G. K. *An Introduction to Fluid Dynamics.* Cambridge: Cambridge University Press, 1967.
3. Carroll, S. B., and M. P. Scott. "Zygotically Active Genes that Affect the Spatial Expression of the *fushi tarazu* Segmentation Gene During Early *Drosophila* Embryogenesis." *Cell* **45** (1986): 113–126.
4. Carroll, S. B., and S. H. Vavra. "The Zygotic Control of *Drosophila* Pair-Rule Gene Expression II. Spatial Repression by Gap and Pair-Rule Gene Products." *Development* **107** (1989): 673–683.
5. Coulter, D. E. R., and E. Wieschaus. "Gene Activities and Segmental Patterning in *Drosophila*: Analysis of *Odd-Skipped* and *Pair-Rule* Double Mutants." *Genes and Dev.* **2** (1988): 1812–1823.
6. Frankel, J. *Pattern Formation.* Oxford University Press, 1989.
7. Gaul, U., and H. Jackle. "Analysis of Maternal Effects Combinations Elucidates Regulation and Function of the Overlap of *Hunchback* and *Krüppel* Gene Expression in the *Drosophila* Blastoderm Embryo." *Development* **107** (1989): 651–662.
8. Gergen, J. P., D. Coulter, and E. F. Wieschaus. "Segmental Pattern and Blastoderm Cell Identities." In *Gametogenesis and the Early Embryo*, 195–220. Alan R. Liss, 1986.
9. Gergen, J. P., and E. F. Wieschaus. "The Localized Requirements for a Gene Affecting Segmentation in *Drosophila*: Analysis of Larvae Mosaic for *Runt*." *Dev. Biol.* **109** (1985): 321–335.
10. Goldschmidt, R. B. "Additional Data on Phenocopies and Gene Action." *J. Exp. Zool.* **100** (1945): 193–201.
11. Goodwin, B. C. *Temporal Organization in Cells.* New York: Academic Press, 1963.
12. Goodwin, B. D. "Structuralism in Biology." *Sci. Progress, Oxford* **74** (1990): 227–244.
13. Goodwin, B. C., and S. A. Kauffman. "Spatial Harmonics and Pattern Specification in Early *Drosophila* Development, Part I. Bifurcation Sequences and Gene Expression." *J. Theor. Biol.* **144** (1990): 303–319.
14. Goodwin, B. C., and S. A. Kauffman. "Bifurcation, Harmonics and the Four Color Wheel Model of *Drosophila* Development." In *Cell-to-Cell Signalling: From Experiment to Theoretical Models*, edited by A. Goldbetter, 213–227. New York: Academic Press, 1989.
15. Goodwin, B. C., and L. E. H. Trainor. "Tip and Whorl Morphogenesis in Acetabularia by Calcium Regulated Strain Fields." *J. Theor. Biol.* **117** (1985): 79–106.

16. Ho, M. W., A. Matheson, P. T. Saunders, B. C. Goodwin, and A. Small-combe. "Ether Induced Segmentation Defects in *Drosophila melanogasler.*" *Roux's Arch. Dev. Biol.* **196** (1987): 511–521.

17. Hunding, A., S. A. Kauffman, and B. C. Goodwin. "*Drosophila* Segmentation: Supercomputer Simulation of Prepattern Hierarchy." *J. Theor. Biol.*, in press.

18. Ingham, P. W., S. M. Pinchin, K. R. Howard, and D. Ish-Horowicz. "Genetic Analysis of the *Hairy* Locus in *Drosophila melanogasler.*" *Genetics* **111** (1985): 463–486.

19. Jackle, H., D. Tautz, R. Schuh, E. Seifert, and R. Lehmann. "Cross-Regulatory Interaction Among The Gap Genes of *Drosophila.*" *Nature* **324** (1986): 668–670.

20. Kauffman, S. A., and B. C. Goodwin. "Spatial Harmonics and Pattern Specification in Early *Drosophila* Developments, Part II. The Four Color Wheel Model." *J. Theor. Biol.* **144** (1990): 321–345.

21. Lacalli, T. C. "Modelling The *Drosophila* Pair-Rule Pattern by Reaction-Diffusion: Gap Input and Pattern Control in a 4-Morphogen System." *J. Theor. Biol.* **144** (1990): 171–194.

22. Lacalli, T. C., D. A. Wilkinson, and L. G. Harrison. "Theoretical Aspects of Stripe Formation in Relation to *Drosophila* Segmentation." *Development* **103** (1988): 105–113.

23. Lohs-Schardin, M. "*Dicephalic*—A *Drosophila* Mutant Affecting Plarity in Follicle Organization and Embryonic Patterning." *Wilhelm Roux's Archives* **191** (1982): 28–36.

24. Macdonald, P. M., P. Ingham, and G. Struhl. "Isolation, Structure, and Expression of *Even-Skipped*: A Second Pair-Rule Gene of *Drosophila* Containing a Homeo Box." *Cell* **417** (1986): 721–734.

25. Matheson, A. "Teratology and The Regulation of Pattern in Drosophila Development." Thesis, Open University, 1991.

26. Matheson, A., and B. C. Goodwin. "Developmental Perturbations Resulting From Permeabilisation of *Drosophila* Embryos and Exposure to Calcium—Disturbing Agents." *Roux's Archives Developmental Biology*, 1991, submitted.

27. Maynard Smith, J. "Evolution and Development." In *Development and Evolution*, edited by B. C. Goodwin, N. Holden, and C. C. Wylie, 33–46. Cambridge: Cambridge University Press, 1983.

28. Meinhardt, H. *Models of Biological Pattern Formation.* London: Academic Press, 1982.

29. Mittenthal, J. E. "Physical Aspects of the Organization of Development." In *Lectures in The Sciences of Complexity*, edited by D. L. Stein, 225–274. Santa Fe Institute Studies in the Sciences of Complexity, Lect. Vol. I. Redwood City, CA: Addison-Wesley, 1988.

30. Murray, J. D. *Mathematical Biology.* Springer-Verlag, 1989.

31. Nagorcka, B. N. *J. Theor. Biol.* **132** (1988): 177–306.

32. Nüsslein-Volhard, C. "Genetic Analysis of Pattern Formation in the Embryo of *Drosophila melanogaster* Characterization of the Maternal Effect Mutant Bicaudal." *Wilhelm Roux's Arch.* **183** (1977): 244–268.

33. Nüsslein-Volhard, C. "Maternal Effect Mutations that Alter the Spatial Co-ordinates of the Embryo of *Drosophila melanogaster.*" In *Determinants of Spatial Organization*, 185–211. New York: Academic Press, 1979.

34. Nüsslein-Volhard, C., and E. Wieschaus. "Mutations Affecting Segment Members and Polarity in *Drosophila.*" *Nature* **287** (1986): 795–801.

35. Odell, G., G. F. Oster, B. Burnside, and P. Alberch. "The Mechanical Basis of Morphogenesis." *Devel. Biol.* **85** (1981): 446–462.

36. Sander, K. "The Role of The Genes in Ontogenesis: Evolving Concepts From 1883 to 1983 as Perceived by an Insect Embryologist." In *A History of Embryology*, edited by T. J. Horder, J. A, Witkowski, and C. C. C. Wylie, 363–395. Cambridge: Cambridge University Press, 1986.

37. Stern, C. "Genes and Developmental Patterns." *Caryologia* **6** (suppl.) (1954): 355–369

38. Stewart, I. *Does God Play Dice?* London: Penguin, 1989.

39. Tautz, D. Personal communication.

40. Thompson, D'Arcy. *On Growth and Form*, 2nd edition. Cambridge: Cambridge University Press, 1942.

41. Turing, A. M. "The Chemical Basis of Morphogenesis." *Phil Trans. R. Soc. London B* **237** (1952): 37–72.

42. Van der Meer, J. M. "Parameters Influencing Reversal of Segment Sequence in Posterior Egg Fragments of *Callosobuchas* (Coleoptera)." *Wilhem Roux's Arch.* **193** (1984): 339–356.

43. Vavra, S. H., and S. B. Carroll. "The Zygotic Control of *Drosophila* Rule-Pair Gene Expression I: A Search For New Pair-Rule Regulatory Loci." *Development* **107** (1989): 663–672.

44. Waddington, C. H. *The Strategy of the Genes.* London: Allen and Unwin, 1957.

45. Weir, M. P., and T. Kornberg. "Patterns of *Engrailed* and *fushi tarazu* Transcripts Reveal Novel Intermediate Stages in Segmentation." *Nature* **318** (1985): 433–445.

46. Wieschaus, E. F., C. Nusslein-Volhard, and H. Kluding. "*Krüppel*, A Gene Whose Activity is Required Early in the Zygotic Genome for Normal Embryonic Segmentation." *Dev. Biol.* **104** (1984): 172–186.

47. Winfree, A. T. "A Continuity Principle for Regulation." In *Pattern Formation*, edited by G. M. Malacinski and S. V. Bryant, 103–124. London: Macmillan, 1984.

48. Winfree, A. T. *The Geometry of Biological Time.* New York: Springer-Verlag, 1988.

Patterns of Constancy and Change: Commentary

Wake and Stork emphasize a principal theme of this book: To analyze the organization of organisms, one must consider a situation-dependent mix of factors—historical contingency, self-organization and selection, emergent features in hierarchies of processes, material processes and genetic regulatory pathways, and intrinsic and extrinsic processes. The following discussion looks at the five preceding chapters to see what contributions these factors can make to structure, within the full temporal spectrum of physiology, development, and evolution.

INTRINSIC PROCESSES

A major problem is to understand the macroscopic structures and transitions of structure to be expected in organisms, given their microscopic organization. Mandell and Selz suggest that phase transitions in allosteric enzyme kinetics may underlie analogous transitions at higher levels of organization. They note that the low-dimensional dynamics used to model molecular transitions may be useful at higher levels, to model the dynamics of order parameters. In a similar vein Goodwin and Kauffman point out, as did DeGuzman and Kelso, that a macroscopic dynamical description can often be made without explicit reference to the underlying microscopic processes. Goodwin and Kauffman investigated the perturbation of a system that normally forms spatially periodic patterns, using methods with which

Principles of Organization in Organisms,
SFI Studies in the Sciences of Complexity, Proc. Vol. XIII,
Eds. J. Mittenthal & A. Baskin, Addison-Wesley, 1992   **297**

Winfree[5] analyzed the perturbation of a temporally periodic system. Their work shows that a macroscopic description not only can ignore the microscopic dynamics of gene regulatory networks, but also can apply in both spatial and temporal domains.

Efforts to characterize a hierarchy of processes often carry, at least implicitly, the assumption of unidirectional determination—that micro-level processes determine macro-level features. Wake argues that macro-micro relations are bidirectional; levels of organization both higher and lower than the level of a feature may influence its development.

As Wake points out, there are many ways to stretch a salamander: A longer, slimmer vertebral column may evolve through an increase in the number of vertebrae or through elongation of the primitive set of vertebrae. Thus structure at a micro-level can sometimes be adjusted in diverse ways to meet performance criteria at a higher level. The existence of alternative microscopic ways to make a macroscopic transition of structure increases the probability of that transition.

Stork's chapter shows explicitly how the kind of changes discussed by Wake occur: Evolution proceeds, in part, through changes in the rules for physiology and development. Changes in the developmental rules for generating the connectivity of a neural network can adjust the speed and flexibility of its evolution: A network with better performance evolves faster if the network can be modified through learning before reproduction occurs, rather than being modified only through selection of variant networks.

THE INTERFACES OF INTRINSIC PROCESSES WITH HISTORY AND ENVIRONMENT DURING EVOLUTION

The chapters point out that the process of evolution has changed through history, because history is cumulative. Early in evolution, generic material and network processes may have operated relatively independently within organisms. However, their interactions with each other and with extrinsic processes in a composite dynamical system have increasingly made them more interdependent. In turn, this interdependence has increasingly constrained the structures available for future exploration.

Observation and conjecture suggest several patterns of increasing constraint. Newman suggests that during evolution material processes have become increasingly canalized through genetic regulation. Similarly, networks of processes retain the imprint of past selection, as Stork's chapter shows. As Riedl[4] pointed out, this imprint can be locked into ontogeny: Later-developing features that depend on a given feature limit its evolutionary variability, and reduce the possibility of its elimination even though the feature itself is irrelevant.

Processes in the environment of an organism, either in its physical environment or its interactions with other organisms, limit the structures that persist through evolution. When many alternative body plans appear at one taxonomic level, as in the Burgess shale, extinctions through chance and selection winnow out all but a few, which then diversify.[2] An elimination of alternatives also occurs when a higher level of organization evolves, as Buss[1] pointed out: Conflict is likely to occur between criteria of selection at the higher level and at lower levels. This conflict must be resolved during subsequent evolution, in ways that constrain processes at all levels. Clarifying and classifying these patterns of constraint is an important task for the future.

Newman suggests that the interface between self-organization and selection does not occur only at the interface between an organism and its environment. Early in evolution generic material processes provided an environment internal to the organism, within which particular genetic networks were preferentially selected. More generally, one should bear in mind that the environment of any part of an organism is the rest of the organism, as well as its external environment; the parts of an organism have coevolved with each other and with the environment.

APPROACHES THROUGH GENERATION AND DESIGN

The chapters suggest that the criteria of performance for organisms have changed through evolution. As an organism proceeds from development to reproduction, it must not only cope with its environment but with its history, which is written into its molecular organization. Early in evolution organisms had little history (geologically speaking), and so may have met different performance criteria than more recent organisms. For early organisms reliability may have been less important than flexibility: In early organisms, presumably, there was little regulation to enhance the faithful preservation of traits between generations. An adequate organization probably would work in diverse environments, despite variations in its details. As Newman points out, as evolution proceeded genetic regulation probably increased the reliability with which material processes generate specific outcomes. Such regulation would stabilize an organism's performance in diverse environments. Selection could sharpen the division of labor within organisms, by favoring a well-defined hierarchy of processes with elaborate regulation. Reliable preservation of these regulatory processes and regulated structures would be favored, while flexibility could be achieved by modifying low-level processes and eliciting new clusters of them.

Evidently the performance criteria for intrinsic processes depend both on the environment of an organism and on its inherited organization. The complexity of organism-environment composite systems makes it unlikely that the optimization of an ultimate fitness function can explain the structure of organisms, although optimality arguments can help to explain specialized aspects.[3] Stork's chapter emphasizes the elusiveness of optimality arguments. The organization of a neural network that evolved through various selection regimes may be suboptimal from the viewpoint of its performance in behavior. However, this organization may provide evolutionary flexibility (evolvability). These conflicts between performance criteria on different time scales may be related to the conflicts between criteria for different levels of organization discussed by Buss,[1] since higher-level processes operate at longer time scales.

REFERENCES

1. Buss, L. *The Evolution of Individuality*. Princeton: Princeton University Press, 1987.
2. Gould, S. J. *Wonderful Life*. New York: W. W. Norton, 1989.
3. Parker, G. A., and J. Maynard Smith. "Optimality Theory in Evolutionary Biology." *Nature* **348** (1990): 27–33.
4. Riedl, R. *Order in Living Organisms*, translated by R. P. S. Jefferies. New York: John Wiley & Sons, 1978.
5. Winfree A. *The Geometry of Biological Time*. New York: Springer-Verlag, 1980.

General Principles of Organization: Introduction

The following four chapters offer arguments from generation and from design for the existence of general principles of organization. The chapters argue that the same principles apply at all levels of organization and at all time scales.

In his chapter (and more directly in Kauffman, 1991), Kauffman surveys the history and present state of efforts to understand the structures of complex adaptive systems. He reiterates the conclusion that arguments based on both self-organization and selection are needed to understand these structures. He discusses the characteristic structures likely to be generated during the evolution of networks of Boolean switches, which he takes as models for networks of genes. With suitable switching rules, a minimal perturbation—changing the state of one gene at one time step—elicits a small cascade of changes in the states of other genes.

Mittenthal et al. call such a cascade—a cluster of processes elicited by a small input—a dynamic module. They suggest that dynamic modules perform many tasks in organisms, and that there is a hierarchy of modules, in which low-level modules are used in various combinations to perform higher-level tasks. They further propose that the structure of an organism mirrors the world in which it acts, in the sense that the pattern of coupling among its intrinsic processes matches the pattern of correlations among the tasks it performs. With a formal model Clarke and Mittenthal support this idea, showing that a hierarchy of unreliable modules can perform a hierarchy of tasks with relatively high reliability.

Principles of Organization in Organisms,
SFI Studies in the Sciences of Complexity, Proc. Vol. XIII,
Eds. J. Mittenthal & A. Baskin, Addison-Wesley, 1992   **301**

Baskin et al. argue that structure can arise from the finiteness of resources. Higher levels of organization are likely to evolve because they can reduce the resources that organisms use to perform tasks, allowing an organism to meet a broader range of constraints with given resources.

Stuart A. Kauffman
Santa Fe Institute, 1660 Old Pecos Trail, Suite A, Santa Fe, NM 87501

The Sciences of Complexity and "Origins of Order"

This article discusses my book, *Origins of Order: Self Organization and Selection in Evolution,* in the context of the emerging sciences of complexity. *Origins,* due out of Oxford University Press in early 1992, attempts to lay out a broadened theory of evolution based on the marriage of unexpected and powerful properties of self-organization which arises in complex systems, properties which may underlie the origin of life itself and the emergence of order in ontogeny, *and* the continuing action of natural selection. The three major themes are: (1) that such self-organized properties lie to hand for selection's further molding; (2) hence, that the order we see is not due to selection alone, but in part reflects the order selection has always acted upon; and finally, (3) that the marriage of natural order and natural selection may inevitably lead living entities to a novel organized state, lying on the edge between order and chaos, as the inevitable evolutionary attractor of selection for the capacity to adapt.

Principles of Organization in Organisms,
SFI Studies in the Sciences of Complexity, Proc. Vol. XIII,
Eds. J. Mittenthal & A. Baskin, Addison-Wesley, 1992   **303**

INTRODUCTION

A new science, the science of complexity, is birthing. This science boldly promises to transform the biological and social sciences in the forthcoming century. My own book, *Origins of Order: Self Organization and Selection in Evolution*,[14] is at most one strand in this transformation. I feel deeply honored that Marjorie Grene undertook organizing a session at the Philosophy of Science meeting discussing *Origins*, and equally glad that Dick Burian, Bob Richardson, and Rob Page have undertaken their reading of the manuscript and careful thoughts. In this article I shall characterize the book, but more importantly, set it in the broader context of the emerging sciences of complexity. Although the book is not yet out of Oxford press's quiet womb, my own thinking has moved beyond that which I had formulated even a half year ago. Meanwhile, in the broader scientific community, the interest in "complexity" is exploding.

A summary of my own evolving hunch is this: In a deep sense, *E. coli* and IBM know their respective worlds in the same way. Indeed, *E. coli* and IBM have each participated in the coevolution of entities which interact with and know one another. The laws which govern the emergence of knower and known, which govern the boundedly rational, optimally complex biological and social actors which have co-formed, lie at the core of the science of complexity. This new body of thought implies that the poised coherence, precarious, subject to avalanches of change, of our biological and social world is inevitable. Such systems, poised on the edge of chaos, are the natural tailsman of adaptive order.

The history of this emerging paradigm conveniently begins with the "cybernetic" revolution in molecular biology wrought by the stunning discoveries[8,9] in 1961 and 1963, by later Nobelists Francois Jacob and Jacques Monod, that genes in the humble bacterium, *E. coli*, literally turn one another on and off. This discovery laid the foundation for the still-sought solution of the problem of cellular differentiation in embryology. The embryo begins as a fertilized egg, the single-cell zygote. Over the course of embryonic development in a human, this cell divides about 50 times, yielding the thousand-trillion cells which form the newborn. The central mystery of developmental biology is that these trillions of cells become radically *different* from one another, some forming blood cells, others liver cells, still others nerve, gut, or gonadal cells. Previous work had shown that all the cells of a human body contain the same genetic instructions. How, then, could cells possibly differ so radically?

Jacob and Monod's discovery hinted the answer. If genes can turn one another on and off, then cell types differ because different genes are expressed in each cell type. Red blood cells have hemoglobin, immune cells synthesize antibody molecules, and so forth. Each cell might be thought of as a kind of cybernetic system with complex genetic-molecular circuits orchestrating the activities of some 100,000 or more genes and their products. Different cell types then, in some real sense, *calculate* how they should behave.

THE EDGE OF CHAOS

My own role in the birth of the sciences of complexity begins in the same years, when I asked an unusual, perhaps near unthinkable question. Can the vast, magnificent order seen in development conceivably arise as a spontaneous *self-organized* property of complex genetic systems? Why "unthinkable?" It is, after all, not the answers which scientists uncover, but the strange magic lying behind the questions they pose to their world, knower and known, which is the true impulse driving profound conceptual transformation. Answers will be found, contrived, wrested, once the question is divined. Why "unthinkable?" Since Darwin, we have viewed organisms, in Jacob's phrase, as *bricolage*, tinkered-together contraptions. Evolution, says Monod, is "chance caught on the wing." Lovely dicta, these, capturing the core of the Darwinian world view in which organisms are perfected by natural selection acting on random variations. The tinkerer is an opportunist, its natural artifacts are *ad hoc* accumulations of this and that, molecular Rube Goldbergs satisfying some spectrum of design constraints.

In the world view of bricolage, selection is the sole, or if not sole, the preeminent source of order. Further, if organisms are *ad hoc* solutions to design problems, there can be no deep theory of order in biology, only the careful dissection of the ultimately accidental machine and its ultimately accidental evolutionary history.

The genomic system linking the activity of thousands of genes stands at the summit of four billion years of an evolutionary process in which the specific genes, their regulatory intertwining and the molecular logic have all stumbled forward by random mutation and natural selection. Must selection have struggled against vast odds to create order? Or did that order lie to hand for selection's further molding? If the latter, then what a reordering of our view of life is mandated!

Order, in fact, lies to hand. Our intuitions, in fact, have been wrong. We may have to revise our view of life. Complex molecular regulatory networks inherently behave in two broad regimes separated by a third-phase transition regime: The two broad regimes are *chaotic* and *ordered*. The phase transition zone between these two comprises a narrow third *complex* regime poised on the boundary of chaos.[3,6,10,12,13,15] Twenty-five years after the initial discovery of these regimes, a summary statement is that the genetic systems controlling ontogeny in mouse, man, bracken, fern, fly, bird,—all—appear to lie in the ordered regime near the edge of chaos. Four-billion years of evolution in the capacity to adapt offers a putative answer: Complex adaptive systems may achieve, in a law-like way, the edge of chaos.

Tracing the history of this discovery, the discovery that extremely complex systems can exhibit "order for free," that our intuitions have been deeply wrong, begins with the intuition that even randomly "wired" molecular regulatory "circuits" with random "logic" would exhibit orderly behavior if each gene or molecular variable were controlled by only a few others. The intuition appears correct. Idealizing a gene as "on" or "off," it was possible by computer simulations to show that large systems with thousands of idealized genes behaved in orderly ways if each gene is directly controlled by only two other genes. Such systems spontaneously lie in the ordered regime. Networks with many inputs per gene lie in the chaotic regime. Real genomic systems have few molecular inputs per gene, reflecting the *specificity* of molecular binding, and use a *biased* class of logical rules, reflecting molecular simplicity, to control the on/off behavior of those genes. Constraint to the vast ensemble of possible genomic systems characterized by these two "local constraints" also inevitably yields genomic systems in the ordered regime. The perplexing, enigmatic, magical order of ontogeny may largely reflect large-scale consequences of polymer chemistry.

Order for free. But more: The spontaneously ordered features of such systems actually parallel a host of ordered features seen in the ontogeny of mouse, man, bracken, fern, fly, and bird. A "cell type" becomes a stable recurrent pattern of gene expression, an "attractor" in the jargon of mathematics, where an attractor, like a whirlpool, is a region in the state space of all the possible patterns of gene activities to which the system flows and remains. In the spontaneously ordered regime, such cell-type attractors are inherently small, stable, and few, implying that the cell types of an organism traverse their recurrent patterns of gene expression in hours not eons, that homeostasis, Claude Bernard's conceptual child, lies inevitably available for selection to mold, and, remarkably, that it should be possible to *predict* the number of cell types, each a whirlpool attractor in the genomic repertoire, in an organism. Bacteria harbor one to two cell types, yeast three, ferns and bracken some dozen, and man about two-hundred and fifty. Thus, as the number of genes, called genomic complexity, increases, the number of cell types increases. Plotting cell types against genomic complexity, one finds that the number of cell types increases as a rough square-root function of the number of genes. And, in parallel, the number of whirlpool attractors in model genomic systems in the ordered regime also increase as a square-root function of the number of genes. Man, with about 100,000 genes should have 370 cell types, but has close to 250. A simple alternative theory would predict billions of cell types.

Bacteria, yeast, ferns, and man, members of different phyla, have no common ancestor for the past 600 million years or more. Has selection struggled for 600 million years to achieve a square-root relation between genomic complexity and number of cell types? Or is this expression of "order for free" so deeply bound into the roots of biological organization that selection cannot *avoid this order?* But if the latter, then *selection is not the sole source of order in biology.* Then Darwinism must be extended to embrace self-organization *and* selection.

The pattern of questions posed here is novel in biology since Darwin. In the Neo-Darwinian world view, where organisms are *ad hoc* solutions to design problems, the answers lie in the specific details wrought by ceaseless selection. In contrast, the explanatory approach offered by the new analysis rests on examining the statistically typical, or generic, properties of an entire class, or "ensemble" of systems all sharing known local features of genomic systems. If the typical, generic features of ensemble members corresponds to that seen in organisms, then the explanation of those features emphatically *does not* rest in the details. It rests in the general laws governing the typical features of the ensemble as a whole. Thus an "ensemble" theory is a new kind of statistical mechanics. It predicts that the typical properties of members of the ensemble will be found in organisms. If true, it bodes a physics of biology.

Not only a physics of biology, but beyond; such a new statistical mechanics demands a new pattern of thinking with respect to biological and perhaps even cultural evolution: Self-organization, yes, aplenty. But selection, or its analogues such as profitability, is always acting. We have no theory in physics, chemistry, biology, or beyond which marries self-organization and selection. The marriage consecrates a new view of life.

But two other failures of Darwin, genius that he was, must strike us. How do organisms, or other complex entities, manage to adapt and learn? That is, what are the conditions of "evolvability"? Second, how do complex systems coordinate behavior?

Consider "evolvability" first. Darwin supposed that organisms evolve by the *successive accumulation of useful random variations.* Try it with a standard computer program. Mutate the code, scramble the order of instructions, and try to "evolve" a program calculating some complex function. If you do not chuckle, you should. Computer programs of the familiar type are not readily "evolvable." Indeed, the more compact the code, the more lacking in redundancy, the more sensitive it is to each minor variation. Optimally condensed codes are, perversely, minimally evolvable. Yet the genome is a kind of molecular computer, and clearly has succeeded in evolving. But this implies something very deep: Selection must *achieve* the kinds of systems which are *able* to adapt. That capacity is not Godgiven; it is a success.

If the capacity to evolve must itself evolve, then the new sciences of complexity seeking the laws governing complex adapting systems must discover the laws governing the emergence and character of systems which can themselves adapt by accumulation of successive useful variations.

But systems poised in the ordered regime near its boundary seem likely to be precisely those which can, in fact, evolve by successive minor variations. The behavior of systems in the chaotic regime are so drastically altered by any minor variation in structure or logic that they cannot easily accumulate useful variations. Conversely, systems deep in the ordered regime are changed so slightly by minor variations that they adapt too slowly to an environment which may sometimes alter catastrophically. Those on the boundary between order and chaos have behaviors which are typically only modified slightly by most mutations, but modified more drastically by other mutations. In a changing world, such systems can adapt gradually in the typical circumstance, but mount massive changes when needed. The evolution of the capacity to adapt might well be expected to achieve poised systems.

How can complex systems coordinate behavior? On several grounds it now appears that complex adaptive entities poised at the edge of chaos may be able to coordinate the most complex behavior.[12,13,14,15] Deep in the chaotic regime, alteration in the activity of any element in the system unleashes an avalanche of changes, or "damage," which propagates throughout most of the system.[17] Such spreading damage is equivalent to the "butterfly effect" or sensitivity to initial conditions typical of chaotic systems. The butterfly in Rio changes the weather in Chicago. Cross currents of such avalanches unleashed from different elements means that behavior is not controllable. Conversely, deep in the ordered regime, alteration at one point in the system only alters the behavior of a few neighboring elements. Signals cannot propagate widely throughout the system. Thus, control of complex behavior cannot be achieved. Just at the boundary between order and chaos, the most complex behavior can be achieved. For example, at this phase transition, a characteristic *power law* distribution of avalanches of damage arises, with many small avalanches and few large avalanches,[17] implying that avalanches on all size scales arise. Sites within a system can therefore communicate with nearby sites often, and distant sites rarely. But order is retained, chaos and the butterfly remain quiescent, because the system remains slightly within the ordered regime.

Finally, computer simulations suggest that natural selection or its analogues actually *can achieve* the edge of chaos. This third regime, poised between the broad ordered regime and the vast chaotic regime, is razorblade thin in the space of systems. Absent other forces, randomly assembled systems will lie in the ordered or chaotic regimes. But let such systems play any of a variety of games with one another, winning and losing as each system carries out some behavior with respect to the others, and let the structure and logic of each system evolve by mutation and selection, and the systems do actually converge towards the edge of chaos! No minor point this, if general: Evolution itself may often bring complex systems, when they must adapt to the actions of others, to an internal structure and logic poised between order and chaos.[13]

We are led to a bold hypothesis: Complex adaptive systems achieve the edge of chaos.

The story of the "edge of chaos" is stronger, the implications more surprising. Organisms, economic entities, and nations, do not evolve; they *coevolve*. Almost miraculously, co-evolving systems, too, can mutually achieve the poised edge of chaos. The sticky tongue of the frog alters the fitness of the fly, and deforms its fitness landscapes—that is, what changes in what phenotypic directions improve its chance of survival. But technological evolution is also coevolution. The automobile replaced the horse. With the automobile came paved roads, gas stations, hence a petroleum industry and war in the Gulf, traffic lights, traffic courts, and motels. With the horse went stables, the smithy, and the pony express. New goods and services alter the economic landscape. Coevolution is a story of coupled deforming "fitness landscapes." The outcome depends jointly on how much my landscape is deformed when you make an adaptive move, and how rapidly I can respond by changing "phenotype."

Are there laws governing coevolution? And how might they relate to the edge of chaos? In startling ways. Coevolution, due to a selective "metadynamics" tuning the structure of fitness landscapes and couplings between them, may typically reach the edge of chaos.[14] *E. coli* and IBM not only "play" games with the other entities with which they coevolve. Each also participates in the very *definition or form* of the game. It is we who create the world we mutually inhabit and in which we struggle to survive. In models where players can "tune" the mutual game even as they play, or coevolve, according to the game existing at any period, the entire system moves to the edge of chaos. This surprising result, if general, is of considerable importance. A simple view of it is the following: Entities control a kind of "membrane" or boundary separating inside from outside. In a kind of surface-to-volume way, if the surface of each system is small compared to its volume, it is rather insensitive to alterations in the behaviors of other entities. That is, adaptive moves by other partners do not drastically deform one partner's fitness landscape. Conversely, the ruggedness of the adaptive landscape of each player as it changes its "genotype" depends upon how dramatically its behavior deforms as its genotype alters. In turn this depends upon whether the adapting system is itself in the ordered, chaotic, or boundary regime. If in the ordered, the system itself adapts on a smooth landscape. In the chaotic regime the system adapts on a very rugged landscape. In the boundary regime the system adapts on a landscape of intermediate ruggedness, smooth in some directions of "genotype" change, rugged in other directions. Thus, both the ruggedness of one's own fitness landscape and how badly that landscape is deformed by moves of one's coevolving partners are *themselves possible objects of a selective "metadynamics."* Under this selective metadynamics, tuning landscape structure and susceptibility, model coevolving systems which interact with one another actually reach the edge of chaos. Here, under most circumstances, most entities optimize fitness, or payoff, by remaining the same. Most of the ecosystem is frozen into a percolating Nash equilibrium, while coevolutionary changes propagate in local unfrozen islands within the ecosystem. More generally, alterations in circumstances send avalanches of changed optimal strategies propagating through the co-evolving system. At the edge of chaos, the size distributions of those avalanches approach a power law, with many small avalanches and few large ones. During such co-evolutionary avalanches, affected players would be expected to fall transiently to low fitness, and hence might go extinct. Remarkably, this size distribution comes close to fitting the size distribution of extinction events in the record. At a minimum, a distribution of avalanche sizes from a common-size small cause tells us that small and large extinction events may reflect endogenous features of coevolving systems more than the size of the meteor which struck.

The implications are mini-Gaia. As if by an invisible hand, coevolving complex entities may mutually "tune" the games they play and thereby attain the poised boundary between order and chaos. Here, mean sustained payoff, or fitness, or profit, is optimized. But here avalanches of change on all length scales can propagate through the poised system. Neither Sisyphus, forever pushing the punishing load, nor fixed unchanging and frozen, instead *E. coli* and its neighbors, IBM and its neighbors, even nation states in their collective dance of power, may attain a precarious poised complex adaptive state. The coevolution of complex adaptive entities itself appears lawful.

If one had to formulate, still poorly articulated, the general law of adaptation in complex systems, it might be this: Life adapts to the edge of chaos.

THE ORIGIN OF LIFE AND ITS PROGENY

This story, the story of the boundary between order an chaos achieved by complex coevolving systems, is but half the emerging tale. A second voice tells of the origin of life itself, a story both testable and, I hope, true.

Life is held to be a miracle, God's breath on the still world, yet cannot be. Too much the miracle, then we were not here. There must be a viewpoint, a place to stand, from which the emergence of life is explicable, not as a rare untoward happening, but as expected, perhaps inevitable. In the common view, life originated as a self-reproducing polymer such as RNA, whose self-complementary structure, since Watson and Crick remarked with uncertain modesty, suggests its mode of reproduction, has loomed the obvious candidate to all but the stubborn.[4] Yet stubbornly resistant to test, to birthing *in vitro* is this supposed simplest molecule of life. No worker has yet succeeded in getting one single-stranded RNA to line up the complementary free nucleotides, link them together to form the second strand, melt them apart, then repeat the cycle. The closest approach shows that a polyC polyG strand, richer in C than G, can in fact line up its complementary strand. Malevolently, the newly formed template is richer in G than C, and fails, utterly, to act as a facile template on its own. Alas.

Workers attached to the logic of molecular complementarity are now focusing effort on polymers other than RNA, polymers plausibly formed in the prebiotic environment, which might dance the still sought dance. Others, properly entranced with the fact that RNA can act as an enzyme, called a ribozyme, cleaving and ligating RNA sequences apart and together, seek a ribozyme which can glide along a second RNA, serving as a template that has lined up its nucleotide complements, and zipper them together. Such an ribozyme would be a *ribozyme polymerase*, able to copy any RNA molecule, including itself. Beautiful indeed. And perhaps such a molecule occurred at curtain-rise or early in the first act. But consider this: a free living organism, even the simplest bacterium, links the synthesis and degradation of some thousands of molecules in the complex molecular traffic of metabolism to the reproduction of the cell itself. Were one to begin with the RNA urbeast, a *nude gene*, how might it evolve? How might it gather about itself the clothing of metabolism?

There is an alternative approach which states that life arises as a nearly inevitable phase transition in complex chemical systems. Life, this hypothesis asserts, formed by the emergence of a collectively autocatalytic system of polymers and simple chemical species.[1,5,10,13]

Picture, strangely, ten-thousand buttons scattered on the floor. Begin to connect these at random with red threads. Every now and then, hoist a button and count how many buttons you can lift with it off the floor. Such a connected collection is call a "component" in a "random graph." A random graph is just a set of buttons connected at random by a set of threads. More formally, it is a set of N nodes connected at random by E edges. Random graphs undergo surprising phase transitions. Consider the ratio of E/N, or threads divided by buttons. When E/N

is small, say .1, any button is connected directly or indirectly to only a few other buttons. But when E/N passes about 0.5, so there are half as many threads as buttons, a phase transition has occurred. If a button is picked up, very many other buttons are picked up with it. In short, a "giant component" has formed in the random graph in which most buttons are directly or indirectly connected with one another. In short, connect enough nodes and a connected web "crystallizes."

Now life according to the new hypothesis. Proteins and RNA molecules are linear polymers built by assembling a subset of monomers, twenty types in proteins, four in RNA. Consider the set of polymers up to some length, M, say 10. As M increases, the number of types of polymers increases exponentially; for example, there are 20^M proteins of length M. This is a familiar thought. The rest of the ideas are less familiar. The simplest reaction among two polymers consists in gluing them together. Such reactions are reversible, so the converse reaction is simply cleaving a polymer into two shorter polymers. Now count the number of such reactions among the many polymers up to length M. A simple consequence of the combinatorial character of polymers is that there are *many more reactions* linking the polymers than there are polymers. For example, a polymer length M can be formed in $M - 1$ ways by gluing shorter fragments comprising that polymer. Indeed, as M increases, the ratio of reactions among the polymers to polymers is about M; hence it increases as M increases. Picture such reactions as black, not red, threads running from the two smaller fragments to a small square box, then to the larger polymer made of them. Any such triad of black threads denotes a possible reaction among the polymers; the box, assigned a unique number, labels the reaction itself. The collection of all such triads is the chemical reaction graph among them. As the length of the longest polymer under consideration, M, increases, the web of black triads among these grows richer and richer. The system is rich with cross-linked reactions.

Life is an *autocatalytic process* where the system synthesizes itself from simple building blocks. Thus, in order to investigate the conditions under which such a autocatalytic system might spontaneously form, assume that no reaction actually occurs unless that reaction is catalyzed by some molecule. The next step notes that protein and RNA polymers can in fact catalyze reactions cleaving and ligating proteins and RNA polymers: trypsin in your gut after dinner digesting steak, or ribozyme-ligating RNA sequences are examples. Build a theory showing the probability that any given polymer catalyzes any given reaction. A simple hypothesis is that each polymer has a fixed chance, say one in a billion, to catalyze each reaction. No such theory can now be accurate, but this hardly matters: The conclusion is at hand, and insensitive to the details. Ask each polymer in the system, according to your theory, whether it catalyzes each possible reaction. If "yes," color the corresponding reaction triad "red," and note down which polymer catalyzed that reaction. Ask this question of all polymers for each reaction. Then some fraction of the black triads have become red. The red triads are the catalyzed reactions in the chemical reaction graph. But such a catalyzed reaction graph undergoes the button-thread phase transition. When enough reactions are catalyzed, a vast web of polymers are linked by catalyzed reactions. Since the ratio of reactions to polymers increases with M, at some point as M increases, at least one reaction per polymer is catalyzed by some polymer. The giant component crystallizes. An autocatalytic set which collectively catalyzes its own formation lies hovering in the now pregnant chemical soup. A self-reproducing chemical system, daughter of chance and number, swarms into existence, a connected collectively autocatalytic metabolism. No nude gene, life emerged whole at the outset.

This theory is now at least well substantiated as a theory. Detailed analytic results and computer simulations[1,5,11] demonstrate that such autocatalytic systems, in principle, can sustain themselves as well as *evolve* in the open space of polymers.

If this new view of the crystallization of life as a phase transition is correct, then it should soon be possible to create actual self-reproducing polymer systems, presumably of RNA or proteins, in the laboratory. Experiments, even now, utilizing very complex libraries of RNA molecules to search for autocatalytic sets are underway in a few laboratories.

If not since Darwin, then since Weisman's doctrine of the germ plasm was reduced to molecular detail by discovery of the genetic role of chromosomes, biologists have believed that evolution via mutation and selection virtually requires a stable genetic material as the store of heritable information. But mathematical analysis of autocatalytic polymer systems suggests this conviction is wrong. As noted, such systems can evolve to form new systems. Thus, contrary to Richard Dawkin's thesis[2] in "The Selfish Gene," biological evolution does not, in principle, demand self-replicating genes at the base. Life can emerge and evolve without a genome. Heresy, perhaps? Perhaps.

Many and unexpected are the children of invention. Autocatalytic polymer sets have begotten an entire new approach to complexity. The starting point is obvious. An autocatalytic polymer set is a *functional integrated whole*. Given such a set, it is clear that one can naturally define the function of any given polymer in the set with respect to the capacity of the set to reproduce itself. Lethal mutants exist, for if a given polymer is removed, or a given foodstuff deleted, the set may fail to reproduce itself. Ecological interactions among coevolving autocatalytic sets lie to hand. A polymer from one such set injected into a second such set may block a specific reaction step and "kill" the second autocatalytic set. Coevolution of such sets, perhaps bounded by membranes, must inevitably reveal how such systems "know" one another, build internal models of one another, and cope with one another. Models of the evolution of knower and known lie over the conceptual horizon.

Walter Fontana came to the Santa Fe Institute and Los Alamos. Fontana had worked with John McCaskill, himself an able young physicist collaborating with Eigen at the Max Planck Institute in Gottingen. McCaskill dreamt of polymers, not as chemicals, but as Turing machine computer programs and tapes. One polymer, the computer, would act on another polymer, the tape, and "compute" the result, yielding a new polymer. Fontana was entranced. But he also found the autocatalytic story appealing. Necessarily, he invented "Algorithmic Chemistry." Necessarily, he named his creation "Alchemy"[7]

Alchemy is based on a language for universal computation called the lambda calculus. Here almost any binary symbol string is a legitimate "program" which can act on almost any binary symbol string as an input to compute an output binary symbol string. Fontana created a "Turing gas" in which an initial stock of symbol strings randomly encounter one another in a "chemostat" and may or may not interact to yield symbol strings. To maintain the analogue of selection, Fontana requires that a fixed total number of symbol string polymers be maintained in the chemostat. At each moment, if the number of symbol strings grows above the maximum allowed, some randomly chosen strings are lost from the system.

Autocatalytic sets emerge again! Fontana finds two types. In one, a symbol string which copies itself emerges. This "polymerase" takes over the whole system. In the second, collectively autocatalytic sets emerge in which each symbol string is made by some other string or strings, but none copies itself. Such systems can then evolve in symbol string space—evolution without a genome.

More, one can study the stability of an autocatalytic set: Inject a few new symbol strings into such a system and it may ignore the strings, which dilute out and disappear, or the input strings can send the set gyrating through string space, perhaps to a new autocatalytic set. Thus, these novel objects are stable or unstable to perturbations of their symbol string contents.

Fontana broke the bottleneck. Another formulation of much the same ideas, which I am now using, sees interactions among symbol strings creating symbol strings as carrying out a "grammar." Work on disordered networks, work which exhibited the three broad phases, ordered, chaotic and complex, drove forward based on the intuition that order and comprehensibility would emerge by finding the generic behavior in broad regions of the space of possible systems. The current hope is that analysis of broad reaches of grammar space, by sampling "random grammars," will yield insight into this astonishingly rich class of systems.

The promise of these random grammar systems may extend from analysis of evolving proto-living systems, to characterizing mental processes such as multiple personalities, the study of technological coevolution, bounded rationality and non-equilibrium market behavior at the foundations of economic theory, to cultural evolution.

Strings of symbols which act upon one other to generate strings of symbols can, in general, be computationally universal. That is, such systems can carry out any specified algorithmic computation. The immense powers and yet surprising limits, known since Gödel, Turing, Church, Kleene, lie before us, but in a new and suggestive form. Strings acting on strings to generate strings create an utterly novel conceptual framework in which to cast the world. The puzzle of mathematics, of course, is that it should so often be so outrageously useful in categorizing the world. New conceptual schemes allow starkly new questions to be posed.

A grammar model is simply specified. It suffices to consider a set of M pairs of symbol strings, each about N symbols in length. The meaning of the grammar, a catch-as-catch-can set of "laws of chemistry," is this: Wherever the left member of such a pair is found in some symbol string in a "soup" of strings, substitute the right member of the pair. Thus, given an initial soup of strings, one application of the grammar might be carried out by us, acting Godlike. We regard each string in the soup in turn, try all grammar rules in some precedence order, and carry out the transformations mandated by the grammar. Strings become strings become strings. But we can let the *strings themselves* act on one another. Conceive of a string as an "enzyme" which acts on a second string as a "substrate" to produce a "product." A simple specification shows the idea. If a symbol sequence within a string in the soup, say 111, is identical to a symbol sequence on the "input" side of one grammar pair, then that 111 site in the string in the soup can act as an enzymatic site. If the enzymatic site finds a substrate string bearing the same site, 111, then the enzyme acts on the substrate and transforms its 111 to the symbol sequence mandated by the grammar, say 0101. Here, which symbol string in the soup acts as enzyme and which is substrate is decided at random at each encounter. With minor effort, the grammar rules can be extended to allow one enzyme string to glue two substrate strings together, or to cleave one substrate string into two product strings.

Grammar string models exhibit entirely novel classes of behavior, and all the phase transitions shown in the origin of life model. Fix a grammar. Start the soup with an initial set of strings. As these act on one another, it might be the case that all product strings are longer than all substrate strings. In this case, the system never generates a string previously generated. Call such a system a *jet*. Jets might be *finite*, the generation of strings petering out after a while, or *infinite*. The set of strings generated from a sustained founder set might loop back to form strings formed earlier in the process, by new pathways. Such *"mushrooms"* are just the autocatalytic sets proposed for the origin of life. Mushrooms might be finite or infinite, and might, if finite, squirt infinite jets into string space. A set of strings might generate only itself, floating free like an *egg* in string space. Such an egg is a *collective identity operator* in the complex parallel processing algebra of string transformations. The set of transformations collectively specifies only itself. The egg, however, might wander in string space, or squirt an infinite jet. Perturbations to an egg, by injecting a new string, might be repulsed, leaving the egg unchanged, or might unleash a transformation to another egg, a mushroom, a jet. Similarly, injection of an exogenous string into a finite mushroom might trigger a transformation to a different finite mushroom, or even an infinite mushroom. A founder set of strings might galvanize the formation of an infinite set of strings spread all over string space, yet leave local "holes" in string space because some strings might not be able to be formed from the founder set. Call such a set a *filligreed fog*. It may be formally undecidable whether a given string can be produced from a founder set. Finally, all possible strings might ultimately be formed, creating a *pea soup* in string space.

In its most general setting, a grammar model considers an *ordered set* of input strings as an *input bundle*, to a second ordered set of strings considered as a *machine*. The machine acts on the input bundle to yield an ordered set of strings as an *output bundle*. Since we are considering symbol strings of unbounded length, the set of all subsets of this denumerably infinite number of strings is, in fact, the *power set* of the infinite set of symbol strings. The mapping of input bundle, via machine, to output bundle is therefore a mapping from the power set via the power set into the power set. This mapping is *not* denumerably infinite. Since algorithms are effectively computable and can correspond to a number, the number of effectively computable mappings is denumerably infinite. It follows that most of the power-set mappings are not effectively computable. In turn, it follows that the effectively computable grammars are denumerably infinite.

The denumerable infinity of computable grammars, in turn, is important since it allows us to construct a grammar space and attempt to establish what physicists might call "universality classes," or regions of grammar space, which correspond to classes of well-defined generic behaviors. For example, jets, mushrooms, eggs, filligreed fogs, pea soups, and other objects might arise with different frequencies as a function of grammar complexity or other features based on diverse measures.

Wondrous stuff, this. But more lies to hand. Jets, eggs, filligreed fogs, and the like are merely the specification of the *string contents* of such an evolving system, not its *dynamics*. Thus, an egg might regenerate itself in a steady state, in a periodic oscillation during which the formation of each string waxes and wanes cyclically, or chaotically. The entire "edge of chaos" story concerned dynamics only, not composition. What will arise in string space?

NEW TERRITORY

Models of mind, models of evolution, models of technological transformation, of cultural succession, these grammar models open new provinces. In *Origins of Order* I was able only to begin to discuss the implications of Fontana's invention. I turn next in this essay to mention their possible relation to artificial intelligence and connectionism, sketch their possible use in the philosophy of science, then mention briefly their use in economics, where they may provide an account, not only of technological evolution, but of bounded rationality, nonequilibrium market, future shock, and perhaps, a start of a theory of "individuation" of coordinated clusters of processes as entities, firms, organizations, so as to optimize wealth production. In turn, these lead to the hint of some rude analogue of the second law of thermodynamics, but here for open systems which increase order and individuation to maximize something like wealth production.

Not the least of these new territories might be a new model of mind. Two great views divide current theories of mind.[16] In one, championed by traditional artificial intelligence, the mind carries out algorithms in which production rules act sequentially on production rules to trigger appropriate sequences of actions. In contrast, connectionism posits formal neural networks whose attractors are classes, categories, or memories. The former are good at sequential logic and action, the latter are good at pattern recognition. Neither class has the strengths of the other. But parallel-processing symbol strings could have the strength of both: Thus, in parallel-processing symbol string systems, each string is a production rule, as in traditional artificial intelligence. Strings acting on strings to produce strings can be computationally universal. But in addition, parallel-processing string systems can be viewed in an extended dynamical system framework. As noted above, an "egg" is a collective identity operator which may reproduce itself in a steady state, a limit cycle, or by chaotic dynamics. Thus, an egg is an attractor in both its string composition and the dynamics by which it reproduces itself. Consider, if only playfully, the image of a stably self-regenerating egg as an "ego," a way of being in the world. Among the remarkable capacities of the human mind is that of multiple personalities; often these are entirely or partially unaware of one another. The image is not so distant from eggs which transform into neighboring eggs upon injection of specific new strings. The potential of grammar models for a theory of mind may warrant serious investigation.

Grammar models may also bear on the philosophy of science. Since Quine we have lived with holism in science, the realization that some claims are so central to our conceptual web that we hold them well nigh unfalsifiable, and hence treat them as well nigh true by definition. Since Kuhn we have lived with paradigm revolutions and the problems of radical translation, comparability of terms before and after the revolution, reducibility. Since Popper we have lived ever more uneasily with falsifiability and the injunction that there is no logic of questions. And for decades now we have lived with the thesis that conceptual evolution is like biological evolution: Better variants are cast up, never mind how conceived, and passed through the filter of scientific selection. But we have no theory of centrality versus peripherality in our web of concepts, hence no theory of pregnant versus trivial questions, nor of conceptual recastings which afford revolutions or wrinkles. But if we can begin to achieve a body of theory which accounts for both knower and known as entities which have coevolved with one another, *E. coli* and its world, IBM and its world, and understand what it is for such a system to have a "model" of its world via meaningful materials, toxins, foods, the shadow of a hawk cast on newborn chick, we must be well on our way to understanding science, too, as a web creating and grasping a world.

Holism should be interpretable in statistical detail. The centrality of Newton's laws of motion compared to details of geomorphology in science find their counterpart in the centrality of the automobile and peripherality of pet rocks in economic life. Conceptual revolutions are like avalanches of change in ecosystems, economic systems, and political systems. We need a theory of the structure of conceptual webs and their transformation. Pregnant questions are those which promise potential changes propagating far into the web. We know a profound question when we see one. We need a theory, or framework, to say what we know. Perhaps like Necker cubes, alternative conceptual webs are alternative grasped worlds. It might be useful to find a way to categorize such "alternative worlds" as if they were alternative stable collective string production systems, eggs or jets by which "knowers" categorize the worlds in which they "live." If holism yields conceptual webs, then scientific progress is often the transformation to a *neighboring web*. What pathways of conceptual change can transform a given conceptual web, and to what "neighboring" conceptual webs? What governs "neighboring webs," each a whole, in this sense? All this, of course, is buried in the actual structure of the web at any point. We know this, but lack a framework to say what we know. I suspect grammar models and string theory may help. Finally, note that conceptual evolution is like cultural evolution. I cannot help the image of an isolated society with a self-consistent set of roles and beliefs as an egg, shattered by contact with our own perhaps supracritical explosive Western civilization.

CLOSING REMARK: A PLACE FOR LAWS IN HISTORICAL SCIENCES

I close this essay by commenting on Burian and Richardsons' thoughtful review of *Origins of Order*. They properly stress a major problem: What is specifically "biological" in the heralded renderings of ensemble theories? This is a profound issue. Let me approach it by analogy with the possible use of random grammar models in economics. The opportunity for the introduction of new goods and services is based on the economic niches afforded by the current goods and services. The computer creates the possibility of the systems engineer. Economists call goods which are used together "compliments," those that replace one another "substitutes." Nut and bolt are complements, screw and nail are substitutes. Economists lack a deep theory of technological evolution, the ways in which the economic web abets its own transformation autocatalytically, because they lack a theory of technological complementarities and substitutes. One needs to know why nuts go with bolts to account for the coevolution of these two bits of econo-stuff. But we have no such theory, nor is it even clear what a theory which gives the *actual* couplings among ham and eggs, nuts and bolts, screws and nails, computer and software engineer, might be. The hope for grammar models is that each grammar model, one of an infinite set of such grammars, is a "catch-as-catch-can" model of the unknown laws of technological complementarity and substitutability. The hope is that vast reaches of "grammar space" will yield economic models with much the same global behavior. If such generic behaviors map onto the real economic world, I would argue that we have found the proper *structure* of complementarity and substitutability relationships among goods and services, and hence can account for many *statistical aspects* of economic growth such as the number of goods typically displaced by introduction of a fundamental innovation, and the avalanche of new goods that innovation brings with it. But such statistical laws will afford no account of the coupling between specific economic goods such as power transmission and the advent of specific new suppliers to Detroit. Is such a theory specifically "economics?" I do not know, but I think so.

Grammar models afford us the opportunity to capture statistical features of historically contingent phenomena ranging from biology to economics, perhaps to cultural evolution. Phase transitions in complex systems may be lawful, power-law distributions of avalanches may be lawful, but the specific avalanches of change may not be predictable. The details may rest on too many throws of the quantum dice or other sources of macroscopic disorder. Thus we confront a new conceptual tool which may provide a new way of looking for laws in historical sciences. Where will the specifics lie? As always, I presume, in the consequences deduced after the axioms are interpreted.

REFERENCES

1. Bagley, R. "A Model of Functional Self Organization." Ph.D. Thesis, University of California, San Diego, CA, 1991.
2. Dawkins, R. *The Selfish Gene.* Oxford, NY: Oxford University Press, 1976.
3. Derrida, B., and Y. Pomeau. "Random Networks of Automata: A Simple Annealed Approximation." *Europhys. Lett.* **1(2)** (1986): 45–49.
4. Eigen, M., and P. Schuster *The Hypercycle: A Principle of Natural Self Organization.* New York: Springer-Verlag, 1979.
5. Farmer, J. D., S. A. Kauffman, and N. H. Packard. "Autocatalyic Replication of Polymers." *Physica* **22D** (1986): 50–67.
6. Fogleman-Soulie, F. "Parallel and Sequential Computation in Boolean Networks." In *Theoretical Computer Science*, vol. 40. North Holland, 1985.
7. Fontana, W. "Algorithmic Chemistry." In *Artificial Life II*, edited by C. G. Langton, C. Taylor, J. D. Farmer, and S. Rasmussen, 159–209. Santa Fe Institute Studies in the Sciences of Complexity, Proc. Vol. X. Redwood City, CA: Addison-Wesley, 1991.
8. Jacob, F., and J. Monod. "On the Regulation of Gene Activity." *Cold Spring Harbor Symp. Quant. Biol.* **26** (1961): 193–211.
9. Jacob, F., and J. Monod. "Genetic Repression, Allosteric Inhibition, and Cellular Differentiation." In *Cytodifferentiation and Macromolecular Synthesis*, edited by M. Locke, 30–64. 21st Symp. Soc. Study of Devel. and Growth. New York: Academic Press, 1963.
10. Kauffman. S. A. "Metabolic Stability and Epigenesis in Randomly Connected Nets." *J. Theor. Biol.* **22** (1969): 437–467.
11. Kauffman, S. A. "Autocatalytic Sets of Proteins." *J. Theor. Biol.* **119** (1986): 1–24.
12. Kauffman, S. A. "Principles of Adaptation in Complex Systems." In *Lectures in the Sciences of Complexity*, edited by D. Stein, 619–712. Santa Fe Institute Studies in the Sciences of Complexity, Lect. Vol. I. Redwood City, CA: Addison-Wesley, 1989.
13. Kauffman, S. A. "Antichaos and Adaptation." *Sci. Amer.* **August** (1991): 78–84.
14. Kauffman, S. A. *Origins of Order: Self Organization and Selection in Evolution.* Oxford: Oxford University Press, in press.

15. Langton, C. "Life at the Edge of Chaos." In *Artificial Life II*, edited by C. G. Langton, C. Taylor, J. D. Farmer, S. Rasmussen, 41–91. Santa Fe Institute Studies in the Sciences of Complexity, Proc. Vol. X. Redwood City, CA: Addison-Wesley, 1991.

16. Smolensky, P. "On the Proper Treatment of Connectionism." *Behav. Brain Sci.* 11 (1988): 1–74.

17. Stauffer, D. "Random Boolean Networks: Analogy with Percolation." *Philosophical Magazine B* 56(6) (1987): 901–916.

Jay E. Mittenthal,* Arthur B. Baskin,† and Robert E. Reinke†
*Department of Cell and Structural Biology, University of Illinois, 505 S. Goodwin St. (and Center for Complex Systems Research, Beckman Institute; and College of Medicine), Urbana, Illinois 61801, U.S.A., and †Department of Veterinary Biosciences, University of Illinois, 1408 W. University Ave, Urbana, Illinois 61801, U.S.A.

Patterns of Structure and Their Evolution in the Organization of Organisms: Modules, Matching, and Compaction

We suggest that organisms tend to evolve a hierarchy of dynamic modules in response to selection for reliable and flexible performance. The clustering of processes in modules tends to match the pattern of correlations among constraints that the modules meet. Modules and matching may be generated and varied through self-organization in networks of processes, and then evolve under selection. Limits on resources favor the evolution of higher-level processes.

INTRODUCTION

Organisms are complex systems with remarkable properties. They are thermodynamic open systems that persist by changing their environments, maintaining themselves, and reproducing their kind. Their competence to persist gives them a quality of self-sufficiency called organic unity.

We are seeking a vantage point from which to understand the material and dynamical structures characteristic of organisms. Here we suggest that a useful

Principles of Organization in Organisms,
SFI Studies in the Sciences of Complexity, Proc. Vol. XIII,
Eds. J. Mittenthal & A. Baskin, Addison-Wesley, 1992   **321**

approach views an organism as a hierarchy of processes.[28] Each process generates outputs when inputs are provided. Processes are coupled, in that outputs of some are inputs to others. A cluster of coupled processes is itself a higher-level process.

An enzyme-catalyzed reaction is the lowest characteristically biological process. Within a cell the molecular processes form networks; the networks of intermediary metabolism, gene regulation, and post-transcriptional modification in the Golgi apparatus are examples. At higher levels multicellular networks operate; the most complex of these mediate neural, endocrine, and immune regulation. At still higher levels, processes involving social interactions occur; these include mating and (unfortunately) paying taxes. In general the processes form a nested hierarchy with bidirectional interactions among levels.

If an organism is a network of processes, what are its characteristics? Why do these characteristics evolve? Here we address these questions for clusters of coupled processes called dynamic modules. We first argue that such modules exist and show a kind of matching: The clustering of processes into modules tends to match the correlation among constraints that the modules meet. We then consider how modules and matching may evolve through self-organization and selection.

MODULES AND MATCHING

THE NETWORK OF PROCESSES CONTAINS MODULES THAT PERFORM ACTIONS

In his classic study of the architecture of complexity, Simon[42] observed that a set of coupled processes tends to form a hierarchy if the couplings among processes are not of equal strength. A cluster of strongly interacting processes may be a higher-level process that is a stable building block, a dynamic module. (A dynamic module is a process, whereas a structural module is a material unit.) A dynamic module can be activated with a small number of inputs from other processes to which it is more weakly coupled. The module can be activated in various contexts; that is, it can act as part of various higher-level modules.

Modules occur: In diverse situations a small set of molecular inputs elicits a cascade of molecular responses that collectively perform an action. At the molecular level, multicomponent complexes of macromolecules implement dynamic modules as they perform DNA replication, transcription of RNA from DNA, splicing of RNA, and translation of messenger RNA into protein. In intermediary metabolism, a branch of a pathway with end-product inhibition is a dynamic module.[39] An extracellular messenger can provoke a single cell to alter its membrane potential, motility, and biosynthetic activities; such a response is also a module. At the multicellular level, a hormone can elicit coordinated changes in the activity of many organs, as adrenalin elicits a fight-or-flight response; this response is a module.

During development a morphogenetic field may be a hierarchy of dynamic modules. Such a field is a control system that generates an organ such as a limb, kidney, or brain. Through an embryonic induction involving a small set of extracellular messenger molecules, signalling cells can induce responding cells to cooperate with them in generating the organ. A relatively simple morphogenetic field uses one lower-level module recursively to generate a branching organ.[29] However, in general the induction of an organ initiates several clusters of processes that act in parallel. Each cluster generates a component of the organ—circulation, innervation, connective tissue, muscle, bone—in a characteristic spatial distribution relative to other components. These clusters may be modules.

During the evolution of development, clusters of processes can sometimes be displaced in space and time, in heterotopy and heterochrony.[35] Dynamic modules offer a parsimonious interpretation of heterotopy and heterochrony: A cluster of processes could easily be displaced if it is a module that can be activated by few inputs. Alternative explanations of such displacements require more complex or unlikely assumptions. For example, the displaced cluster of processes might require many inputs, all of which happened to be available at the new site as well as the old one. (This seems rather unlikely unless most of the inputs are systemically available factors, such as hormones). In a particularly implausible alternative, a limb might be generated at a new time or site through a new cluster of processes, unrelated to those occurring in ancestors. From this argument, the occurrence of heterochrony supports the hypothesis that there are developmental modules.

Apparently some processes are not modular. For example, the molecular organization of a high-level developmental process, segmentation in the fruit fly *Drosophila*, does not reveal a hierarchy of modules. Segmentation involves many genes, each of which has many inputs.[16,18] Thus, in segmentation, a network of modules is not clearly evident.

A PRINCIPLE OF MATCHING: THE COUPLING AMONG PROCESSES MAY MATCH THE CORRELATION AMONG CONSTRAINTS

We have argued that modules occur and shown examples of them. We now argue that the clustering of processes matches the correlation among constraints. Organisms must perform many tasks to meet constraints on the persistence of their lineage. These constraints include universal laws, resource limitations, and conditions particular to the organisms and their environment.[28] It is likely that, just as one module often performs one task, distinct modules will perform tasks that are dissociable, or variably associated. That is, the pattern of coupling among processes may match the pattern of correlation among constraints that the processes meet. This principle of matching has been proposed, and examples have been presented, for diverse adaptive complex systems.[4,5,6,13,33,42,36,40]

As an example of this principle of matching at the cellular level, consider aggregation of amebae of the slime mold *Dictyostelium discoideum* to form a slug. This process is a model for the development of a multicellular embryo. Because the amebae are scattered individual cells that can aggregate on a petri dish, the distribution of extracellular messenger molecules and the movements of individual amebae have been characterized more extensively than comparable variables in embryos.

Aggregation can be regarded as a constraint in the life history of *Dictyostelium*. To aggregate the amebae must migrate and signal to each other; thus migration and signalling are subsidiary constraints. An ameba could migrate without signalling, or signal without migrating; so these constraints are dissociable.

The pathways mediating signalling and migration are also dissociable. Amebas signal by means of the first messenger cyclic adenosine monophosphate (cAMP). In response to a pulse of cAMP from a localized source, an ameba secretes a pulse of cAMP—that is, relays the signal—and migrates toward the source. A cluster of intracellular processes performs each of these activities. In the cluster that mediates relaying, extracellular cAMP binds to a receptor protein and stimulates the activity of membrane-bound intracellular adenylate cyclase. This enzyme produces cAMP that the ameba secretes. In the cluster of processes that mediates migration, extracellular cAMP binds to a receptor protein and stimulates production of the second messenger inositol phosphate. This messenger activates a cascade of molecular processes that mediate elongation and locomotion of the ameba toward the source. These processes include polymerization of microtubules and of actin, and phosphorylation of myosin. There are mutations that affect relaying but not migration, and *vice versa*.[12,30] Thus the pathways mediating relaying and migration are dissociable clusters of processes, activated by a single input, cAMP. These pathways seem to be dissociable modules that meet the dissociable constraints of signalling and migration.

Note that cAMP activates a module that produces aggregation; this module contains the subsidiary modules that produce relaying and migration. Thus a hierarchy of modules meets a hierarchy of constraints. In fact, aggregation occurs after many iterations of relaying and migration; lower-level modules are active repeatedly within a higher-level module. These iterated responses to cAMP not only bring the amebae together, but insert molecules on their surfaces so that they adhere and communicate after they touch.

Thus matching occurs at the molecular level: A pattern of structure in modules matches a pattern in constraints. Processes match constraints at levels higher than the molecular. Examples considered elsewhere in this volume include the entrainment of an oscillator to a periodic driver,[8] as occurs in circadian rhythms, and the development of visual receptive fields that match visual input during a critical period.[32,45] Matching is also evident in the organization of vision in adult primates. Away from human-dominated environments, an object is usually illuminated mainly by a single light source, so its shape is strongly correlated with its shading. Correspondingly, our brains interpret patterns of illumination as shaded objects using the assumption of a single light source, as visual illusions show.[34] Shape and shading are processed together in the nervous system as they are correlated in the external world. However, human and monkey visual systems have dissociable pathways for processing some dissociable aspects of visual input. A faster-conducting (magnocellular) pathway provides information about the orientation, motion, and depth of objects, whereas a slower-conducting (parvocellular) pathway provides data on color and details of form.[26] Additional dissociations occur in the input (e.g., between form and color) but are not necessarily used in visual processing. The principle of matching suggests only that when dissociable pathways are used, they meet dissociable constraints.

THE EVOLUTION OF MATCHING AND MODULES

Evolution occurs through a dialogue between self-organization and selection. Variants of a structure are generated through the intrinsic dynamics, or self-organizing activity, of a system. Once diverse variants exist, they typically differ in stability in the environments where they occur. That is, the variants can undergo natural selection.

Here we consider the circumstances under which self-organization and selection tend to produce modules and matching. As we shall see, networks of processes have intrinsic dynamics that can generate modules. Finite resources limit the size and diversity of the modules that can evolve. Matching evolves: Modules can perform tasks to meet constraints. Selection can refine modules and matching that arise through self-organization. Selection can also favor the evolution of a hierarchy of modules that increases the reliability and flexibility with which organisms perform tasks.

NETWORKS OF PROCESSES HAVE INTRINSIC DYNAMICS THAT CAN GENERATE MODULES

In a multivariate dynamical system, large-scale or global structures arise through local dynamics.[24] In organisms large-scale structures such as modules are associated with patterns of coupling among molecular processes. These patterns will change during evolution because random variations in the genome will tend to scramble the coupling. The rules governing these variations are the local dynamics of evolutionary change. The variations are inevitable because the sequence of nucleotides in the input and output regions of each gene can change through diverse processes, including point mutation, recombination, gene conversion, gene amplification, and the insertion or deletion of transposons. These changes can add, delete, or modify transcription factors and cis-regulatory elements, thereby changing the coupling among genes.[1,19,37]

The clustering of processes in the organismal network may change in several ways. If two coupled processes become uncoupled, or dissociated, one of them may be deleted from a higher-level process. Adding coupling can have diverse consequences: The addition of inhibitory coupling can allow a context-dependent dissociation between two processes. Additional excitatory coupling can add a process to a higher-level process, or produce a new higher-level process.

The global consequences of local dynamics in the genome are not well understood for organisms, but some consequences have been explored in models for Boolean networks of binary genes.[22,23] The self-organizing evolution of Boolean networks with suitable parameters can produce modules. With a low connectivity among genes or with certain biases in the Boolean switching rules, a minimal perturbation—a transient switch in the state of one gene—elicits a small cascade of changes in the states of other genes. Such a cascade, initiated by a minimal perturbation, corresponds to a dynamic module.

If a model network has parameter values such that it operates at the edge of chaos, minimal perturbations will elicit many small cascades and few large cascades, with a power-law distribution of cascade sizes. Qualitatively, developing systems often show few large cascades (corresponding to major developmental transitions, such as fertilization, gastrulation, and induction) and many small cascades.[27,38] Thus organismal networks may operate at the edge of chaos.

FINITE RESOURCES LIMIT THE VARIETY OF MODULES THAT CAN EVOLVE

The resources available in a network limit the size and diversity of the modules that can evolve. To see this, suppose the activity of macromolecules implements low-level modules, such as individual enzyme-catalyzed reactions. Additional modules can evolve through variations and combinations of these modules, as follows: (1) Variations in the genome, discussed above, can generate new low-level modules that are variants of existing ones. (Variant modules are more likely to evolve than are wholly new types of modules.) (2) Macromolecules may associate in various combinations to form higher-level modules. Useful combinations of macromolecules can form through random collisions in solution, but this process is slow and unreliable. (3) Higher-level processes may evolve that generate specific combinations of lower-level modules reliably.

Ultimately, the finiteness of resources limits the extent to which variation and combination of macromolecules can generate new modules. Such limits include the coding capacity of the genome, the error rate for syntheses of macromolecules, and the pool sizes of monomers. The coding capacity of the genome is limited, in that the probability of an error in copying a linear macromolecule (DNA or RNA) increases with its length. The accuracy of copying (error rate) is also limited, by the error-correcting processes that are available. The pool sizes of monomers are limiting in that syntheses of different types of macromolecules compete for shared pools. All these limits on resources limit the diversity and quantity of macromolecules that can be synthesized.

These limits imply two further ways in which new modules can evolve. (4) More resources may become available, so a limit increases. For example, a new error-correcting process may evolve. (5) Existing resources may become available for new modules if a compaction occurs—if higher-level processes evolve that reduce the resources needed for existing modules.

SELECTION CAN REFINE MODULES AND MATCHING THAT ARISE THROUGH SELF-ORGANIZATION

The preceding arguments suggest that the evolutionary dynamics of networks with suitable parameters will tend to generate modules, though the available resources will limit the spectrum of modules. It seems likely that a newly formed module will have some rudimentary capacity to perform a task, as a polypeptide is likely to have a rudimentary capacity to catalyze some reaction.[21] That is, a rudimentary matching of processes to constraints must occur spontaneously. Once a sufficient set of tasks are performed sufficiently well, selection can refine the performances, since variants of the original modules that perform better persist more stably through evolution. The natural outcome of this refinement may be a match between the coupling among processes and the correlations among constraints.

Note that finiteness limits are among the large-scale features subject to selection. For example, selection for a short life cycle and a high rate of reproduction has favored a relatively small genome, with a correspondingly limited coding capacity, in some viruses, in bacteria, and in *Drosophila*.[9,20,44] If selection for rapid reproduction is relaxed, variations in the copying of DNA may enlarge the genome, increasing its coding capacity.

As one might expect, self-organization and selection may work at crossed purposes. A module that occurs spontaneously may have disadvantageous consequences for the species using it. For example, arms races can evolve through strong coupling between the sexes within a species in sexual selection, or through strong coupling among coevolving species as in predator-prey or host-parasite interactions. An arms race may be an evolutionary module. Within an arms race, selection may generate specialized features that hinder many activities of the carrier, such as the antlers of the Irish elk[17] or the plumage of a bird of paradise.[7]

SELF-ORGANIZATION AND SELECTION HAVE GENERATED MODULES IN A SPECTRUM OF LEVELS OF ORGANIZATION

We now suggest that self-organization and selection have tended to generate the spectrum of possible modules, beginning with low-level modules implemented by small macromolecules. As Bonner[4] noted, organisms will proliferate in the niches they occupy, and so will tend to use up the environmental resources in those niches. Thus limits on resources favor the evolution of higher-level processes that perform additional tasks and so give access to new niches.

The repertoire of processes has probably increased in all of the ways considered above—through variation, combination, changes in finiteness limits, and compaction. According to the exon theory of genes,[1,3,9,10,11,14,15] early in evolution statistical fluctuations produced mini-genes, individual exons, that encoded peptides of about 15-20 amino acids. Some of these peptides had catalytic activity and so could perform a low-level task. (1) Variations in the genome produced variant peptides. (2) Multicomponent complexes of peptides could perform sequences of tasks faster and with fewer side reactions than if substrates diffused among the peptides in solution. Such complexes could also perform additional tasks, as the association of peptides changed their configurations and made new interfaces between them. (3) The evolution of RNA splicing allowed the synthesis of proteins containing several of the older peptides in sequence. Thus particular associations of catalytic and regulatory activities could be produced reliably, while diversity was maintained by recombination and alternative splicing. (4) The evolution of error correction in replication and in translation[25] increased the variety of proteins that could be synthesized reliably.

Now consider a possible example of compaction. (5) A compaction may have increased the reliability with which splicing of heterogeneous nuclear RNA excises introns and makes messenger RNA. Early in evolution each intron may have acted as an RNA enzyme, a ribozyme, to catalyze its own excision and the splicing of the adjacent exons.[41] Probably a considerable part of the nucleotide sequence in each intron was needed for this autocatalysis. However, at present in eukaryotes a multicomponent complex of small nuclear ribonucleoproteins, a spliceosome, performs splicing. For the activity of a spliceosome short intervals of maintained DNA sequence suffice; these include the coding regions for the components of the spliceosome and the recognition elements of each intron that bind the spliceosome.[43]

The probability of an error in transcription increases with the length of the transcribed RNA. Since the length of RNA needed for spliceosome-mediated splicing is appreciably less than the length needed for self-splicing of all introns, spliceosomal splicing is more reliable than is self-splicing. Here a compaction occurred—a higher-level process, splicing by spliceosomes, evolved and reduced the amount of a resource, RNA sequence, needed to splice all genes. Within introns the RNA that formerly contributed to self-splicing (or the DNA that encodes it) became available to help with alternative tasks, such as binding proteins that regulate transcription or alternative splicing.

Thus organisms can perform more diverse tasks not only by varying and combining modules, but also by raising finiteness limits and by compaction. Baskin et al. [2] discuss these processes further.

SELECTION FOR FLEXIBLE AND RELIABLE PERFORMANCE FAVORS THE EVOLUTION OF MODULES AND MATCHING

Finally, we shall consider selection pressures that may especially favor the evolution of modules and matching. Selection can favor the evolution of organismal networks that meet diverse combinations of performance criteria. In general terms these criteria are flexibility, reliability, economy, and speed.

Modular organization favors some of these criteria but disfavors others. It increases the reliability of a network of unreliable processes, the probability that the network generates an output that meets constraints.[5,6,42] Modular organization also favors flexibility: As Simon[42] emphasized, if dissociable modules perform dissociable tasks, then a module can be modified without altering other modules so long as it accepts appropriate inputs and provides needed outputs. A vertebrate heart can be partitioned into two, three, or four chambers, as long as it pumps blood through the circulation. Heterochrony and heterotopy are aspects of flexibility associated with displacing the activity of modules relative to one another in time and space. Recalling that model Boolean networks at the edge of chaos generate a spectrum of module sizes, it is noteworthy that such networks also adapt flexibly to a changing environment.[22]

These considerations suggest that modules and matching are to be expected when selection favors flexibility and reliability. However, a network with modular organization may produce output more slowly than a special-purpose, nonmodular network. That is, flexibility and reliability may be sacrificed for speed of performance, particularly if nonmodular networks can be produced at low cost and at a high rate. As prokaryotes evolved they may have lost introns, sacrificing the flexibility associated with introns (through alternative splicing and recombination) for the speed associated with replicating a smaller genome.[9] A tradeoff between speed and reliability may also occur in the development of the fruit fly *Drosophila*.[20]

DISCUSSION

We have suggested that organisms tend to evolve a hierarchy of dynamic modules in response to selection for reliable and flexible performance. The clustering of processes in modules may tend to match the pattern of correlations among constraints that the modules meet. Modules and matching may be generated and varied through self-organization in networks of processes, and then evolve under selection. (Similarly, Newman[31] and Goodwin and Kauffman[16] argue that the generic dynamics of the materials in embryos played a major role in the evolution of morphogenesis.) Limits on resources favor the evolution of higher-level processes.

These proposals imply a program of research. The preceding arguments are qualitative or are based on models considerably abstracted from the realities of biology. It is necessary to see whether the conclusions emerge from more realistic models. Do such models predict the observed clustering of processes into modules? What modules are likely to occur in an organism, and how are resources likely to be allocated among them? Models that approach these questions should provide a deeper insight into the characteristic structures of organisms.

ACKNOWLEDGMENTS

This work benefitted substantially from discussions with many colleagues in Britain (where J. E. Mittenthal began it) and in the U.S. We appreciate discussions and comments on draft manuscripts by John Campbell, Brian Goodwin, Lloyd Gold-wasser, David Lambert, Marc Mangel, Yoshi Oono, Michael Stryker, and partic-ipants in a workshop on Development and Evolution at the Santa Fe Institute, November, 1989. J. E. Mittenthal appreciates the support of Jim Murray during a sabbatical at the Center for Mathematical Biology. Preparation of this paper was supported in part by U.S.P.H.S. grant HD16577 and grant GR/D/13573 from the Science and Engineering Research Council of Great Britain.

REFERENCES

1. Alberts, B., D. Bray, J. Lewis, M. Raff, K. Roberts, and J. D. Watson. In *Molecular Biology of the Cell*, 2nd edition, Chapters 3, 5, 9, 10. New York: Garland, 1989.
2. Baskin, A. B. B., R. E. Reinke, and J. E. Mittenthal. "Exploring the Role of Finiteness in the Emergence of Structure." This volume.
3. Blake, C. "Exons—Present From The Beginning?" *Nature* **306** (1983): 535–537.
4. Bonner, J. T. *The Evolution of Complexity*. Princeton: Princeton University Press, 1988.
5. Clarke, B., and J. E. Mittenthal. "Modularity and Reliability in the Organi-zation of Organisms." *Bull. Math. Biol.*, **54** (1992):1-20.
6. Clarke, B., and J. E. Mittenthal. "Reliability of Networks of Genes." This volume.
7. Dawkins, R. *The Blind Watchmaker*. New York: Norton, 1986.
8. DeGuzman, C. G., and J. A. S. Kelso. "The Flexible Dynamics of Biological Coordination: Living in the Niche Between Order and Disorder." This vol-ume.
9. Doolittle, W. F. "Genes in Pieces: Were They Ever Together?" *Nature* **272** (1978): 581–582.
10. Doolittle, W. F. "What Introns Have to Tell Us: Hierarchy in Genome Evo-lution. *Cold Spring Harbor Symposia on Quantitative Biology* **52** (1987): 907–913.
11. Dorit, R. L., L. Schoenbach, and W. Gilbert. "How Big is the Universe of Exons?" *Science* **250** (1990): 1377–1382.
12. Drummond I. A. S., and R. L. Chisholm. "Precocious Developmental In-duction of cAMP Regulated Genes in a Signal Transduction Mutant of *Dic-tyostelium discoideum*." *J. Cell Biol.* **107** (1988): 55a.
13. Gelernter, D. "The Metamorphosis of Information Management." *Sci. Am.* **261**(2) (1989): 66–73.
14. Gilbert, W. "Why Genes in Pieces?" *Nature* **271** (1978): 501.
15. Gilbert, W. "The Exon Theory of Genes." *Cold Spring Harbor Symposia on Quantitative Biology* **52** (1987): 901–905.

16. Goodwin, B., and S. Kauffman. "Deletions and Mirror Symmetries in *Drosophila* Segmentation Mutants Reveal Generic Properties of Epigenetic Mappings." This volume.

17. Gould, S. J. "The Misnamed, Mistreated, and Misunderstood Irish Elk." In *Ever Since Darwin*, 79–90. New York: W. W. Norton, 1977.

18. Ingham, P. "The Molecular Basis of Embryonic Pattern Formation in *Drosophila*." *Nature* **335** (1988): 25–34.

19. John, B., and G. Miklos. *The Eukaryote Genome in Development and Evolution*. London: Allen & Unwin, 1988.

20. Karr, T. L., and J. E. Mittenthal. "Adaptive Mechanisms that Accelerate Embryonic Development in *Drosophila*." This volume.

21. Kauffman, S. A. "Autocatalytic Sets of Proteins." *J. Theor. Biol.* **119** (1986): 1–24.

22. Kauffman, S. A. "Antichaos and Adaptation." *Sci. Amer.* **265**(2) (1991): 78–84.

23. Kauffman, S. A. "The Sciences of Complexity and 'Origins of Order.'" This volume.

24. Kelso, J. A. S., M. Ding, and G. Schoner. "Dynamic Pattern Formation: A Primer." This volume.

25. Kirkwood, T. B. L., R. F. Rosenberger, and D. J. Galas. *Accuracy in Molecular Processes*. New York: Chapman and Hall, 1986.

26. Livingstone, M., and D. Hubel. "Segregation of Form, Color, Movement, and Depth: Anatomy, Physiology, and Perception." *Science* **240** (1988): 740–749.

27. Loomis, W. F. "Essential Genes for Development of *Dictyostelium*." *Progress in Molecular and Subcellular Biology* **11** (1989): 159–183.

28. Mittenthal, J. E. "Physical Aspects of the Organization of Development." In *Complex Systems*, edited by D. Stein, 225–274. Santa Fe Institute Studies in the Sciences of Complexity, Lect. Vol. I. Redwood City, CA: Addison-Wesley, 1989.

29. Nelson, T. R. "Biological Organization and Adaptation: Fractal Structure and Scaling Similarities." This volume.

30. Newell P. C., G. N. Europe-Finner, N. V. Small, and G. Liu. "Inositol Phosphates, G-proteins and *ras* Genes Involved in Chemotactic Signal Transduction of *Dictyostelium*." *J. Cell Science* **89** (1988): 123–127.

31. Newman, S. A. "Generic Physical Mechanisms of Morphogenesis and Pattern Formation as Determinants in the Evolution of Multicellular Organization." This volume.

32. Obermayer, K., H. Ritter, and K. Schulten. "A Model for the Development of the Spatial Structure of Retinotopic Maps and Orientation Columns." This volume.

33. Olson, E. C., and R. L. Miller. *Morphological Integration*. Chicago: University of Chicago Press, 1958.

34. Ramachandran, V. S. "Perception of Shape From Shading." *Nature* **331** (1988): 163–166.

35. Raff, R. A., and T. C. Kauffman. *Embryos, Genes, and Evolution*. New York: Macmillan, 1983.
36. Riedl, R. *Order in Living Organisms*. Translated by R. P. S. Jefferies. London: John Wiley, 1976.
37. Roth, V. L. "The Biological Basis of Homology." In *Ontogeny and Systematics*, edited by C. J. Humphries, 1–26. New York: Columbia University Press, 1988.
38. Sargent, T. D., and I. B. Dawid. "Differential Gene Expression in the Gastrula of *Xenopus laevis*." *Science* **222** (1983): 135–139.
39. Savageau, M. A. *Biochemical Systems Analysis*. Reading, Massachusetts: Addison-Wesley, 1976.
40. Schmalhausen, I. I. 1949. *Factors of Evolution*. Chicago: University of Chicago Press, 1986.
41. Sharp, P. A. "On the Origin of RNA Splicing and Introns." *Cell* **42** (1985): 397–400.
42. Simon, H. "The Architecture of Complexity." *Proc. Am. Phil. Soc.* **106** (1962): 467–482.
43. Steitz, J. A. "Snurps." *Sci. Am.* **258(6)** (1988): 56–63.
44. Stryer, L. *Biochemistry*, 3rd edition. New York: W. H. Freeman, 1988.
45. Stryker, M. "Activity-Dependent Reorganization of Afferents in the Developing Mammalian Visual System." This volume.

Bertrand Clarke† and Jay E. Mittenthal‡
†Department of Statistics, Purdue University, 1399 Mathematical Sciences Building, West Lafayette, IN 47907-1399, U.S.A., and ‡Department of Cell and Structural Biology, University of Illinois (and Center for Complex Systems Research, Beckman Institute; and College of Medicine), 505 South Goodwin Street Urbana, IL 61801, U.S.A.

Reliability of Networks of Genes

The probability that an organism persists depends on the reliability of the processes underlying its activities and on the coupling among those processes. In a problem that abstracts salient features of genetic networks, we evaluated the reliability of alternative networks that meet a set of constraints, to see whether highly reliable networks have a characteristic organization. Among the networks with high reliability is a modular network in which the coupling among processes matches the correlation among constraints. This report summarizes the model and conclusions presented more fully elsewhere.[1]

For mathematical convenience we deal with simplified genetic networks in which each gene has a regulatory region with one cis element that binds one messenger protein, and an output region that makes the gene product, a protein. When the cis element binds a messenger, the output region makes its product with probability p. Two genes are coupled when the product of one can bind to the cis element of the other. Many genes can be coupled to form a network. Thus one messenger can elicit the formation of several proteins that cooperate in performing an activity.

We are interested in alternative networks of processes that lead to synchronous synthesis of proteins A, B, C, and D. These four monomers associate to form dimers AB and CD, which associate to form the tetramer $ABCD$.

Principles of Organization in Organisms,
SFI Studies in the Sciences of Complexity, Proc. Vol. XIII,
Eds. J. Mittenthal & A. Baskin, Addison-Wesley, 1992   **333**

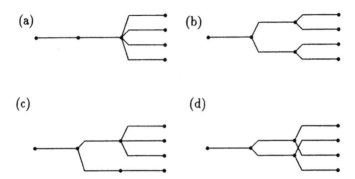

FIGURE 1 Four patterns of coupling that can produce the monomers synchronously. Each arrow represents a gene; the arrowhead points to the gene product, and the tail represents the cis element. However, messenger proteins are not shown explicitly. The contact of an arrowhead with a tail represents a coupling between two genes, in which a messenger can bind to a cis element. The constraints, which are the same for all the networks, are only shown explicitly for the modular network. Networks: (a) shotgun; (b) modular; (c) pseudomodular; (d) nonmodular.

We define the reliability of a network as the steady-state probability that it makes a tetramer in response to binding an initial messenger. The reliability depends on the pattern of coupling among genes. Four patterns of coupling are shown in Figure 1. They are the most reliable of more than fifty alternative networks that meet the constraints of the model.

We calculated the reliability $R(p)$ with which the shotgun, modular, pseudo-modular, and nonmodular networks meet the constraints on forming a tetramer. We modelled the activity of the networks as a discrete-time Markov chain. The monomers last one time step; the dimers last two time steps. The persistence of dimers for more than one time step makes this problem nontrivial.

Figure 2 shows that for $0 < p < 1$,

$$R_{shotgun}(p) > R_{modular}(p) > R_{pseudomodular}(p) > R_{nonmodular}(p).$$

In the modular network a module produces each dimer, and these two modules are coupled to form a higher-level module that produces the tetramer. The shotgun network is a degenerate case of the modular network, in the sense that both dimer-forming modules are activated by the same messenger. Thus the networks that make the tetramer with highest reliability have a modular organization that matches the pattern of interactions among the monomers and the dimers.

The maximal difference in reliability between modular and nonmodular networks occurs at approximately $p = .85$. Other results (not shown) suggest that the corresponding maximal difference occurs at a higher value of p in a more complex network. Thus the high reliability associated with molecular processes (replication, transcription, translation) may maximize the benefit of organizing these processes in a modular way.

We conclude that a network with modular organization can meet constraints with relatively high probability.

(a)

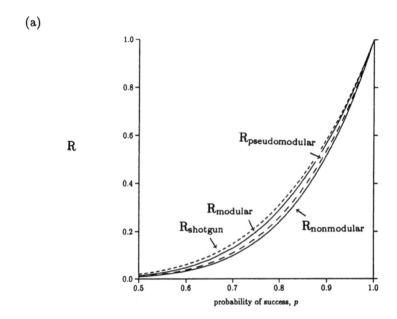

(b)

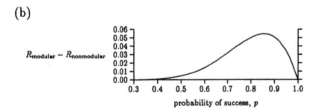

FIGURE 2 The reliability of shotgun, modular, pseudomodular, and nonmodular networks for synthesizing the tetramer as functions of the probability p of success. (a) Reliability, R, of four networks. (b) The difference $R_{\text{modular}} - R_{\text{nonmodular}}$

REFERENCES

1. Clarke, B. and J. E. Mittenthal. "Modularity and Reliability in the Organization of Organisms." *Bull. Math. Biol.*, in press.

Arthur B. Baskin,† Robert E. Reinke,† and Jay E. Mittenthal‡
†Department of Veterinary Biosciences, 1408 W. University Avenue, University of Illinois, Urbana, Illinois 61801; and ‡Department of Cell and Structural Biology, 505 South Goodwin St. (and Center for Complex Systems Research, Beckman Institute; and College of Medicine), University of Illinois, Urbana, Illinois 61801

Exploring the Role of Finiteness in the Emergence of Structure

The fact that complex adaptive systems have existed in the biological world for millennia suggests that there are forces which foster or reinforce such systems. The development of a normative theory of organization for these complex adaptive systems may make it easier to understand biological systems and will, certainly, assist in the design of man-made systems. This paper explores the role of finiteness in the emergence of structure by relating biological structures to a formal mathematical model, investigating structural properties of the model, and relating those properties back to their biological equivalents. Based on the emerging theory, we conclude that a clash between requirements (for reliability, accuracy, and speed) and finiteness limits (on control states and building blocks) provides both an impetus for and a pressure to sustain higher levels of structure.

Principles of Organization in Organisms,
SFI Studies in the Sciences of Complexity, Proc. Vol. XIII,
Eds. J. Mittenthal & A. Baskin, Addison-Wesley, 1992   **337**

1. INTRODUCTION

Living organisms are complex systems of interacting processes organized according to a highly structured pattern. Indeed, much of physics and biology is a quest for understanding the structure underlying the physical and biological worlds. The fact that highly structured biological systems arose and persist suggests that there are forces which foster, or at least reinforce, such systems. The purpose of this paper is to explore the hypothesis that higher degrees of structure arise from the imposition of finiteness limits. As part of a broader effort to develop a normative theory of complex organizations, this paper attempts to identify the types of finiteness limits which operate to foster increased structure. This new theory could be used to summarize what is currently known about the organization of organisms and to make predictions about ways in which other complex organizations might be constructed.

In this paper, we will essentially develop four major points about the emergence of structure in complex adaptive systems:

1. The Degree of structure can be quantified and measured along a relative scale.
2. The shape of the space of structural alternatives can provide a bias for increased structure.
3. Increasing capability requirements coupled with resource limits provide a systematic pressure for increased structure.
4. Increases in structure which result from expanding capability requirements colliding with finiteness limits will promote modular structure which matches patterns of decomposability in the capability requirements.

In the conclusion section, we will return to these points and provide elaboration based on the formalism we develop and the examples we explore below.

The first step in the development of any theory is the identification of basic concepts which will be used in the construction of the theory. The intuitive concepts described in the next section point to a general direction for the theoretical development and serve as a check on the final theory. Our theory rests primarily on the notion of organisms as problem solvers. We evaluate the degree of structure of the organism by comparing the nature and amount of structure in the organism to the maximum structure that the problem allows. In the sections which follow, we identify intuitive notions of structure, develop a formal mathematical model which embodies them, explore properties of the mathematical model, and then apply these results to biological systems.

2. SOME INTUITIVE NOTIONS OF STRUCTURE

The behavior of a biological system can be thought of as the process of solving a mathematical function, i.e., as a mapping from inputs to outputs. For instance, a protein synthesis process can be thought of as a function where mRNA and chemical building blocks are the inputs and the synthesis product is the output. Such a function can be characterized by its set of inputs, its set of outputs, and the mapping between inputs and outputs.[1]

Notions of problem size, complexity, and degree of structure underlie much of our proposed theory. We begin by noting that size and complexity are intrinsic properties of a problem, while degree of structure is a property of a particular solution to the problem. For a given problem with a particular intrinsic complexity, there may be a number of solutions with widely varying degrees of structure. Thus, in this section, we explore four major points:

1. Operation of a biological system can be modeled as evaluating a function.
2. Problem size can be computed as the size of the function.
3. Complexity is a property of the problem itself, not a specific solution.
4. Structure is a property of a solution and not the problem.

We have already pointed out the correspondence between the behavior of a biological system and the evaluation of a function. The remaining subsections briefly discuss the remaining points.

2.1 PROBLEM SIZE IS A FUNCTION OF INPUTS AND OUTPUTS

For widely differing problems, the difference in problem size is easily determined intuitively: the weather forecasting problem is much larger than the (still large) problem of counting cards in blackjack. Alternatively, the problems of two-digit multiplication and two-digit addition are similar in size because they can each be fully enumerated with a table containing 10,000 rows.[2]

Perhaps the most natural representation of the size of a problem is given by the size of a look-up table for the problem. The rows of such a table list the situations and the corresponding responses. Even when we do not know how to assign the actions for each possible situation, we can frequently at least determine how many situations might occur. Since the size of the problem increases directly with the number of input situations, we can determine the size of a problem even when we do not know enough about it to determine its complexity.

In mathematical terms, a function is defined as a subset of the cross product of the input set and the output set. In keeping with the intuitive notions above, we define problem size as the cardinality of such a set representing the function. Where a function can meaningfully have an infinite input set or an infinite output set, we take the problem size to be infinite.

[1]For now, we will assume that there is always the same output for a given input. Extensions of this formalism to the situation where the output is a stochastic function of the inputs is likely to be straightforward, but is beyond the scope of this paper.

[2]Each row would contain the pair of input numbers along with the product or sum respectively.

2.2 PROBLEM COMPLEXITY MEASURES THE IRREDUCIBLE CORE OF A PROBLEM

The notion of problem size is easier to define and measure than either problem complexity or structure of the solution. Unfortunately, we are generally not as interested in the size of a problem as we are in its underlying complexity or the structure of the solution. The difference between problem size and complexity shows up clearly in the calculation of Pi. Despite the fact that there are an infinite number of digits in the output, the calculation of Pi is not at all complex (essentially just a single division step, given the inputs). In a similar way, the process of multiplication is intuitively more complex than addition because multiplication can be decomposed into successive additions.

It seems clear that the problem of balancing a checkbook is less complex than developing a balanced national budget. Similarly, the game of tic-tac-toe can be easily demonstrated to be less complex than checkers, which has, in turn, been demonstrated to be less complex than chess. Comparing the intrinsic complexity of problems helps us to predict those which will be easily solved and those which will be difficult to solve.

We intuitively associate problems where the solution is easy to describe as not very complex. In a similar way, solutions which are complicated and lengthy are generally taken to be complex. For example, consider the addition vs. multiplication example above. Because the algorithm for multiplication must contain the entire algorithm for addition as a component, the algorithm for multiplication will be longer and more complicated than the algorithm for addition. In this way, we identify intrinsic complexity with the length of the minimum algorithm needed to describe the behavior of the solution to the problem.

The notion of complexity has been extensively studied and there is considerable agreement that the complexity of a mathematical function is associated with the shortest possible description of the function. A more mathematically precise version of this notion is generally called Kolmogorov complexity. We leave a more detailed analytical definition to section 3 below.

2.3 STRUCTURE IS A PROPERTY OF THE SOLUTION, NOT THE PROBLEM

Structure can be naturally thought of as the opposite of randomness. If the solution to a problem is accomplished by truly random activity, then that solution is not at all structured. Conversely, if the solution is accomplished by an intricate sequence of steps, then the solution has been structured into those same steps. The degree of structure can be taken to be the distance away from random behavior toward a more structured solution to the problem.

The definitions of problem size and complexity above are intrinsic to the problem and do not depend on the way in which the problem is solved, i.e., the structure of the solution. The notion of structure, on the other hand, is usually thought of as a property of the solution—not only of the problem. Thus, one solution may be more or less structured than another solution to the same problem. In this way of looking at structure, it makes little sense to compare the degree of structure of two solutions to two entirely different problems of widely differing complexity.

2.3.1 STRUCTURE IS CHARACTERIZED BY ORDER AND ORGANIZATION. Orderly behavior and the existence of organizations are clearly intuitively associated with highly structured systems. For example, a military parade clearly has more order than a rioting mob of an equal number of people. The actions of the people in the military parade are more ordered, more easily described, and more predictable than those of the members of the mob. Clearly, any theoretical model of structure in complex systems should reflect this intuitive distinction.

In a similar example, consider the organization of a colony of ants in which a relatively small number of castes exist, each with its own set of duties. Such a colony is even more highly organized (and, thus, structured) than the military parade. The ant colony is more structured because a greater portion of the activity of the ants is covered by the caste system and because the overall variety of activities among the ants is less than for the soldiers. The degree of structure of the ant colony is in stark contrast to the comparatively random behavior of a similar number of unicellular creatures in a drop of water.

2.3.2 STRUCTURE IS CHARACTERIZED BY CONCISENESS OF DESCRIPTION. The intuitive notion of "not random" as a definition of structure contains embedded within it the notion of concise description. The notion of randomness is essentially that there is no commonality of activity among the parts. The activity and state of each part must be described separately along with all of its interactions. This lack of commonality means that the description cannot be shortened by summarizing the behavior of similar parts. Consider, for a moment, the difference between a good lecture and a rambling, unfocused presentation of a collection of material. Most good lectures begin with a concise capsule of the material to be covered (sometimes using simplifying assumptions which must be relaxed later) in order to concisely convey the essence of the material. Armed with this organization in a concise form, the listener has "a place to put" the material.

In a more mathematical example, consider the number Pi. This number is thought to have an infinite number of digits where there is no repeating pattern. Such a description would appear to be completely unstructured. By contrast, consider the situation when Pi is described not as an infinite set of non-repeating digits but as the circumference of a circle divided by its diameter. Given this relationship and a precisely measured circumference/diameter pair, Pi can be calculated to any desired degree of accuracy. Such a description of Pi is clearly more concise, and therefore more structured, than the equivalent string of digits.

2.3.3 STRUCTURE IS CHARACTERIZED BY COMPRESSION. The order, organization, and conciseness invoked above to capture an intuitive notion of structure in complex systems can all be summarized as some form of compression. The compression may come in the variety of roles as in the military and ant examples, or the compression may come in the size of the description, as in the case of Pi.

Most would agree that a diamond is more structured than the lump of coal from which the diamond was formed by compression. In forming a diamond, physical compression acted to drive the relatively unstructured lump of coal toward a more highly structured crystalline state. The regularities of the crystalline structure allow the diamond to take up less space than the corresponding coal and, thus, to meet the externally imposed compression constraint.

2.3.4 STRUCTURE CAN BE DEFINED IN TERMS OF PROBLEM SIZE AND COMPLEXITY. The interval between problem size and intrinsic problem complexity defines the interval of possible degrees of structure. We have already defined a solution that is expressed as an exhaustive enumeration of all situations and their associated outputs as equal to the problem size. We now add that such a solution has minimal structure. Similarly, a solution which has already maximally taken advantage of potential structure in the solution to arrive at the minimal encoding of the solution is maximally structured. We define such maximally structured machine to have a structure of 1.0.

When a biological system is organized in such a way that the function is explicitly enumerated as a look-up table, then the size of the description is equivalent to the size of the problem. We define the degree of structure to be minimal at that point.

2.4 PUTTING THE INTUITIVE CONCEPTS TO WORK

The intuitive concepts presented above may be better understood in terms of a concrete example. Consider the evaluation of a function F which is of the form:

x	$F(x)$
0	-4
1	0
2	8
⋮	⋮

where this tabular (and, thus, minimally structured) version of the function may involve an infinite table and could be rewritten as an algebraic formula:

$$F(x) = 2x^2 + 2x - 4 \tag{1}$$

for as many values of x as are allowed. Clearly, the formula summarizes the information contained in the much larger table by capturing the algebraic "shape" of the function. Now consider an alternate structure for the solution to this same problem:

$$F(x) = (2x + 4)(x - 1) \tag{2}$$

which is a factored version of the formula above for as many values of x as are allowed. Like the original formula which captured structure implicit in the table, this formula captures structure implicit in the previous formula. While the intrinsic complexity of the problem is constant (because the problem is the same), the two formulas do not have the same structure. At the end of the next section, we will argue that formula (2) has more structure than formula (1).

In summary, during this section we have explored four major points:

1. By appropriate choice of inputs and outputs, biological operations can be modeled as functions.
2. Structure is a property of a solution to a problem and different solutions to the same problem can be compared on a relative scale.
3. An exhaustive enumeration of the solution to the function defines the problem's size and corresponds to minimal structure.
4. The complexity of a problem defines "how much room" there is for structure, i.e., more complex problems can have more complex solutions than simple problems.

The material in the next section will explore each of these major points in a more formal way and in greater detail.

3. MAKING THE INTUITIVE DEFINITIONS MORE PRECISE

The intuitive definitions of problem size, intrinsic problem complexity, and degree of structure can be made more precise by adopting a single representation for all functional descriptions. A useful representation for functions, drawn from theoretical computer science, is the *Turing machine*. A Turing machine is an abstract computational device with a precise mathematical definition. Turing machines compute the class of recursively enumerable functions; it is believed that this class is equivalent to the class of functions that are computable (by any computing device). For our purposes, we will consider a class of simple Turing machines in which there is a potentially infinite tape containing data, a read/write head, and a controller that oversees the operation of the machine. The controller specifies tape movement operations, read operations that sense the contents of the tape, write operations that modify the tape, and termination conditions for the calculation.

The symbols on the tape are chosen from a specified alphabet of possible symbols. The minimum alphabet consists of {01B}. The one and zero symbols are used to encode data, and the B symbol is a blank used as a separator. The controller for the Turing machine is a finite-state machine where transitions between the states are determined by the original input data or data read from the tape. The controller can be characterized by its number of states, with a starting state and a final state being required. Thus, the simplest Turing machine consists of a two-state controller and a two-element alphabet (every alphabet must have a separator, so the B is not counted). Such a simple machine cannot perform interesting calculations, but by adding alphabet symbols and/or additional states, a Turing machine can be constructed for any function.

In the remainder of this section, we will develop four formal concepts which parallel each of the four intuitive concepts explored in the previous section:

1. Turing machines are a formal model of computing a function.
2. Turing machine size is based on alphabet size and number of states.
3. Complexity is defined by the size of the smallest Turing machine for the function.
4. Relative structure is measured by comparing reduced Turing machine sizes.

In drawing a parallel between biological systems, mathematical functions, and Turing machines, we do not intend to suggest that biological systems are actually implemented as Turing machines. Rather, we intend to take advantage of the constrained formalism for the functions and machines to explore issues of structure and organization which would be too complicated to explore with full biological details in tow.

3.1 PROBLEM SIZE IS A FUNCTION OF STATES AND ALPHABET

The so-called Universal Turing machine is a Turing machine that reads a description of another Turing machine on its tape and simulates the machine described on the tape. When the Universal Turing machine starts, the tape contains a state transition table and initial data for the machine being simulated. A Universal Turing machine can be thought of as a machine that interpretively executes a "program" stored on the tape.

The size of a problem can be defined in terms of a Turing machine which exhaustively enumerates the solution as either a look-up table or a hard-coded machine to solve the problem for each input. We will associate the size of the problem and, thus, minimum structure with the length of the state transition table and the initial data on the tape of a Universal Turing machine that emulates the Turing machine for solving the problem. Such an exhaustive enumeration, whether encoded as an input symbol/output symbol pair in a table on the tape or in the state transition behavior of the controller, makes the minimum use of any structure which may be available in the solution in order to reduce the size of the enumeration.

The simplest way to think about problem size is to use the size of a look-up table stored on the tape and a simple two state controller that merely searches the tape for a single input character. In such a system, each distinct input combination and output result will be encoded in one symbol pair on the tape. The size of the maximum irredundant alphabet needed to completely encode the function as a look-up table on the tape in this way will, thus, define the size of the problem.

The number of states of the controller can also be used to describe the size of the problem. The result is not quite as intuitively clear, but is actually functionally identical to encoding the function in a look-up table on the tape. If the tape alphabet is minimal, then the maximal information about the function must be encoded in the state behavior for the controller. When there is exactly one output state for each input combination, and otherwise there is a minimal number of states, then the resulting number of states determine the size of the problem in much the same way that the size of the alphabet determined the size before.

For any biological system, the number of control states corresponds to the (usually genetically encoded) control machinery and the size of the alphabet corresponds to the variety of chemical building blocks in the immediate vicinity of the control machinery. In general, it may be possible to identify states and alphabet symbols in more than one way for a given system. For example, for a set of proteins, each protein might be identified as an alphabet symbol, or each of the 20 amino acids might be an alphabet symbol, or each of four nucleotides could be used. In this example, as the size of the alphabet is reduced, additional processing machinery is required to assemble strings drawn from smaller alphabets in order to encode the full set of proteins. In these cases, there may be interesting trade-offs between extent of control machinery and the variety of molecular building blocks that must be maintained.

Figure 1 shows the rectangle describing the size of an arbitrary function in terms of either the maximum irredundant number of states or alphabet symbols. The precise size of the function depends on the details of what the function computes, but the overall shape will always be similar to Figure 1. The curve linking two corners of the size rectangle corresponds to the available trade-off of states vs. alphabet size for the function in question. The curve defines the maximum irredundant number of states and alphabet symbols for computing the function. The region near the origin bounded by the curve corresponds to the set of Turing machines which cannot compute the function due to a lack of states, alphabet size, or both. The region outside the curve-bounded region near the origin corresponds to redundancy—the system has more than the maximally irredundant alphabet and/or set of states.

We have already defined the degree of structure to be minimal when the solution is expressed in a maximally irredundant form. We now extend that definition to say that all Turing machines in Figure 1 which can compute the function have a minimal degree of structure. Those machines which correspond to the curved line have a maximally irredundant combination of states and alphabet. Those machines off of the line in the unshaded area correspond to machines with minimal structure and either redundant states, redundant alphabet symbols, or both.

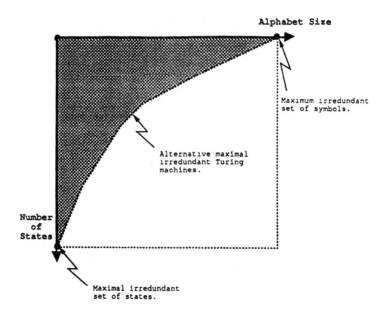

FIGURE 1 Problem size is defined as the area of the rectangle formed by the product of the maximum irredundant alphabet and the maximum irredundant set of states for Turing Machines that implement the solution to the problem. In this plane, the region between the curve and the origin contains machines that are not redundant and cannot solve the problem (this plan is the bottom plane of Figure 3).

3.2 PROBLEM COMPLEXITY MEASURES THE SIZE OF THE IRREDUCIBLE TURING MACHINE

The Kolmogorov complexity of a function, mentioned above, is usually defined as the size of the smallest Turing machine (over the minimal alphabet) that computes the function (see Li and Vitanyi[7] for a survey). We have used the number of states, S, for the size of a machine or for the size of the solution it generates. Thus, the Kolmogorov complexity is defined by the most concisely described controller for the smallest possible alphabet.

It will be useful to extend the definition of Kolmogorov complexity to alphabets that are not the smallest possible. The Kolmogorov complexity is essentially a measure of the smallest representation of the solution to the problem counting both alphabet and control states. In order to achieve the smallest statement of the solution, a maximum amount of structure must be used. As was the case before, it is possible to trade off the number of states and alphabet size without altering the actual amount of structure in the solution. Such solutions are equivalently concise

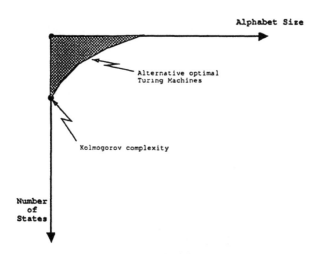

FIGURE 2   Problem complexity as defined by the Turing Machine with a minimum alphabet and a minimum number of states. The region between the curve and the origin corresponds to Turing Machines too small to solve the problem (this plane is the top plane in Figure 3).

and by our definition are equally (and maximally) structured. Thus, in Figure 2, our complexity measure traces out a curve in the plane of states vs. alphabet size which intersects the structure axis at the Kolmogorov complexity. This plane represents a plane of maximal structure. As was the case in Figure 1, the region near the origin bounded by the curve is unreachable and corresponds to Turing machines which cannot compute the function in question. Again, the region beyond the minimal alphabet and states corresponds to redundancy.

3.3 STRUCTURE IS A PROPERTY OF THE TURING MACHINE NOT THE PROBLEM

What remains in defining the degree of structure is to specify the behavior of the measure between the two end points of the scale already established. In our definitions thus far, we have essentially defined any group of Turing machines (and therefore the solutions they represent) as equally structured if they are equally close to the Kolmogorov complexity for the problem at hand. Two alternative representations of the solution which differ only in the relative mix of alphabet size and states define a plane of equal degree of structure like those in Figures 1 and 2. The plane in Figure 1 describes the alternative ways to build a Turing machine to solve the problem by using no structure to encode an exhaustive solution to the problem.

By gradually using more structure, we pass from this exhaustive (therefore maximally irredundant) representation for the solution of the function to the minimally irredundant description of the function as defined by the Kolmogorov complexity.

The degree of structure defines a third axis along which two solutions to the same problem can be compared. We call the resulting 3-space a structure space. We can represent both a problem and its solution as Turing machines corresponding to a point in this space. Figure 3 shows the region between Figures 1 and 2 in terms

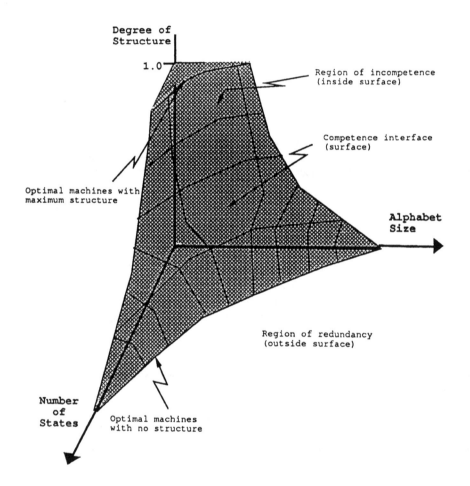

FIGURE 3 The relationship between problem complexity and problem size. If the problem is structured, the complexity is different than the size. Optimal machines with a given amount of the possible structure lie on a curve on the shaded surface. Machines under the surface cannot solve the problem.

of the optimal (irredundant) machines for solving a given problem. We can define the degree of structure of a solution as K/S, where S is the size of the solution (ignoring redundancy) and K is the Kolmogorov complexity (the size of the optimal solution). Combining the degree of structure axis with the previous axes (number of states and alphabet size) defines a space of possible structures for solutions.

3.4 PUTTING THE FORMAL NOTIONS TO WORK: A SIMPLE EXAMPLE

As an example, recall the simple function from the previous section:

$$F(x) = 2x^2 + 2x - 4. \tag{3}$$

The tabular version of the function could simply be encoded on the (potentially infinite) tape as pairs of values. The size of the alphabet required is the size of the set of input symbols. If the variable x is allowed to range over all the integers, then the size of the function is infinite. The Turing machine with its infinite tape can represent such function, but the more structured version in Eq. (3) is clearly more compact and uses fewer alphabet symbols.

While Eq. (3) above uses fewer alphabet symbols to represent the function, it uses more complex control machinery in order to capture the structure in the solution. To illustrate this point, consider the following code for the control behavior of the Turing machine which implements Eq. (3):

Step	Command	Comment
1	Push 2	Push first coefficient onto stack
2	Push x	Prepare to square x
3	Push 2	Push power of x onto stack
4	power	Compute x to the power 2
5	*	Multiply the top stack entries ($2*x$ power 2)
6	Push 2	Push second coefficient onto stack
7	Push x	Prepare to calculate $2x$
8	*	Calculate $2x$ and leave result on stack top
9	+	Add the first two terms and leave result top
10	Push -4	Push last coefficient onto stack
11	+	Add top stack entries and leave result on stack

where the Turing machine's tape will be used to hold the (potentially infinite) stack. The stack machine code above is similar to that used in some calculators and constitutes a high-level description of the Turing machine's state behavior. In this example the alphabet consists of the symbols for $\{\text{Push}, x, 2, \text{power}, *, -4, +\}$ for an alphabet of seven symbols and there are 11 steps in our description of the control machinery.

Now consider the factored form of Eq. (3):

$$F(x) = (2x + 4)(x - 1) \tag{4}$$

which can also be encoded for our Turing machine-based stack machine:

Step	Command	Comment
1	Push 2	Push first coefficient onto stack
2	Push x	Prepare to calculate $2x$
3	*	
4	Push 4	Prepare to calculate $(2x+4)$
5	+	Leave $(2x+4)$ on stack top
6	Push x	Prepare to calculate $(x$-$1)$
7	Push -1	
8	+	Leave $(x$-$1)$ on stack top
9	*	Calculate result and leave on stack top

where the alphabet contains $\{\text{Push}, *, 2, x, 4, -1, +\}$ or seven symbols as before. In this case, there are nine steps in the controller for the machine. According to our definition of structure, Eq. (4) above has a more concise machine which implements it and, thus, it reflects more structure than Eq. (3). Both of these formulas reflect more structure than an exhaustively enumerative look-up table.

In this section, we have refined the four intuitive concepts from the previous section:

1. A universal Turing machine is a model of a function which is, in turn, a model of a biological system solving a problem or meeting external constraints.
2. The size of the machine description on the universal Turing machine tape (states and alphabet) determines the size of the machine and, thus, the size of the problem.
3. The Kolmogorov complexity is the size of the minimal Turing machine that computes the function. This index of complexity defines the maximally structured set of machines.
4. Relative structure between exhaustive enumeration and the Kolmogorov complexity can be measured by comparing sizes of machines.

In the sections that follow, we will introduce the notion of finiteness limits, explore properties of the space of possible structures, and, finally, relate the theoretical results back to biological systems.

4. FINITENESS CAN PROMOTE THE EMERGENCE OF STRUCTURE

We identify a particular adaptive system with a single point in the structure space, i.e., a particular alphabet size, number of states, and degree of structure. Such an adaptive system must have the ability to move to an alternate point in structure space in order to survive. If the system moves outward at a constant level of structure, then adaptation introduces redundancy into the system. If the system moves along the boundary between incompetence and redundancy at a constant structure, then adaptation is just trading states and alphabet symbols. If the system moves inward at a constant level of structure, then adaptation is removing redundancy. Finally, adaptation may increase or decrease the amount of structure in the solution.

In this section, we will explore both static and dynamic properties of the structure space defined in the previous section. In particular we will explore the following points:

- Static properties of structure space influence the evolution of structure.
- Constrained regions of structure space limit structural options.
- Competence requirements and finiteness limits promote an increase in structure.
- Structure space is non-uniform. Dynamic properties of structure space govern adaptive change.
- Structure can arise from collisions with competence and finiteness boundaries.
- Adaptive systems tend to converge on "attractive" points in structure space.
- Modularity is a natural consequence of increasing structure.

In the remainder of this section, we will explore each of these six points in terms of our formalism. In the next section, we will revisit these points in terms of their biological equivalents.

4.1 STATIC PROPERTIES OF STRUCTURE SPACE INFLUENCE THE EVOLUTION OF STRUCTURE

While it is not possible to fully separate the static and dynamic properties of structure space, it is useful to explore the ways in which the shape of the structure space can play a role in defining structure for adaptive systems. In section 5, we will make a similar distinction when exploring the biological equivalents of the formalism below.

4.1.1 DEFINED REGIONS OF STRUCTURE SPACE LIMIT STRUCTURAL OPTIONS.
The space of possible structures contains a surface defined by possible trade-offs between minimal number of states and the corresponding minimal alphabet size for a given degree of structure. The surface is shaded in Figure 3 and intersects the planes in Figures 1 and 2 at their boundaries between shaded and unshaded areas. The surface defines an interface between a region of incompetence enclosing the origin and a region of redundancy which lies outside the surface. The region of incompetence corresponds to problem solvers (whether Turing machines or organisms) which lack either states or alphabet (or both) required to compute the solution. The surface itself defines the minimal problem solver for the given level of structure and a trade-off between states and alphabet at the given level of structure

which can compute the function. The region of redundancy extends arbitrarily far from the origin and corresponds to problem solvers that compute the function with unnecessary states and/or alphabet symbols.

The competence interface (the shaded surface in Figure 3) represents the minimal possible solutions to the given problem. Trading number of states for alphabet size moves us along the surface in a plane representing a fixed degree of structure. But, we may also change the amount of structure by changing S (the overall size, which is a function of both number of states and alphabet size). For instance, we may begin with a look-up table (maximum alphabet, minimum states) and reduce the alphabet by merging inputs that have common outputs. This moves us along the competence interface (in the degree of structure/alphabet size plane) until we reach the curve representing the Kolmogorov complexity of the problem. At this point, to continue reducing the alphabet we must add states, which moves us along the top curve in Figure 3 until we reach the minimum alphabet. Now adding states will reduce the degree of structure, which will move us down the competence interface to the opposite extreme from our starting point (minimum alphabet, maximum states).

We defined the region of redundancy as the portion of structure space extending outward from the competence interface. We now add another constraint, namely a finiteness limit, to the definition of the region of redundancy. A finiteness limit on states and/or alphabet is an externally imposed limitation on the size of the alphabet or the number of states in the finite-state machine controller. Finiteness limits essentially constrain the outer limits of the region of redundancy.

For problems with a small problem size, it is possible to actually construct a solution to the problem which is exhaustive. That is, the maximally irredundant number of states or alphabet symbols is obtainable. As the size of the problem grows, or the context in which the problem must be solved intrudes, it may no longer be possible to build the maximally irredundant machine to solve the problem.

We now refine our definition of the region of redundancy to only include the region between the competence interface and the finiteness limits. If the finiteness limits are beyond the problem size, then they limit only the available amount of redundancy. If, on the other hand, they are within the problem size, then they actually limit the allowable structures. In effect, they cut off the tails of the competence interface and define a minimum required structure. Whenever the competence interface and the finiteness limits intersect, we speak of a "collision" between the competence interface and finiteness limits.

4.1.2 IMPOSITION OF COMPETENCE REQUIREMENTS AND FINITENESS LIMITS PROMOTE STRUCTURE. The shape of the region of redundancy promotes increased structure. The competence interface is a surface (not necessarily continuous) composed of the minimal problem solvers with a given degree of structure for structures ranging from exhaustively enumerative to machines at the Kolmogorov complexity. As shown in Figure 4(a), when the distance between the competence interface and the finiteness limits is large, the shape of the competence interface plays little role in defining the portion of the space which is accessible.

As shown in Figure 4(b), when the competence interface and a finiteness limit intersect, then the shape of the competence interface will mean that machines with low structure will be excluded from the region of redundancy first. In a situation where the competence interface advances or the finiteness limit becomes more stringent, the alternative structures become more limited as the space between the competence interface and the finiteness limits is reduced. Clearly, the first alternatives to go are the regions near the boundaries. Due to the generally upward and inward shape of the competence interface, there will be more viable structures at higher levels of structure as the space is constrained. This relative preference for higher structures as competence and finiteness collide provides a static preference for increased structure.

4.1.3 STRUCTURE SPACE IS NONUNIFORM. The region of redundancy has a nonuniform density of points and numbers of alternative problem solvers. In the general case, there may be several distinct problem solvers with the same number of states, alphabet symbols, and degree of structure. In this case, we speak of a point in structure space as a set of machines. The region of incompetence described above corresponds to a region where the set of machines that solve the function is empty due to a lack of states or alphabet symbols. In a similar way, the finiteness limits bound regions where there are no available machines that both solve the function and meet the finiteness limits. On the competence interface and within the region of redundancy, there will generally be a number of machines at each structure point that solve the problem, although in potentially different ways. As an example, consider that, subject to a finiteness limit on number of states, any number of pairs of instructions such as:

Step	Command	Comment
n	Push A	Waste time Pushing arbitrary value
$n+1$	Pop	Pop back off stack, so no lasting effect

could be inserted between any two instructions in the examples in Section 3. Notice that a single push or pop operation could NOT be inserted without causing a distortion in the calculation of the function. Thus, not all potential forms of redundancy are compatible with continuing to solve the function. In our example, paired Push/Pop operations would mean that only redundant points that differ by pairs of redundant instructions would solve the function. Points that would have an odd number of redundant instructions would generally not compute the function and, thus, the region of redundancy would contain a nonuniform distribution of numbers of states.

Clearly, points along the competence interface and in the region of redundancy differ in other ways as well. They may require differing amounts of time to compute a result, they may compute some results faster than others, and they may consume varying amounts of the Turing machine's tape in performing the calculation. We will have more to say about these points in the next section when we discuss biological interpretations of the formalism.

(a)   (b)

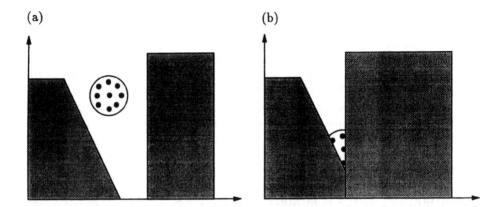

FIGURE 4   (a) A cohort of nine organisms located centrally in the region of redundancy away from both the shaded competence interface and finiteness limits. Such a cohort of organisms will be within structure space due to local variations in structure space but will not be influenced by the boundaries. (b) A cohort of individuals which has been caught in a collision between the sloping competence interface and the vertical fitness limit. Because of the shape of structure space in the region of the cohort, there will be a bias for increased structure which will tend to drive the cohort to higher structure in order to survive.

4.2 STRUCTURAL ADAPTATION IS DETERMINED BY THE DYNAMICS OF MOVEMENT IN STRUCTURE SPACE

Complex adaptive systems such as biological organisms "explore" the structure space as they adapt. Thus, a structure point in our formalism may undergo change as a result of a structural reorganization. When the reorganization results in a different number of alphabet symbols, states in the controller, or degree of structure, then the adaptation results in a change from one point in structure space to another. We will argue below that this exploration is essentially a random process which continually results in small structural changes.

In addition, the boundaries in structure space are not stable. The competence interface and the finiteness limits may vary with time. The competence interface can vary due to changes in the environment of the system or due to actions by the system itself. For example, an organism which seeks to adopt a new environmental niche will usually be changing the competence interface from that defined by the current niche to that defined by the new niche. Alternately, environmental constraints can change finiteness limits as in the case of depletion of a nutrient from the environment (thereby reducing the available alphabet).

A cohort of individuals defines a cluster in structure space. In order to explore the dynamics of movement in structure space, we need to introduce the concept of a "cohort" of adaptive systems. By cohort, we mean a group of adaptive systems each undergoing adaptive change according to the dynamics we will explore below. The members of the cohort may represent multiple copies of identically the same machine, alternative structures within a single point in structure space, or a cluster of points in structure space, or any combination of all three.

A cohort of adaptive systems is taken to be solving essentially the same problem, but with potentially different degrees of structure or redundancy. The cohort is said to have a "centroid," a single point in structure space, which is simply the frequency weighted average position of the group. The centroid is essentially the "center of gravity" of the group. Rather than trying to track the dynamics of structure for a single adaptive system, we will track the centroid for a cohort in the remainder of this section.

4.2.1 STRUCTURE CAN ARISE FROM COLLISIONS WITH BOUNDARIES. The discussion of local dynamics for movement in structure space above did not involve finiteness limits. Figures 5(a) and 5(b) illustrate the collision between a cohort of machines and a finiteness limit and a competence interface respectively. In each case, some of the potential structure points are precluded by the boundary. Notice that due to the shape of the competence interface (up and in toward the origin), it is more likely that structure points with low structure will be precluded by the

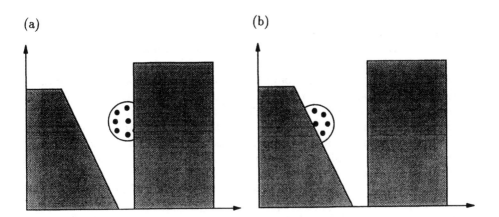

FIGURE 5 (a) A cohort of individuals colliding with a finiteness interface. The cohort will tend to move away from the competence interface and back into the region of redundancy. (b) A cohort of individuals colliding with an expanding competence interface. Because of the shape of the competence interface, the cohort will tend to move off the competence interface and out into the region of redundancy. Due to the fact that there are more structural alternatives at higher levels of structure, the cohort will be slightly more likely to be up in structure than down.

boundary than higher structure points. In the collision between the cohort and the finiteness limit (Figure 5(a)), structural alternatives with larger numbers of states and alphabet symbols are precluded by the finiteness limit. Such a restriction on states and alphabet, if stringent enough, will require increased structure in order to continue to solve the function. In both these situations, the boundaries can create a preference for increased structure.

We can make the intuitive notions above more precise using the notion of a centroid for both the cohort of organisms and a region of space. We defined the centroid of a cohort of organisms above as a weighted average of points in structure space, weighted by the number of organisms from the cohort at the given point. This weighting produces the center of gravity of the cohort. By performing a similar calculation, we can determine if the "shape" of the structure space itself provides a pressure for change of structure.

Instead of weighting points in structure space by number of organisms present, we can weight points by the number of different machines possible at each point. If we perform this calculation for the region of structure space containing the cohort, we can again calculate a centroid. In this case, it is a centroid of structural alternatives. If the centroid of structural alternatives is different from the centroid of the cohort of organisms, then the cohort will tend to move toward that centroid because there are more structural alternatives for the cohort there. Thus, we can now see that systematically chopping off structure points (as happens with competence and finiteness limits) will distort the centroid of structure space in the neighborhood of the limits.

Using a similar line of reasoning, unstable boundaries can be shown to repel organism points. When a competence interface or a finiteness limit changes with time, organism points near the boundaries must either develop extra structure to allow them to arrest growth and development until the boundary recedes or the organisms must reorganize so as to move away from the boundary. Figure 6 illustrates the trade-off in cost of responding to a changing competence interface vs. the cost of maintaining states and alphabet not essential to the task at hand (redundancy). As can be seen from the graph, there is an optimal region away from the competence interface where the problem solver is poised to use otherwise redundant machinery to solve the new problem(s) presented by the expanding competence interface. A similar analysis of the collision between a cohort and a finiteness limit would show the finiteness limit precluding some structures for periods of time when the finiteness limit is unstable (as is the case when marsh land is periodically dry and inundated). Again, either the organism must evolve structure to allow it to arrest growth and development until the finiteness limits are relaxed or the organisms must reorganize in such a way as to reduce resource requirements.

**4.2.2 ADAPTIVE SYSTEMS WILL TEND TO CONVERGE ON "ATTRACTIVE" STRUC-
TURES.** As pointed out in section 3.1 above, structure space is nonuniform. Figure 7
shows a cohort of adaptive systems in a central region in the space. If each of the
alternative points in structure space contain an equal number of actual machines,
then the centroid for the cohort will be the point in the center of the group in
Figure 7. As discussed in the previous section, we can now compute the center of
gravity of a section of structure space by essentially taking a weighted average of
the structural alternatives, weighted by the number of alternative problem solvers
within each structure point. If the number of potential points in structure space
is also equal at each point, then the centroid of the region of structure space will
coincide with the centroid of the cohort of machines.

If the number of alternatives is unequal at different points in the structure
space, then the centroid of the available structural alternatives in a region of structure space will not always be the centroid of the area of structure space. Due to
nonuniform distributions of actual machines in a cohort and potential machines in
the same region of structure space, it is possible for the centroid of the cohort to

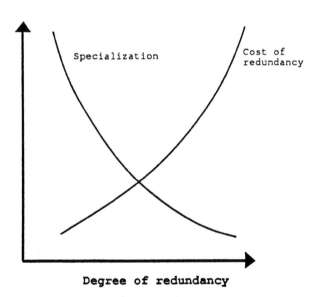

FIGURE 6 The relationship between cost of redundancy and specialization. The
degree of redundancy is the distance from the shaded surface in Figure 3. As
degree of redundancy increases, the amount of specialization decreases and so
the cost of conforming to change (weighted by the probability of change occurring)
decreases. Conversely, as degree of redundancy increases, the cost of maintaining that
redundancy increases. Organisims should settle where the two cost lines intersect.

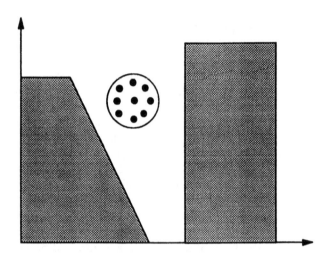

FIGURE 7 A cohort of individuals located centrally in the region of redundancy. The central point within the group coincides with the centroid of the cohort when each black dot represents the same number of organisms. The centroid of structure space in the same region is calculated by using available structure points rather than individual organisms.

be different from the centroid of the available machines. All other factors being equal, we hypothesize that a cohort of adaptive systems will gradually converge on the centroid of available machines when these two are different. We term this convergence based on availability.

In the world of Turing machines as well as the real world of biology, not all structural transformations are equally possible and not all structure points are equally easy to reach. Even for two different machines within a single point in structure space, neighboring points may not be equally accessible. (This differential attractiveness is equivalent to the notion of attractors in dynamical systems and differential accessibility corresponds to movement along a given trajectory in structure space.) Thus, we must now expand our calculation of the centroid of structural alternatives to include the weighing based on the probability of a transformation as well as number of available alternatives. Structure points that are "difficult" to reach will get a lower weight when computing the centroid of the alternative machines. Machines that are more adaptable will have more nearby machines to which structural transformations are easy. We term this convergence based on accessibility.

In a similar way, not all machines are equally efficient or reliable. Again, we generalize the weighing function from a simple count weighted by accessibility to one weighted by performance criteria as well. Unlike the distortions of the uniformity of structure space above, a particular performance criterion will have little effect without selection pressure for it. We term this convergence based on performance.

The three types of convergence presented above—availability, accessibility, and performance—characterize the operation of two nested dynamical systems—organisms, and the composite of organisms and their environment. Availability captures the notion that not all structures of organisms are possible. Accessibility captures the notion that, among the possible structures, only a subset is accessible from any given one. Performance captures the notion that different structures meet performance criteria associated with the environment to differing degrees.

Using the qualitative analyses above, we can predict patterns of movement in structure space away from the competence and finiteness boundaries. We have assumed that small-scale structural alteration is a constant property of adaptive systems. In this way, a cohort of organisms will "explore" a nearby region of structure space and tend to move toward "attractive" points within the space. Absent pressures from limits, the probability of transition from one structure point to another can be assumed to fall off with the extent of the structural reorganization required. In the region of redundancy, we see that structure space is "warped," i.e., it contains regions of nonuniform attractiveness. A cohort of organisms will tend to converge on an area in structure space which is analogous to a gravitational well or an attractor in dynamical systems terms.

In quantifying the pressure to move within the region of redundancy, we define a "basket of criteria" which can be used to evaluate each point in the structure space and determine its attractiveness. By averaging over adjacent points, we can again compute a centroid as described above. The important thing to note about the basket of criteria is that there are no universal weights for the components: number of structural alternatives allowed by material properties (availability), self-organizing alternatives to which an adaptive system is likely to converge (accessibility), and better performing systems which can provide a selection advantage (performance). In our notion of the basket, the relative weights of each member of the basket can only be computed once the details of the space are known. Thus, it is not possible to explain the general patterns of structure that we see without including the impact of each element of the basket of criteria with its own situation-specific weight. This notion of a basket of criteria is a generalization of the notion of a balance of structural stability and adaptability note by Crutchfield.[3] In this context, Crutchfield observes that: "The balance itself is not static: the driving force underlying dynamic creation of complexity derives from an individual's limited metabolic, computational, and observational resources." We will return to this concept in section 5.2.2 below.

4.2.3 MODULARITY IS A NATURAL CONSEQUENCE OF INCREASING STRUCTURE.

For a truly random function, there is no simpler version of the function than a lookup table. Thus, no compression in the solution is possible; the size of the problem and the intrinsic complexity are equal. For such a system, we would say that the degree of structure is minimal and that it represents the maximum possible degree of structure.

In order to produce a more concise, and by our definition more structured, solution to a problem, we must make use of commonality within the solution to the function. As pointed out above, where there is no commonality, there is no compression, and there is no structure other than exhaustive enumeration available. Where multiple inputs can be mapped to a single output or the overall problem can be decomposed into a collection of reusable components, then the overall size of the solution can be reduced by capitalizing on this structure in the solution. In this way, modularity should be expected to emerge in situations where there is pressure for a more concise solution (such as the collision between a competence interface and a finiteness limit).

Our definition of degree of structure has the pleasing property that it provides a more precise definition for the intuitive relationship between levels of organization and structure.[8] Mittenthal et al. have pointed out that structure in biological systems often takes the form of levels of regulation. The curve in structure space traced by reducing alphabet and climbing the competence interface corresponds to adding levels of organization to a system; up to a certain point, this will increase the amount of structure (that point being where the number of levels matches the problem's intrinsic complexity), after which adding levels will decrease the amount of structure by proceeding to an exhaustive enumeration of the function using states. In the next section, we illustrate these intuitive notions about the relationship between structural levels and modularity with a concrete example.

4.3 PUTTING THE FORMAL PROPERTIES OF STRUCTURE SPACE TO WORK: A CONTINUED EXAMPLE

In the algebraic example of evaluating a quadratic, we identify the competence interface with the ability to solve the function with the minimal number of states (program steps in our example) and the minimum alphabet for a given level of structure. We can increase the alphabet by, for instance, adding additional constants other than 1, 2, and 4 already used. Alternately, we can introduce redundancy as was done in section 4.1.3 by adding a sequence of states (program steps) that add bulk to the machine but do not fundamentally change its operation.

Recall that Eq. (3) above required 11 steps in the description of the control machinery and Eq. (4) required nine steps. As long as the finiteness limit on steps in the program is larger than 11, both structures are allowed. If the finiteness limit is reduced to ten steps, then the 11-step solution is no longer viable and only the nine-step solution will suffice. In this case, the competence interface is defined by the overall shape of the function to be solved (a quadratic polynomial) and the finiteness limit could be imposed on the number of program steps (equivalent to states) or the alphabet of symbols.

We now generalize this example to include the notion of modularity. If we generalize the specific problem from the preceding sections to a generalized polynomial, we obtain a problem of the form:

$$Ax^2 + Bx + C \tag{5}$$

which can also be expressed as:

$$(Dx + E)(Fx + G) \qquad (6)$$

where each capital letter corresponds to some constant. The general version of the problem in Eq. (5) requires essentially the same 11 steps as in the previous version with explicit constants. By inspecting a traditional computer compiler parse tree for Eq. (6) or by just by inspecting the formula itself, it is possible to identify two instances of one module—a constant times x plus another constant. The following simple "program" illustrates the definition of the "module" and its use to solve the function:

Step	Command		Comment
1	Module	Factor	Begin definition of reusable module
2	Push	x	x is common to all terms
3	*		Multiply x by is coefficient
4	+		Add constant to "x" term
5	EndModule		Close up the module definition
6	Push	E	Prepare constant for use by module
7	Push	D	Prepare coefficient of x for module
8	Call	Factor	Invoke module
9	Push	G	Prepare constant for use by module
10	Push	F	Prepare coefficient of x for module
11	Call	Factor	Invoke module
12	*		Multiple the two module-derived terms

where the generalized version of the problem requires 12 steps for its solution. It is rather common that for simple versions of a problem, the more structured, i.e., more modular, version will require more steps. As the problem becomes more complex, the size of the modular version will become significantly smaller than the non-modular version. The modular version of an nth-order polynomial, which in general has $n + 1$ terms, is a product of n factors, or modules. It requires six steps to evaluate any nonlinear term; for example, Ax^n requires Push A, Push x, Push n, Power, *, and then a final + to add the term onto the other terms. In the modular version, an arbitrary factor requires four or fewer steps: two Push commands, a Call, and a final *. Thus, as the degree of the polynomial increases, the difference in number of steps to evaluate the modular and non-modular versions increases. The modular version simply recycles machinery it already has while the less structured version must add a whole set of machinery for each different power of x encountered.

TABLE 1

Input abc	Output abc
0 0 0	0 0 1
0 0 1	0 1 0
0 1 0	0 1 1
0 1 1	1 0 0
1 0 0	1 0 1
1 0 1	1 1 0
1 1 0	1 1 1
1 1 1	0 0 0

4.4 STRUCTURE SPACE AND BOOLEAN NETWORKS: A SIMPLE EXAMPLE

Boolean networks have been explored as a model for biological systems.[4,5] By analyzing these networks, Kauffman has presented a strong case that biological systems live "on the edge of chaos." In the remainder of this section, we will apply our formalism to families of boolean networks and explore the relationship between finiteness and the edge of chaos.

We first associate solving a function with presenting one set of inputs to the system and determining the output from the next state of the system. In effect, the inputs select one state and the state transition table for the network determines the next state (and thus the output). For example, consider a simple three-digit binary counter. This counter counts from zero up to seven and then rolls over back to zero. The behavior of the counter can be summarized in Table 1 where each of the eight rows in the table above corresponds to a situation and its resulting output.

The size of the function is eight because there are eight rows. The size of the alphabet for the problem is clearly eight because there are eight distinct input combinations. As is always the case, we may trade alphabet symbols or states without affecting the amount of structure. In this case, we may assume that we have an alphabet size of eight in which case we need only one eight-valued input state and one such output state. In the more traditional binary interpretation, we would have an alphabet of two symbols but now we need three states in order to encode the function.

TABLE 2

Input abc	Output abc
0 0 0	0 0 1
0 0 1	0 1 0
0 1 0	0 1 1
0 1 1	1 0 0
1 0 0	1 0 1
1 0 1	1 1 0
1 1 0	1 1 1
1 1 1	0 0 0

TABLE 3

Input abc	Output abc
0 0 0	0 0 0
0 0 1	0 1 0
0 1 0	0 0 1
0 1 1	1 1 1
1 0 0	0 1 1
1 0 1	0 1 1
1 1 0	0 1 1
1 1 1	1 1 1

By inspection of the table, we see that no two outputs are the same. This corresponds to a function where there is no opportunity to use structure to reduce the size of the table. As we discussed in section 2 above, this corresponds to the relatively uninteresting situation where the size of the function and the minimum length encoding of the function are the same. In this situation, the most structured version of the function is simply the exhaustive enumeration of the function.

The finite state machine which implements the counter above is a simple ring composed of a state for each counter value, where a state is associated with a row in Table 1. In Kauffman's terms, this corresponds to a boolean system with only one circuit which is of maximal length. A single traversal of the circuit passes through each of the states and no other circuits are possible. Kauffman characterizes networks by the number of inputs to each state required to determine the next state of the system. For our simple counter, this parameter, K, is three (the maximum possible value) because the next state of the counter can only be determined by looking at all three inputs at each step in the counting process.

As we have already observed, structure in a solution is possible when multiple inputs give rise to a common output. In the simple boolean function shown in Table 2, there are only five distinct output combinations where K is still three, but now the function can be expressed in a more compact form by capturing the structure which is present in the solution. For instance, the following list of five rules summarizes the table in a more structured way:

If b=1 and c=1	then 1 1 1
Else if a=1	then 0 1 1
Else if c=1	then 0 1 0
Else if b=1	then 0 0 1
Else	0 0 0

where the rules must be checked in sequence. This replacement of a complete tabular representation with a more compact rule-based representation is similar to the multiplication table example in section 2.

Now consider an example also used by Kauffman[4] in which the output "a" is given by the AND of inputs "b" and "c"; outputs "b" and "c" are given by the OR of the inputs "b" and "c." This simple network has the tabular representation in Table 3.

Clearly, in this case K is two because each output only depends on two inputs. In this situation, we require three states at a minimum to encode the outputs but we only require an alphabet of at most four symbols. (One symbol suffices for each

TABLE 4

Input abc	Output abc
0 0 0	0 0 0
0 0 1	0 1 1
0 1 0	0 1 1
0 1 1	1 1 1
1 0 0	0 1 1
1 0 1	0 1 1
1 1 0	0 1 1
1 1 1	1 1 1

of the four input combinations based on B and C.) In fact, if we are fortunate enough to have the proper encoding machinery already, we can optimally code this function using three input symbols (one for "b"=0 and "c"=0, one for "b"=1 and "c"=1, and one for other) and the three binary states "a," "b," and "c."

In the examples above, we have discussed degree of structure of various functions and their solutions. We have not discussed the imposition of finiteness limits. In Kauffman's analysis of boolean networks more complicated than those shown here, he has identified three very interesting types of networks. For small values of the connectivity K, Kauffman notes that the networks have rigid behavior. That is, small perturbations to the network do not really change its behavior. At the other extreme, highly interconnected networks (with large K) have state cycles which tend to be long and extremely sensitive to initial conditions. Kauffman suggests that a value of K between two and three represents an optimal trade-off for boolean networks between the "frozen" behavior at $K = 1$ and the "chaotic" behavior for large K.

A limit on K corresponds to a limit on the size of the alphabet (in this case the alphabet is $2k$). Thus, we have a reason to speculate that a limit on alphabet size may be inherent in adaptive networks. If we follow this speculative line further, we arrive at the conclusion that, as we increase the number of states in the system, we cannot increase the connectivity of the network (and thus the alphabet) at the same rate. From this analysis, we propose that for complex systems the finiteness limit on alphabet size will generally be more restrictive than a finiteness limit on number of states of the control machinery.

Summarizing the material presented in this section, we have explored the following points:

- Static properties of structure space influence structure.
- Regions of competence, incompetence, and redundancy limit structural options.
- The shape of the region of redundancy promotes increased structure by proscribing more simple structures than complicated ones.
- Points in the region of redundancy correspond to different numbers of machines with differing internal compositions but equivalent levels of structure.
- Dynamic properties of structure space govern adaptive change.
- Structure can arise from collisions with competence and finiteness boundaries by proscribing simpler alternatives.
- Adaptive systems will tend to converge on "attractive" points in structure space based on number of alternatives, stability, and accessibility of alternatives.
- Increasing levels of organization and modularization naturally arise from the need to meet finiteness limits on control machinery (states) or variety of building blocks (alphabet) or both.

In the next section, we will explore many of these aspects of the emergence of structure in biological contexts which parallel the formalism presented in this section.

---

5. EXTENDING THE MODEL TO COVER GROUPS OF ORGANISMS AND LINEAGES

Armed with the formal system defined in the preceding section, we can see that finiteness limits and the shape of the competence interface constrain where and how structure will appear. Though our formalism has been defined in terms of Turing machines, we shall use it to discuss paths of evolution for lineages of organisms. This shift of focus is plausible in that Turing machines and organisms generate outputs that depend on the state of the system and on inputs. (Recall that Turing machines can calculate an input-output relation of arbitrary complexity.) While we believe that our formalism applies to entire organisms, it is easier to construct examples of activities within organisms, such as the synthesis of protein or RNA, rather than to analyze entire organisms. In order to construct more concrete examples, we will generally confine ourselves to these simpler examples.

In this section, we will explore the following points:

- Static properties of alternative biological structures influence what structures evolve.
- Incompetence, competence, redundancy and finiteness have biological equivalents.
- The distribution of competent structures can promote increased structure.
- Alternative biological structures are not uniformly available. The dynamics of variation and selection govern adaptive change.
- Structure can arise from collisions with competence and finiteness boundaries.
- Adaptive systems will converge on "attractive" biological structures.
- Finiteness limits foster levels of organization and modularity that match constraints within the problem.

5.1 STATIC PROPERTIES OF ALTERNATIVE BIOLOGICAL STRUCTURES INFLUENCE WHAT STRUCTURES EVOLVE

A point in structure space represents the variety of building blocks used (alphabet size), the size of the control machinery (number of states), and the structure of a type of organism; this will be called an organism point. A lineage of identical organisms would all correspond to the same organism point. A cluster of organism points represents a population of organisms. As an organism persists, fluctuations at the molecular level change its structure. The heritable changes range from point mutations to recombination of chromosomes that rearrange and restructure genes. Correspondingly, the organism points for a population diffuse in structure space.

Adjacent points in structure space differ in size of alphabet, number of states, and/or amount of structure. Variations at an organism point correspond to alternative choices of alphabet symbols and control states, without changes in their numbers. It is important to note that a small displacement in structure space can correspond either to very small or very large changes in the structure of an organism. In a protein, one amino acid can be substituted for another while its action is preserved. However, if the base sequence in a cis-regulatory element of a regulatory gene changes, the change can alter the use of that gene in development, initiating a cascade of consequences that markedly changes the structure of the adult.

5.1.1 INCOMPETENCE, COMPETENCE, REDUNDANCY, AND FINITENESS HAVE BIO-LOGICAL EQUIVALENTS. As an organism survives in an ecological niche, it meets certain externally imposed constraints. The constraints specify a competence interface in the structure space; only organisms with structures on the interface or in the region of redundancy beyond it can meet the constraints. During evolution the constraints can vary unpredictably in time; the position of the competence interface varies correspondingly. In fact, each organism in a population lives in a slightly different niche and solves slightly different problems. This variety of constraints corresponds to a family of moving competence interfaces. However, for convenience we will refer to individual organism points and individual competence interfaces in the following discussions.

An organism point that is above a competence interface in structure space solves a problem larger than the one defining the interface, or contains redundancy that does not directly contribute to solving the present problem. An organism for which the organism point is located above a competence interface can be regarded as preadapted, in that it is at least partially prepared to solve a problem that is a superset of the problem it is actually faced with.

It is important to keep in mind that each organism point is characterized by the number of alphabet symbols and the number of states—not by the particular ones used. The choice of which particular alphabet and state variables are used determines the detailed internal structure of the organism point. For example, there are many ways to adjust regulatory variables associated with protein synthesis. Choosing a particular mix of regulatory sites, messenger RNAs, and polypeptides would define a particular solution within an organism point. There may be several distinct ways to pick a mutually compatible set of such states and alphabet symbols. As long as the number of states and alphabet symbols are constant, the organism point does not change. To increase reliability or to respond to a wider variety of situations, an organism point might move out into the region of redundancy and utilize additional alternative regulatory sites, messenger RNAs, or polypeptides in order to make the same protein in a different way. In a similar way, duplicated (and therefore redundant) machinery can be used to increase the speed of synthesis for a needed quantity of synthesis product.

Finiteness limits can be identified in protein synthesis as well. In order to survive, an organisms must be able to make a variety of proteins that play essential roles in its inner workings. Two basic finiteness limits constrain these syntheses. Chemistry limits the alphabet, i.e., the variety of nucleotides and amino acids that can be polymerized to make nucleic acids and proteins. Natural fluctuations limit the reliability of the processes involved in replicating and expressing genes. Through fluctuations, errors can occur; e.g., inappropriate monomers are sometimes placed in position for bonding into an elongating polymer. Once formed, a polymer may be modified inappropriately by further reactions or may fold with an inappropriate tertiary structure. While error-correcting systems can repair some of these errors, others remain. These limits constrain the complexity of reliable control machinery (and thus the number of states) that can operate the process of synthesis.

5.1.2 THE DISTRIBUTION OF ALTERNATIVE STRUCTURES CAN PROMOTE INCREASED STRUCTURE. The alphabet and the number of states are discrete variables. Hence, the variables on which the degree of structure depends—the available pathways of chemical reaction and the number of molecules available for each reaction—are also discrete variables, although the variation may occur at such a fine scale as to approximate a continuum. The discrete character of the state variables implies that the competence interface is rough. Furthermore, its position varies unpredictably in time, though large variations are rare.

With increasing constraints, the structure space defined by the region of redundancy will decrease. If the competence interface exceeds the finiteness limits, the organism cannot persist. In some cases, it will be possible for an organism point to respond to an expansion of the competence interface by an increase in structure that moves the finiteness limit outward. An example of this is the incorporation of surface proteins in cell membranes. This additional structure allows the cell to respond specifically to a larger variety of environmental signals (inputs) than would otherwise be possible. The additional membrane proteins increase the limit on effective input alphabet, giving the cell more room in the region of redundancy to explore alternative structures.

Similarly, the addition of DNA repair mechanisms has increased the limit on genome length (the number of control states). Because repair mechanisms can allow more reliable replication of a larger genome, the organism may maintain additional redundant (possibly variant) control states, allowing it to cope with further increases in constraints.

At the level of nucleotides and amino acids, limits on the maximum alphabet size and the maximum number of states may have been reached early in evolution. One can imagine that at some point, the molecular machinery in cells was so highly organized that it could not tolerate the introduction of a new type of nucleotide or amino acid at random positions in nucleic acids and proteins. In this case the alphabet would be essentially frozen at the existing size. Improvements in the error-correcting processes may now be so costly or improbable as to preclude further increases in the maximum number of states.

If the finiteness limits are fixed, organism points near these limits can generally only move away from an expanding competence interface by increasing the degree of structure—moving upward in structure space, parallel to the axis for degree of structure. When the organism point moves up, it is located in the region of redundancy and has an increased (potential) capability through exploitation of increased structure.

5.1.3 ALTERNATIVE BIOLOGICAL STRUCTURES ARE NOT UNIFORMLY AVAILABLE.
As we saw previously in the case of abstract Turing machines, there is structure in the region of redundancy. Not all combinations of numbers of alphabet symbols and numbers of states are equally possible. The nonuniformity of structure space results most directly from the fundamental laws of physics and chemistry. Certain reactions will occur and others will not; some compounds are stable and others are not.

In a dynamical system the states are not all equally stable. An attractor is a point in structure space that is fundamentally different from the neighboring points in the space. The attractor is more accessible than neighboring points, in that almost any adaptive organism near it will converge on it.

The nonuniformity of structure space also results from generic properties of materials[9,10] and networks.[4,6,8] Natural selection can operate to optimize material properties or to stabilize a structure to better exploit pre-existing material properties such as surface tension or differential density. Generic properties of networks are beginning to emerge that operate to constrain structure in much the same way as dynamical system attractors. There appear to be favored patterns of connectivity or control which exploit a boundary between rigidly stable behavior and chaotic behavior. Again, the alternatives are not proscribed, but they sufficiently less likely to be useful that they can be considered to be unavailable. Finally, historical contingency operates to limit the availability of alternative structures. In our formalism, historical contingency establishes a trajectory through structure space and defines the particular organismal machine(s) that actually exist within a given structure point. Clearly, the availability of successor states with differing numbers of states, alphabet symbols, or structure depends on how the particular structures are related. Structural transformations that involve smaller numbers of changes in which alphabet symbols or states are used will be generally be more likely than transformations which involve wholesale replacement of the former structure with a new structure of similar size. We will return to some of these points in section 5.2.2 below.

5.2 IN AN EVOLVING SYSTEM, THE DYNAMICS OF STRUCTURAL ALTERNATIVES GOVERN ADAPTIVE CHANGE
Biological systems rarely, if ever, exhibit the precisely defined structure and reproducible behavior we associate with a fine Swiss watch. By their very nature, biological systems behave stochastically and embody an inherent imprecision. In our model we see biological systems exploiting this inherent stochastic behavior to experiment constantly with variations of structures

and processes. This variation is not a deficiency, but an integral component of adaptation. We have called this variation "noodling" in structure space. It is possible to identify structural noodling on physiological, developmental, and even evolutionary time scales.

The competence interface can vary due to changes in the requirements; that is, the definition of the problem to be solved has changed. These changes can result (among other things) from changes in the environment, competition among organisms, or adopting a new niche. In each case, an organism or lineage will be required to exist on the competence interface or in the region of redundancy in order to meet the imposed constraints. If a constraint increase does not move the competence interface beyond existing finiteness limits, an organism can cope by increasing its alphabet size, number of states, or its degree of structure.

The finiteness limits on alphabet size and number of states may also change with time. Each such limit corresponds to a plane parallel to the structure axis and perpendicular to the axis of the limit. Changes in the limit correspond to moving the plane along the limiting axis. These changes can correspond to the removal or introduction of a nutrient by the environment of the problem solver, to modification of an error rate by error-correcting machinery, or to a combination of change in both state and alphabet restrictions.

The alphabet itself may change as chemical building blocks are added to or deleted from the repertoire of molecules used as fundamental building blocks. Such changes may have occurred many times early in the evolution of life. For example, deoxyribonucleotides may have supplemented an earlier use of ribonucleotides (and amino acids?) as monomers for making polymers that transmit information about the sequence of monomers in other polymers. An increase in the alphabet might occur if the environment provides trace elements which are useful for such operations as oxygen transport, and if these elements were not previously used.

An increase in the number of states might occur as new genes are generated by recombination. For example, duplication of a gene followed by divergence of the two genes produces a kind of overlapping redundancy and augments the capabilities of the organism. The number of states may increase with the maximum number of genes that can reliably be maintained by processes of DNA repair. Thus, the number of states increases with the reliability of the processes that replicate, transcribe, and translate genetic information.[2] An increase in the degree of structure can meet a set of constraints with fewer genes (e.g., by reusing machinery described in a more modular fashion), leaving genes available for meeting additional constraints.

In the discussion above, we arrive at a picture in which all the features of structure space—the location of the point corresponding to an organism, the shape and location of the competence interface, and the limits on alphabet size and number of states—may change during evolution. To make use of this picture, we must understand the dynamics of processes through which these features evolve. Where will organism points tend to be located in structure space? What determines the directions in which the competence interface and organism points tend to move? Under what circumstances will the finiteness limits change? For these changes, what relative rates may we expect? In the remainder of the paper, we will begin to address these questions.

5.2.1 STRUCTURE CAN ARISE FROM COLLISIONS WITH COMPETENCE AND FINITE-NESS BOUNDARIES.

If a constraint on competence becomes chronically more severe, the competence interface will move farther from the origin, and the organisms near the interface will either be pushed away from the surface into what was the region of redundancy or they will not be able to meet the more stringent constraint. Our theoretical framework predicts that a lineage of organisms will respond to increasingly severe constraints by moving either out into the region of redundancy if finiteness limits allow or up to a more structured form. This can occur through an increase in the degree of structure or an increase in the number of states or alphabet symbols used. Changes in the competence interface which are relatively small but overall steadily increasing will provide continuous pressure for increased structure as finiteness limits are reached. Such a process is thought to have occurred as vertebrates evolved from wholly aquatic organisms (fishes) to partially aquatic organisms (amphibians) under the constraint of intermittent drying-up of ponds and streams. On the other hand, abrupt (and large) changes in finiteness limits or competence interfaces can spell extinction for a group of organism points overtaken by the boundary.

As the competence interface moves outward, an organism point that persists must remain away from it (Figure 6). Fluctuations in the structure of organisms may produce two kinds of changes that provide an organism point with space above the competence interface:

1. A change in the structure of the organism may change a finiteness limit, increasing the volume of the structure space. This makes additional space in the region of redundancy available, allowing the system to increase the alphabet size or number of states.
2. An organism point that has not reached the limits of alphabet size, number of states, or degree of structure can increase on one or more of these variables.

Organism points will generally tend to remain near the competence interface (but not too near), in a zone of minimum cost. Organism points will not be inside the competence interface because organism points in the region of incompetence cannot meet the externally imposed constraints that define the survival problem(s) to be solved.

Because the shape of the competence interface varies in a complex way in structure space and in time, organism points will tend to remain in the region of redundancy, away from the interface. An organism point that remains on or near the interface is likely to be overtaken by changes in the externally imposed constraints and forced to enter the region of incompetence and perish. The further an organism point is from the minimal surface, the less is the cost it may have to pay to protect itself if conditions change relatively quickly. (For animals in a climate where large unpredictable fluctuations in temperature occur, the redundant protection of houses, central heating, and sweaters avoids the cost of a sudden migration to a more equable climate.) Clearly, there is a cost of responding to change that decreases with increasing redundancy (utilizing available, presently unused capability has almost no cost) and increases with increasing specialization (changing an organization specialized for a particular function has high cost). However, the redundancy itself costs resources. As Figure 6 shows, there is a trade-off between the costs of responding to change and the cost of maintaining non-essential structure. We suggest that the sum of these costs defines a zone of minimum cost, within the region of redundancy, in which organism points are likely to be found.

5.2.2 ADAPTIVE SYSTEMS WILL CONVERGE ON "ATTRACTIVE" BIOLOGICAL STRUCTURES.
In the region of redundancy away from the finiteness limits, structural adaptation will be more governed by the "local dynamics" than the "boundary conditions" described above. In section 4.2.2 we developed the notion of a basket of criteria which could be used to evaluate the relative "biological attractiveness" of alternative points in structure space. This notion forms the basis of our hypothesis that adaptive systems will converge on attractive structures.

As has been pointed out elsewhere in this volume,[10] the explanation of why we see the particular pattern of biological structure we see lies in the melding of a number of perspectives and not predominantly in one perspective. The interaction between natural selection and dynamical self-organization is a cooperative dialogue rather a competition for primacy. We attempt to capture this notion in our basket of criteria. The criteria fall into three major categories: availability, accessibility, and performance. Within each category, there are several sub-components. We assert that no *a priori* assignment of weights to these different members of the basket of criteria will work in all situations. One must calculate the relative weights for elements of the basket in each particular context.

The basket of criteria is a generalization of the notion of calculating a centroid for a region in structure space. Given the appropriate weights for the situation to be evaluated, the cluster of structure points inhabited by a cohort of organisms can be evaluated to determine the centroid of attractiveness for the region (essentially a weighted average of the points using the basket of criteria to calculate the weight for each point). If we also calculate the centroid of the actual organisms within the space, we can now compare these relative positions. If the centroid of attractiveness of the space is close to the centroid of the group, then the cohort of organisms will not move. If, however, the centroids do not roughly coincide, then the cohort of organisms can be expected to migrate toward the centroid of the region of structure space.

In this analysis, motion in structure space well away from the boundaries can be regarded as the response to a selection pressure for fitness. The essential insight of the basket of criteria is that it is meaningless to try to speak of a universal fitness measure for all situations. In our basket of criteria, we are essentially computing a dynamically reconfigurable fitness measure which must be tuned to reflect the situation at hand.

5.2.3 FINITENESS LIMITS WILL FOSTER THE EVOLUTION OF HIGHER-LEVEL MOD-
ULES. The degree of structure increases by combination of units at one level to form a new level of structure. Such increases have occurred many times. Early in evolution, small molecules polymerized to make catalytically active polymers, including nucleic acids and proteins. Polymers have associated to form multi-component complexes, including replication complexes, RNA polymerase, spliceosomes, ribosomes, nucleosomes, and complexes that perform sequences of steps in intermediary metabolism. Complexes associate to form higher-order complexes. For example, in eucaryotic cells, organelles with the interior topologically equivalent to the exterior of the cell—endoplasmic reticulum, Golgi apparatus, endosomes, lysomes—may have evolved through the association of molecular complexes in a patch of the plasma membrane that enlarged, invaginated, and detached from the cell surface.[1] A cell is an association of complexes higher in the hierarchy of associations. Cells associate to form levels of organization in multicellular organisms, which form conspecific populations and inter-specific assemblages. We hypothesize that these levels of structure emerge due to underlying finiteness limits that constrain structural alternatives.[3]

In some cases, one can argue that a new level of structure allows the meeting of old constraints with less cellular machinery. With the new degree of structure, additional redundancy is then available; this redundancy can be committed as reliably maintained machinery to meet new constraints. For example, for the synthesis of proteins, a system that required direct recognition of each amino acid would require a bigger alphabet and provide less redundancy than the triplet code, which provides 64 codons to code for 20 amino acids. The redundant codons have become used to regulate protein synthesis and to increase the reliability of coding. The reduction of alphabet from 20 elements to four requires the additional (modular) machinery to recognize the triplet code and translate it to particular amino acids.

An increase in the degree of structure may provide additional degrees of freedom for a cell's activities in other situations. Consider a speculative example: After a gene has been transcribed, some machinery must prevent ribosomes from initiating the translation of the RNA transcript to protein, in order to allow splicing—excision of non-coding sequences (introns) from the transcript, and ligation of the adjacent coding sequences (exons). Translation of an unspliced transcript would abort soon after the first exon was translated, and complete proteins would not be synthesized. Translation of unspliced transcripts could be prevented in a non-modular way by negative translational control, in which a repressor of translation bound to a regulatory site on each transcript.[1] However, such a strategy would require maintenance of the regulatory sites on all of the genes, requiring a large number of states. Evolution has used an alternative more modular strategy: Enclosure of the unspliced transcripts within the nucleus has made their splicing a prerequisite for exit from the nucleus to the cytoplasm, where ribosomes mediate translation of the spliced messenger RNAs.[1] At least initially, the formation of a primitive nucleus that partially sequestered unspliced transcripts away from the cytoplasm may have involved less maintained machinery than maintenance of regulatory sites for negative translational control. In a similar way, spliceosomes may represent a smaller set of maintained machinery (and, thus states) than the regions of introns that performed self-splicing early in evolution.[8] Perhaps analogous arguments can be made for the evolution of other higher levels of structure.

5.3 A BIOLOGICAL EXAMPLE ILLUSTRATING THE FORMALISM

We can illustrate the major points in our formalism in the context of an interesting biological example although space and the state of the theoretical formalism do not allow a detailed treatment of this example. In this section, we are primarily illustrating the major components of our formalism, not providing a detailed treatment of the extensive literature in this area. We will proceed by comparing two human senses: smell and vision.

It is widely believed that the sense of smell was more important than vision early in evolution. Recent results[2] suggest that the ability to detect up to 10,000 different smells is rather directly encoded in the human genome by the presence of up to 1000 genes. This situation corresponds to the notion of an exhaustive encoding of this perception function for smell by using what amounts to an alphabet of thousands of different symbols. Each alphabet symbol is detected by a separate smell receptor. As our formalism would predict, with so extensive a vocabulary, we would expect to see relatively few processing states required for smell because so much of the problem has been solved by having such a large alphabet.

The availability of such a large alphabet of smells depends on the availability of a large number of smell receptors which, in turn, depend on the availability of genetic coding machinery to pass the structure to successive generations. The direct encoding of the structure of machinery for recognizing such a large vocabulary should be extensive compared to a more structured solution to the problem. This result is what we would expect when there is considerable space between the competence interface and the finiteness limits. In this case, we would say that the competence interface was defined by the ability to identify around 10,000 smells and that there was chemical machinery and coding space available to solve this problem in what is essentially a table look-up fashion. Our formalism indicates that when the structure space is fairly uniform in density and when the competence interface and finiteness limits are well separated, then there will be no systematic pressure for increased structure. We hypothesize that these situations pertain for smell.

The human sense of vision is of a different character from smell. There is no clear alphabet of visual items comparable to the 10,000 smells. Even if we adopt 10,000 distinct colors and intensities, we are left with orders of magnitude of visual information with no representation before we even attempt to deal with motion. Clearly, an exhaustive solution to visual perception like that possibly used for smell is infeasible. In our formalism, we would say that the finiteness limits on both alphabet and number of states intersect the competence interface for visual perception, eliminating low structure solutions from consideration.

The biological solution to this problem is at least consistent with our formalism. Human visual perception is a highly structured process when compared to smell.

There is significantly more processing (states as opposed to alphabet symbols) in vision. Indeed, the "visual alphabet" has been relatively well identified in contrast to the details of the extensive processing machinery. The visual alphabet consists of approximately 32 shades of gray and a similar number of intensities of three primary colors. This limitation of the visualization alphabet is in stark contrast to the enumerative approach taken for smell. Again, this result is consistent with what would be predicted from our theory where the finiteness limits cut off many of the less structured alternatives.

For both vision and smell, it is reasonable to postulate competitive advantage for organisms which improved smell or vision. These improvements correspond to an expanding competence interface which now supports more varied perception. In the case of smell, where the competence interface and finiteness limits are widely separated, we would expect non-modular increases in capability by simply directly encoding another smell. In the case of vision, the comparatively tight space between the competence interface and the finiteness limits means that as the competence interface increases, additional low structure alternatives will be excluded. Thus, we would expect to see modularization and reuse of component parts rather than direct encoding of new alphabet symbols. Indeed, some researchers have advanced the hypothesis that perception of the color blue is a relatively recent acquisition. This hypothesis is consistent with recycling existing machinery in a modular way in order to avoid a finiteness limit.

6. SUMMARY AND CONCLUSIONS

It is desirable to devise a theory that will summarize the organization of organisms and show how this organization tends to evolve. We have sketched the rudiments of such a theory. Our analysis rests on a relatively well-known model: the input-output behavior of a Turing machine, the equivalence of recursively enumerable functions and Turing machines, the complexity of a function defined in terms of its corresponding Turing machine, and the relationship between alphabet size and controller size for a given Turing machine. The complexity of the Turing machines as measured by their minimum length encoding, their number of controller states, and their alphabet size determines the three axes of our space of possible structures for the solution to a given problem. This space is bounded: With a controller of minimal structure, the solution to the problem can always be exhaustively enumerated using the maximum irredundant alphabet. With a minimal alphabet, the solution can be analogously exhaustively encoded in the finite-state machine behavior of the controller. Finiteness limits bound the space by defining limits on states and alphabet symbols that can be used for any particular solution. The degree of structure for a given problem ranges from a minimum for exhaustive enumeration to unity for solutions which achieve the Kolmogorov complexity limit (of conciseness of statement of the solution).

Within the structure space, a region near the origin represents Turing machines incompetent to solve the problem. The rest of the space represents Turing machines that can solve the problem with redundancy. Separating the region of incompetence from the region of redundancy is a surface that represents the Turing machines with the minimal irredundant alphabet and/or states sufficient to solve the problem.

We propose that an organism meeting constraints is analogous to a Turing machine solving a problem. This analogy, in the context of structure space, suggests that an evolving lineage of organisms should eventually respond to increasingly severe constraints in ways which can be explained by analyzing properties of our formalism.

The degree of structure can be quantified and measured along a relative scale by looking at the number of alphabet symbols and the number of states (or steps) in the control machinery for solving a problem. When comparing the amount of structure in two systems, completely useless states and unnecessary alphabet symbols should be ignored. The number of such states and symbols that have to be ignored is a measure of the amount of redundancy in the solution.

As we have seen, when a competence interface and a finiteness limit collide, the resulting groove at the bottom of the region of redundancy means that more highly structured systems will be favored. This shape of the structure space can provide a systematic pressure for developing and sustaining complex structures in organisms. Such increases in structure will promote modular structure which matches patterns of decomposability in the capability requirements. This concept offers an interpretation for the increase in levels of organization that has occurred during biological evolution. We hope that the theory sketched here will stimulate further inquiries into the organization of complex systems that evolve and will provide a framework within which some aspects of the development of structure can be understood.

REFERENCES

1. Alberts, B., D. Bray, J. Lewis, M. Raff, K. Roberts, and J. D. Watson. *Molecular Biology of the Cell*, 2nd edition. New York: Garland, 1989.
2. Buck, L., and R. Axel. "A Novel Multi-Gene Family May Encode Odorant Receptors: A Molecular Basis for Odor Recognition." *Cell* **65** (1991): 175–187.
3. Crutchfield, J. P. "Evolutionary Mechanics: Towards a Thermodynamics of Evolution." Presented at the Second Artificial Life Workshop, Sante Fe Institute, Santa Fe, NM and personal communication, 1990.
4. Kauffman, S. A. "The Sciences of Complexity and 'Origins of Order.'" This volume.
5. Kauffman, S. A. "Antichaos and Adaptation." *Sci. Am.* (1991): 78–84.
6. Kelso, J. A. S., D. G. DeGuzman, and T. Holroyd. "The Flexible Dynamics of Biological Coordination: Living in the Niche Between Order and Disorder." This volume.
7. Li, M., and P. M. B. Vitanyi. "Two Decades of Applied Kolmogorov Complexity: In Memoriam of A. N. Kolmogorov 1903–1897." In *Proc. 3rd IEEE Conference on Structure in Complexity Theory.* 1988.
8. Mittenthal, J. E., A. B. Baskin, and R. E. Reinke. "Patterns of Structure and Their Evolution in the Organization of Organisms: Modules, Matching, and Compaction." This volume.
9. Newman, S. A. "Generic Physical Mechanisms of Morphogenesis and Pattern Formation as Determinants in the Evolution of Multicellular Organization." This volume.
10. Wake, D. B. "Homoplasy: The Result of Natural Selection, or Evidence of Design Limitations?" This volume.

General Principles of Organization: Commentary

The preceding four chapters argue that characteristic phenomena in organisms are likely to emerge from the dynamics of organism-environment composite systems. The interaction of intrinsic processes with history and environment brings this about, in ways we now consider.

INTRINSIC PROCESSES

Kauffman argues that simple models of networks can generate behaviors characteristic of biological networks. In a Boolean network of genes that obey suitable rules for transitions of state, dynamic modules are likely to occur. With these rules the network is also likely to operate at the edge of chaos, at a phase transition between order and chaos.

Furthermore, in games where interacting Boolean networks coevolve and the state transition rules can change, the rules evolve toward those that make the system operate at the edge of chaos. As Stork demonstrated, selection can change the rules of development and physiology. Kauffman's coevolution games go beyond this general conclusion to suggest that a particular class of rules evolves.

Modular organization and operation at the edge of chaos are macro-features of networks—network generic processes—that emerge from the lower-level dynamics of interacting genes. These generic processes are analogous to the material generic

Principles of Organization in Organisms,
SFI Studies in the Sciences of Complexity, Proc. Vol. XIII,
Eds. J. Mittenthal & A. Baskin, Addison-Wesley, 1992   **379**

processes of development discussed by Newman, and to multicellular physiological oscillators considered by DeGuzman and Kelso (for limb movement) and by Nelson (for heart rate). Most analyses of network processes have used simulation. Clearly formal methods are needed to infer the characteristic structures generated by network generic processes, as the dynamics of material processes can generate vortices and deforming droplets. Baskin et al. sketch a kinematic argument that a cloud of points in a space of organismal structures moves toward a favorable configuration; perhaps formal implementation of such a scheme could generate modules and behavior at the edge of chaos.

THE INTERFACES OF INTRINSIC PROCESSES WITH HISTORY AND ENVIRONMENT

Baskin et al. suggest that the finiteness of resources exerts a systematic pressure for increasing structure. The limited resource may be located in the environment or within the organism; an example of the latter is the limited coding capacity of nucleic acids, set by the error rates for transcription and translation. Resource limitations may favor the evolution of a hierarchy of dynamic modules, because an organism using modules combinatorially in clusters may be likely to perform more demanding tasks, given constraints of finiteness, than a non-modular organization.

It seems likely that during the development of an embryo each organ is generated by a module, a morphogenetic field. If alternative modules are available for generating each organ, many kinds of organisms can be generated, as many meals can be selected from the sections of a Chinese menu. Gould[1] proposed this metaphor; as he pointed out, it offers an interpretation for a common trend in evolution. After a new module or set of modules evolves, many new kinds of organisms appear, as diverse combinations of old and new modules are tried. Chance and selection eliminate many of these kinds of organisms, and smaller-scale variations of the remaining ones evolve.

These arguments suggest that, as observed, higher levels of structure should evolve, but only a small fraction of the variants that could be generated by combining lower-level processes should persist.

APPROACHES THROUGH GENERATION AND DESIGN

Several of the preceding features—dynamic modules, operation at the edge of chaos, transcendence of resource limitations by evolution of higher-level processes—not only are generated by organism-environment composite systems, but also provide combinations of reliability, flexibility, economy, and speed. Networks operating on the edge of chaos have several potentially desirable performance characteristics. Kauffman finds that they combine stability with evolutionary flexibility (evolvability): Most minimal changes in connectivity of the network have little effect on its pattern of attractors, but a few such mutations produce larger alterations. Furthermore, a network on the edge of chaos can both transport information (within itself and relative to the environment) and store information. By contrast, if the behavior of a network is too ordered it can store information well but transport it poorly. If the behavior is too chaotic or noisy, it can transport information but not store it well.[2]

The preceding discussion suggests that selection may favor the evolution of modules for reliability and flexibility. The dissociability of modules in a hierarchy underlies much of evolvability, as Simon[3] argued. Furthermore, Clarke and Mittenthal show in a model problem that, among alternative networks of gene activity that meet constraints, networks having dynamic modules perform with high reliability.

Baskin et al. argue that organisms can survive and reproduce with more severe and diverse performance criteria if they respond to limits on the availability of resources by evolving higher levels of organization. An increase in structure can bring economy: A relatively small investment of resources in a higher-level process can produce a relatively large reduction in the resources required to perform lower-level processes. Furthermore, limits on the availability of storage and control elements favor the evolution of modular organization.

REFERENCES

1. Gould, S. J. *Wonderful Life*. New York: W. W. Norton, 1989.
2. Langton, C. G. "Computation at the Edge of Chaos." *Physica D* **42** (1990): 12–37.
3. Simon, H. A. "The Architecture of Complexity." *Proc. Am. Phil. Soc.* **106** (1962): 467–482.

General Conclusions About the Principles and Theory of Organization in Organisms

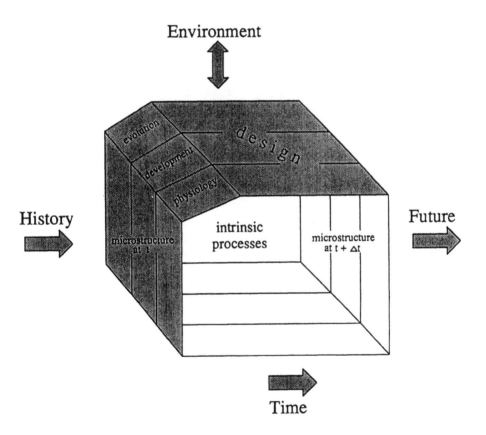

General Conclusions about the Principles and Theory of Organization in Organisms

1. INTRODUCTION

Our experiences at the workshop and in preparing this book strengthened our belief that there are principles and a theory of organization for complex systems. We are optimistic that relevant mathematical and computational tools are sufficiently developed to make such a theory attainable. A sketch of the principles and theory has made it possible to recognize additional parts of the theory, to see how specific content fits into it, and to appreciate how aspects of content that initially appeared distinct fit together.

1.1 AIMS

The object of the theory is to explain patterns of structure in living or past organisms, or patterns to be expected in future organisms. Here structure means both the physical structure of an organism and the network of processes that its physical structure supports. The theory may not specify features of specific biological organisms because the structures realized by historical contingency may be a small and idiosyncratic sample of the possibilities.[3] However, the theory should offer tools for understanding the variety of organisms we see, for evaluating possible modifications of biological organisms, and for designing artificial organisms.

Principles of Organization in Organisms,
SFI Studies in the Sciences of Complexity, Proc. Vol. XIII,
Eds. J. Mittenthal & A. Baskin, Addison-Wesley, 1992   **385**

Our aim in this concluding essay is to provide a broad overview of the progress toward principles and theory that was made through the workshop.

1.2 A SKETCH OF THE THEORETICAL FRAMEWORK

Processes intrinsic to organisms interact with historical and environmental processes to limit structures that can arise. (The interaction with prior processes is unidirectional, whereas organisms and environment can affect each other.) The constraints on the occurrence of a structure are its availability, accessibility, stability, and performance. By analyzing the dynamics of processes intrinsic to a hypothetical organism, one can assess whether its structure is available—that is, whether its occurrence is compatible with the laws of nature. Such compatibility with natural dynamics is often called self-organization. If a structure is available, it may be possible to estimate its accessibility—the probability that historical contingency will produce it. Given an available and accessible structure, its stability in an ecological niche—whether it can persist under natural selection—can be assessed by testing whether it adequately meets performance criteria characteristic of the niche, or whether it can persist in a model for the niche. Trade-offs among these constraints—changes in some constraints allowed by changes in others—can influence the structures that occur. For example, structures of low stability but high accessibility might be as likely to occur as structures of higher stability but lower accessibility.

The treatment of self-organization, contingency, and selection advocated here is a synthesis of the separate conceptual frameworks used at present for these subjects. According to neo-Darwinism, evolution proceeds through natural selection among heritable variations in organisms. This theory does not deal explicitly with differences in the availability and accessibility of structures. Treatments of historical contingency also do not deal explicitly with the availability of structures. This book argues that no subset of these views suffices to explain structure, but that a synthesis of them can do the job. Self-organization, contingency, and selection all limit what structures occur.

It may be possible to develop dynamical arguments to analyze the consequences of these three constraints. However, the task will be difficult because the mix of the three has varied through evolution. For example, Newman suggests that early in evolution self-organization was most important because there was little previous history of life and the environment was rich. Later, selection and contingency increasingly constrained evolution. In addition to this long-term trend, the mix of constraints that limits structure has varied with situations.

Thus there is not a unique objective function, fitness, in terms of which the performance of organisms can be evaluated. Rather, a theory for the organization of organisms should use a multi-objective function, with the weighting of the objectives depending on intrinsic processes, contingency, and selection in a situation-sensitive trade-off. Possibly a stochastic dynamics that gives the probability for various paths of evolution can incorporate availability, accessibility, stability, and performance in a sufficiently flexible mix.

This framework for the theory is illustrated in Figure 1, the diagram proposed as a framework for the book in the General Introduction. Processes intrinsic to an organism (represented as a box) occur on physiological, developmental, and evolutionary time scales. Prior to the current time t, the lineage of the organism has persisted through intrinsic processes and interacted with extrinsic processes in the physical and biological environment. The molecular structure of the organism—its microstructure—at any time embodies its history. The intrinsic and extrinsic processes occuring during a time interval Δt modify the microstructure. As in typical physical theories, dynamical rules generate the time course of the system's behavior (as described by a set of state variables) when initial and boundary conditions are provided. Here the state variables and the dynamic will include those relevant to intrinsic processes of interest. (However, to study coevolution it will be useful to analyze the dynamics of composite systems of intrinsic and extrinsic processes.) The initial conditions represent histories of state variables. The boundary conditions specify interactions of intrinsic and extrinsic processes. At least in some cases, these interactions can be regarded as projected into a set of performance criteria, which represent the action of environmental processes on the organism.

2. INTRINSIC PROCESSES INTERACTING WITH THE ENVIRONMENT AND DEPENDING ON HISTORY LIMIT STRUCTURE THROUGH A BASKET OF CONSTRAINTS.

The preceding framework defines the agenda for this essay. The first section examines the impact of historical contingency on intrinsic processes. We then look at interactions among intrinsic processes, and next characterize selection through interactions between intrinsic and extrinsic processes. The following section examines relations between self-organization and selection.

2.1 THE IMPACT OF HISTORICAL CONTINGENCY ON INTRINSIC PROCESSES

History provides initial conditions for intrinsic processes. In this way historical contingency limits the accessibility of structures that, from the viewpoint of dynamics, are available and stable. Natural processes of variation may or may not generate a structure; the probability of its occurrence depends, in part, on the structures already present. For example, the probability of generating a particular sequence of nucleotides in a chromosome depends on the initial sequence of nucleotides and on the processes of variation that may change the sequence, such as mutation, recombination, and insertion of transposons. Unpredictable eventualities, from viruses to cataclysms, can extinguish structures. It is difficult to predict whether historical contingency will give access to a specific structure, although variation may generate some of the structures in a class eventually if suitable environmental conditions persist long enough. (In this sense, environmental conditions provide a selection pressure, to which organisms can respond with a flux of structure.) For example, the evolution of multicellular organisms became likely after the level of oxygen in the atmosphere increased sufficiently.[9] It may be possible to make dynamical arguments about the probability that structures will evolve (given sufficient information about the dynamics of change within a genome) or become extinct (given sufficient information about the dynamics of populations and the environment).

2.2 ANALYZING SELF-ORGANIZATION IN TERMS OF INTERACTIONS AMONG INTRINSIC PROCESSES

Self-organization describes the dynamics of processes intrinsic to organisms and of processes in organism-environment composite systems. The laws of physics and chemistry limit the processes and states that are potentially available. Some processes and states are available but so unstable, or stable over such a narrow range of parameters, that they are not useful in organisms. Others are stable attractors of dynamical systems. (The Appendix illustrates these concepts.)

The chapters in this book present examples of structures that are probably at least partially self-organizing on all three time scales. Nelson suggests that in physiology a kind of structural turbulence may be relatively likely in organs, since it can be generated by the iterated operation of a few processes. This fractal structure can be spatial, as in the morphology of a branching organ, or temporal, as in variations of the heartbeat rate. On a developmental time scale, Newman notes that in an early embryo the cells probably make shapes natural to the material properties of the tissues they constitute. Early morphogenesis makes rounded cavities, pockets, and edges—shapes natural for motile and adhesive cells to make, since they form a fluid-like mass. The formation of sharp edges later, as in the skeleton, is associated with the secretion of a solid-like extracellular matrix. On an evolutionary time scale, some structures are easier to evolve because natural modes of variation tend to generate them. New kinds of proteins tend to evolve by duplication and divergence of genes, or by exon shuffling, within a family of proteins rather than through the invention of new families.

Higher-level intrinsic processes arise from component processes in several ways. The structure of an organism is a conjunction of material transformations that occur as an organism develops and moves, and a network of processes that mediates its passage through the life cycle. Diverse interactions between material and network processes occur. At one extreme, morphogenesis can proceed without ongoing genetic intervention; in the unicellular alga *Acetabularia*, a cap can regenerate independently of the nucleus.[4] At the opposite extreme, pattern formation may proceed through a continuing dialog between cytoplasmic factors and genes in nuclei, as in segmentation of a *Drosophila* embryo.[6] Newman suggests that during evolution, network processes have increasingly regulated the material transformations of morphogenesis and pattern formation; Goodwin and Kauffman develop this theme for segmentation in *Drosophila*.

An organism embodies a hierarchy of processes, in which interacting processes at a lower level form a higher-level process. As Wake emphasizes, in general the macro-micro interactions are bidirectional: Although processes at a lower level cooperate or compete to generate higher-level dynamics, higher-level processes also constrain those at a lower level. This is clearly evident in the development of mammalian visual systems, as Stryker shows: Interaction among neurons generates their collective dynamics in receptive fields, but synaptic transmission among neurons also affects the morphology of individual neurons.

During development and evolution, interactions among lower-level processes generate processes at a higher level. An increase in coordination occurs during development, as progressively higher levels of organization are generated from a fertilized egg: Interactions among cells produce tissues, which interact to generate organs; organs come under organism-level coordination by the nervous, endocrine, and immune systems. During evolution the intrinsic processes can change through variations and novel combinations of existing processes. However, the resources available to an organism limit the variety of new processes, since resources constrain the size of the set of processes and the rates at which they are active. Resources may become available for new processes if a compaction occurs—if higher-level processes evolve that reduce the resources needed to perform the current repertoire of processes. Mittenthal et al. and Baskin et al. argue that resource limitations exert a selection pressure for the evolution of higher-level processes, since compaction increases the variety of environments in which a lineage of organisms can persist. These authors focus on the evolution of macromolecules, but also treat examples from higher levels of organization.

Processes at different time scales can also interact and trade off. For example, Keesing and Stork[8] show a trade-off between changes in behavior through physiological modification of neural pathways (learning) and through heritable modification of the genome.

2.3 SELECTION OCCURS THROUGH INTERACTIONS BETWEEN INTRINSIC AND EXTRINSIC PROCESSES

Given available and accessible structures, selection based on the performance of organisms constrains the kinds of organisms that may evolve. In many situations the environment acts mainly on the organism and is weakly influenced by it; that is, the interactions are mainly unidirectional. For example, this occurs when the features undergoing selection enable an organism to perform in a physical environment that it does not alter appreciably. A fish needs a visual system that can see objects in water, and a locomotor system that can propel it through water; its vision and locomotion have a negligible effect on the optical and mechanical characteristics of the water, at least on a short time scale.

In such cases the criteria of performance can often be expressed in terms of reliability, flexibility, economy, and speed. Each of these criteria can be specified at different time scales. As an example, consider flexibility: An organism shows responsiveness (controllability) if its physiological performance responds to input from the environment. Such a case is the interaction of a physiological oscillator with an external periodicity, discussed by DeGuzman and Kelso. Development is adaptable if its course changes in response to input; this occurs as the shape of the lung conforms to the thoracic cavity (Nelson) and as visual pathways are modified by experience (Stryker, Obermayer et al.). A structure is evolvable if it readily changes during evolution, in ways that enable a lineage to survive in diverse niches. The salamanders that Wake considers show such evolutionary flexibility.

Diverse combinations of performance criteria are relevant in particular ecological niches, and limitations on resources sometimes necessitate trade-offs to meet particular criteria. For example, selection for a high rate of reproduction favors rapid development and early reproduction. This is the case for the fruit fly *Drosophila*; as Karr and Mittenthal discuss, its development is rapid because many processes occur in parallel and because the embryo has a spatial organization that reproducibly clusters cooperating structures. Given the finiteness of resources, streamlining for speed may tend to sacrifice reliability, flexibility, and economy.

There are other likely trade-offs. Flexibility may be sacrificed for reliability: Stabilizing the production of a particular output may make resources less available for change. Economy may be attained at the price of flexibility: If the minimal set of structures necessary for survival is maintained (as may be the case in some viruses), variants cannot be modified to do an altered task. Physiological reliability can be traded for evolutionary speed: Should an organism invest resources in repairing itself and so tend to last longer, or invest in making progeny and so be less repairable? Blackstone shows that this dilemma applies to colonies of organisms as well as to individuals. Trade-offs are not limited to intrinsic processes; recently humans have evolved less by changing their bodies than by changing the environment.

Performance criteria may suffice to characterize selection when the environment affects an organism but is negligibly changed in the interaction. However, bidirectional interactions are important in coevolution, when there are feedback loops among interacting organisms. In this book the term design has been used to characterize all interactions between organisms and environment, though doing so extends the conventional usage of the term to include coevolution.

For features of organisms such as language and social behavior, the environment of conspecific organisms is crucial. In such cases the composite system of organisms and environment can be modelled explicitly. Alternatively, by analogy with physical theories, it may be possible to integrate over processes in the environment, and to represent the outcome of the integration as performance criteria that provide a model of the environment (including other organisms) as an organism represents it. For example, when two organisms play a game, each can use its previous experience of the other's play in selecting a response to the opponent's next move. The distillation of experience can be as rudimentary as a look-up table ("Do this when she does that") or as sophisticated as a system of strategy and tactics. A human ethical system can be regarded as a set of performance criteria distilled from the history of humans' interactions with each other and with the nonhuman environment.

2.4 SELF-ORGANIZATION AND SELECTION ARE ASPECTS OF THE DYNAMICS OF COMPOSITE ORGANISM-ENVIRONMENT SYSTEMS

As Goodwin[2] has noted, selection can be interpreted in terms of dynamical systems theory: The lineages of organisms that persist through selection are relatively stable components of organism-environment composite systems. The dynamics of organism-environment interactions make it likely that particular kinds of intrinsic processes will evolve in composite dynamical systems. For example, Kauffman says that coevolving networks of genes may tend to evolve input-output rules for individual genes that generate network dynamics at the boundary between order and chaos. DeGuzman and Kelso note that an oscillator behaves with relatively high reliability and flexibility at the edge of entrainment, which can be regarded as a boundary between order and chaos. Others have observed analogous phenomena in cellular automata.[10,11]

In characterizing the dynamics of organism-environment composite systems, one must understand the boundary conditions at the interfaces between processes. These conditions can be regarded as a matching of performances through interaction among intrinsic processes, and between intrinsic and extrinsic processes. Such matching may be a fundamental property of physical systems in general, not only of complex adaptive systems.[5] In organisms, structures perform processes to meet constraints. For example, an embryo develops a specialized surface such as the lining of a gut to digest and absorb nutrients from food, and a circulation to transport nutrients within the body. Typically there is structure-function matching, a correspondence between components of constraints (digestion, transport) and material structures (gut, circulatory system).

Usually structure-function matching refers to the competence of a material structure, such as the heart, to perform actions that meet constraints. Mittenthal et al. suggest that a kind of structure-function matching also characterizes the architecture of the network of intrinsic processes: The coupling among processes matches the correlation among constraints that the processes meet. This principle of matching summarizes the matching of generative processes to design constraints that is an overarching theme of the book. Within an organism intrinsic processes present constraints to each other and are matched. Regulatory pathways match capabilities of the organs they regulate. For example, the contractile properties of muscle fibers are matched to the patterns of discharge in the motoneurons innervating them.[12]

Matching of intrinsic and extrinsic processes also occurs at several levels of organization. At the molecular level, macromolecules are matched in shape and affinity to the ligands they bind. For example, Mandell and Selz show that a cholinergic receptor and peptide snake venoms that bind to it have similar power spectral transformations for runs of hydrophobicity of amino acids. At a physiological level, as Stryker and Obermayer et al. discuss, generally the receptive fields that develop in visual cortex match the visual stimuli that a developing mammal receives during a critical period. When extrinsic processes vary, the matching is dynamically adjustable: DeGuzman and Kelso note that matching of the capacities for coordination of intrinsic processes to extrinsic information—to task performance criteria and physiological constraints—determines the relative stability of alternative intrinsically generated patterns, and results in pattern selection.

3. PRINCIPLES OF ORGANIZATION EXIST, SUMMARIZE PATTERNS OF STRUCTURE, AND SUGGEST DIRECTIONS FOR FURTHER WORK.

In summary, our aim in organizing the workshop was to seek principles of organization in organisms and a theory that could generate these principles, as Newtonian mechanics generates Kepler's laws of planetary motion. The chapters propose principles of organization that are independent of time scale and level of organization, and that make predictions about structure without recourse to micro-level details. Among these candidates are principles of coordination, evolution to the edge of chaos, the matching of processes to constraints, and the evolution of higher-level processes as a way to surmount resource limitations. Such general principles, which may be characteristic of any evolving complex system, must be used in conjunction with properties of the specific materials and processes in biological organisms to understand biological structure. The fluidity of tissues is a property of embryos but not of semiconductor-based computers. The matter matters.

The following theoretical framework emerged from the workshop:

- The availability, accessibility, stability, and performance of structures constrain the likelihood that particular structures will occur.
- These constraints can be evaluated from the dynamics of composite systems of organisms and environment, on physiological, developmental, and evolutionary time scales. Aspects of these dynamics appear as historical contingency and selection.
- There is no comprehensive procedure for evaluating what limitations the constraints impose on structure. However, in some cases arguments based on restricted aspects of a composite system's dynamics are possible.

The emerging principles and theory offer exciting opportunities for further work. The process of testing and interrelating them, through modeling and experimentation, promises a lively ferment in the near future. Some of the concepts in this book originated in the study of complex systems other than organisms, through engineering and computer experiments in artificial life, robotics, parallel computation, and computational neuroscience. The evident utility of these concepts encourages increasing integration and cross-fertilization between these fields and biology, to achieve a comprehensive theory of complex adaptive systems.

REFERENCES

1. Buss, L. W. *The Evolution of Individuality*. Princeton, NJ: Princeton University Press, 1987.
2. Goodwin, B. C. "Evolution and the Generative Order." In *Theoretical Biology: Epigenetic and Evolutionary Order from Complex Systems*, edited by B. Goodwin and P. Saunders, 89–100. Edinburgh University Press, 1989.
3. Gould, S. J. *Wonderful Life*. New York: W. W. Norton, 1989.
4. Hämmerling, J. "Nucleo-Cytoplasmic Interactions in Acetabularia and Other Cells." *Ann. Rev. Plant Physiol.* **14** (1963): 65–92.
5. Hübler, A. "Modeling and Control of Complex Systems: Paradigms and Applications." In *Modeling Complex Phenomena*, edited by L. Lam. New York: Springer-Verlag, 1992.
6. Ingham, P. W. "The Molecular Genetics of Embryonic Pattern Formation in *Drosophila*." *Nature* **335** (1988): 25–34.
7. Kauffman, S. A. "Antichaos and Adaptation." *Sci. Am.* **265(2)** (1991): 78–84.
8. Keesing, R., and D. G. Stork. "Evolution and Learning in Neural Networks: The Number and Distribution of Learning Trials Affect the Rate of Evolution." In *Advances in Neural Information Processing Systems - 3*, edited by R. P. Lippmann, J. E. Moody and D. S. Touretzky, 804–810. Morgan Kaufmann, 1991.
9. Knoll, A. H. "End of the Proterozoic Eon." *Sci. Am.* **265(4)** (1991): 64–73.
10. Langton, C. G. "Computation at the Edge of Chaos." *Physica D* **42** (1990): 12–37.
11. Packard, N. H. "Adaptation Toward the Edge of Chaos." Technical Report CCSR-88-5, Beckman Institute, University of Illinois at Urbana-Champaign, 1988.
12. Salmons, S. "Functional Adaptation in Skeletal Muscle." *Trends in Neuroscience* **3(6)** (1980): 134–137.

Appendix

J. A. S. Kelso,† M. Ding,† and G. Schöner‡
†Program in Complex Systems and Brain Sciences, Center for Complex Systems, Florida Atlantic University, Boca Raton, FL 33431; ‡Present address: Institut für Neuroinformatik, Ruhr-Universität, Bochum, Germany

Dynamic Pattern Formation: A Primer

1. INTRODUCTION

Morphology, embryology, evolution, developmental biology, neurobiology, physiology, and behavior are separate fields that now function largely in isolation of each other. Yet all deal, in one form or another, with structure or pattern formation processes. And all deal, in one form or another, with the fundamental issues of stability and change. Thus, separation of the disciplines does not necessarily negate the possibility that common principles may exist that underlie pattern formation and change. Although the details may differ, *events* on any chosen scale of description rely only upon interactions among components: words like *coordination* or *cooperation* express this fact. One may argue that rather than "things" (material ingredients), the coordinative relations or dynamical actions among things should

Principles of Organization in Organisms,
SFI Studies in the Sciences of Complexity, Proc. Vol. XIII,
Eds. J. Mittenthal & A. Baskin, Addison-Wesley, 1992   **397**

be the primary focus of science.[1] If the universe were composed of non-interacting units that do not speak to each other, then there would be no need to understand pattern formation in physical, chemical, or biological systems. We would not be seeking principles of biological organization because, without interaction, chemistry and biology would simply not exist.

Given, then, the presence of interactions, what form do they take, i.e., what are the laws of coordination in complex biological systems? How do we find these laws, i.e., which, if any, strategies are useful? And what language is appropriate to express these laws? If structures or patterns represent coordinated states of affairs, how do we identify and define these collective states and specify the conditions guaranteeing their persistence and change? Unlike many physical and chemical systems where the important variables and equations are well known, in living systems we have to find key observables and their dynamics (equations of motion).

In this elementary primer we provide a language and a strategy for attacking the above questions. Our approach relies upon, and is inspired by, theories of spontaneous (self-organized) formation of patterns in nonequilibrium systems, especially Haken's[21] synergetics, the qualitative theory of dynamical systems,[19] and the theory of stochastic dynamics.[10] For a long time, biologists have looked to concepts like self-organization and cooperative phenomena for insights into complexity, but it is only recently that the necessary theoretical tools have become available to place these concepts on a firm foundation. Intuitions like "the whole is greater than, or different from the sum of its parts" may thus be seen in a new light when the "whole" is characterized as a self-organized structure whose dynamics are capable of generating enormous behavioral complexity.

Before proceeding, we should mention that the application of dynamical concepts did not occur in a vacuum or because of James Gleick's best seller *Chaos*,[16] but rather was motivated in the context of specific experimental studies of biological coordination. We view this theoretical cum experimental analysis of coordination as a necessary *window*, even an essential step, into uncovering general principles. In

---

[1] Two unlikely bedfellows shared this viewpoint. Witness Henri Poincaré[58]: "...the aim of science is not things themselves, as the dogmatists in their simplicity imagine, but the relations among things; outside these relations there is no reality knowable." And D'Arcy-Thompson[72]: "The *things* which we see in the cell are less important than the *actions* we recognize in the cell." This latter view was considered so outdated that it was omitted in J. T. Bonner's[73] abridgement of Thompson's *On Growth and Form*.[26] A still broader viewpoint is that the classical dichotomy between structure and function may be one of appearance only. A unified treatment would treat structure and function on the same terms, viz. as dynamical processes separated only by the numerous time scales on which they live. Although a detailed mathematical description of such a unified view is beyond the scope of this primer, elsewhere we have established an analogous linkage between the (relatively) slow time scale dynamics of learning and the faster time scale of behavior itself.[67]

this we complement other approaches to coordinative complexity[2] (cf. Goodwin and Kauffman[18] Newman,[55] Mittenthal,[52] and Baskin[3]).

Where appropriate, we will mention experimental examples as a means of illustrating the concepts with the aim of establishing a linkage between analytic tools and experimental data. But our main goal is to review some elementary concepts and techniques pertinent to the nonlinear dynamics of pattern formation. By and large, the primer is conceptual rather than mathematical, and is aimed at communicating with a biological audience. Before beginning we wish to stress that in the present approach, "understanding" is sought, not through some privileged scale of analysis but within the more abstract level of essential, biologically relevant variables and their dynamics *regardless of scale* or *material substrate*. There is no ontological priority of one observational scale over another. What is "macro" at one level can be "micro" for another. Molecules, for example, are macroscopic structures for the particle physicist and microscopic structures for the typical cell biologist. Here, it is the methodological strategy and the reduction to (dynamically expressed) principles across levels of investigation that is the focus of our attention. Insight, within the present approach, is not necessarily gained by increasingly precise quantitative analysis (important though that may be) or by using increasingly complicated equations. Rather the aim is to account for a larger number of empirically observable features with a smaller number of theoretical concepts.

2. DYNAMIC PATTERN FORMATION: GENERAL REMARKS

Why should an understanding of structurally complex biological systems and their corresponding pattern complexity be sought in terms of organizational principles? One reason is that the same, reproducible patterns can be produced by many different material substrates and mechanisms. Even in physical systems a given pattern (e.g., Bénard convection cells) need not relate to a unique mechanism: examples in biology are nearly too numerous to mention.[34] To cite one instance, an extensive review in the well-developed field of invertebrate pattern generation reveals that there is a uniform *lack* of common neuronal mechanisms, despite similarities between the patterns generated.[69] This fact, that many physical mechanisms may instantiate the same pattern, hints strongly of *universality*, that some underlying law(s) or rule(s) govern pattern formation. At the same time, biological structures

---

[2]For example, Kauffman[29] asks: "How a complex system of interacting genes *coordinating* one another's activities behaves and what must be supposed in order to obtain anything like plausible *coordinated* behavior" (italics ours). He then idealizes the activity of the element, the gene in this case as an on/off binary device and proceeds to some surprising conclusions regarding order formation (*sans* selection) in model genetic systems. We choose a particular experimental model system and level of description[30,31] in order to pursue the possibility that principles of coordination lie at the level of dynamic patterns (or forms, cf. von Holst[76]; Goodwin[17]). Once the laws at the pattern level are found, they can be derived as self-organized stable states of coupled nonlinear dynamics among the individual components.

are clearly multifunctional: the same set of anatomical components may support a variety of functions. Again, recent evidence from invertebrate central pattern generators shows that the same neural circuit can switch flexibly among different "functional states," and can reconfigure itself according to current conditions.[49] Minimally, then, any principle(s) of pattern formation should handle *compositional complexity* on the one hand (e.g., how a given pattern is constructed from the interactions among a very large number of heterogeneous components) and *pattern complexity* on the other (e.g., multiple dynamical behaviors). Words familiar to biologists, such as (multi-) *stability* (ability to persist under various environmental conditions) and *flexibility* (ability to adjust to changing internal or external conditions) should be fundamental features of any organizational principles. The "forces," or generic processes (e.g., competition), that lead to pattern *selection* should also be clear.

In the last decade or so, tremendous progress in understanding pattern formation in open, nonequilibrium physical, chemical, and biochemical systems has been made (see, e.g., Collet and Eckmann,[6] Kuramoto,[45] Nicolis and Prigogine,[56] Babloyantz[1]). In particular, synergetic construction principles[20,21] have established the concepts of instability, fluctuations, and slaving as crucial to understanding and predicting the spontaneous (self-organized) occurrence of order in complex systems. Synergetics deals typically with equations of the following form:

$$\dot{\mathbf{q}} = \mathbf{N}(\mathbf{q}, \text{parameters}, \text{noise}) \tag{1}$$

where the dot denotes the derivative with respect to time, $\mathbf{q}$ is a potentially high-dimensional state vector specifying the state of system Eq. (1), and $\mathbf{N}$ is a nonlinear function of the state vector and may depend on a number of parameters (including time) as well as random forces acting on the system.

In general, when parameters in Eq. (1) change continuously, the corresponding solutions of Eq. (1) also change continuously. However, when a continuous change in the control parameter crosses a critical value (or critical point), the system may change qualitatively, or discontinuously. *These qualitative changes are frequently associated with the spontaneous formation of ordered spatial or temporal patterns.* This process of self-organization always arises via an *instability.* The emergence of pattern and pattern switching or change arise solely as a result of the cooperative dynamics of the system (the function $\mathbf{N}$ in Eq. (1)) with no specific ordering influence from the outside and no homunculus-like pattern generator (note the noun form) inside. Examples include the formation of convection rolls or hexagons in the Bénard system, vortex formation in the Taylor system, the onset of lasing in the laser, the formation of concentration patterns in certain chemical systems such as the Belousev-Zhabotinski reaction, reaction-diffusion systems, and the well-studied Turing instability that has served as a model of morphogenesis. In all these cases, near the vicinity of critical regions (i.e., near an instability), the system's macroscopic behavior is dominated by just a few collective modes, the so-called *order parameters.* The latter are the only variables needed to describe the evolving self-organized state exhaustively. This compression of degrees of freedom (**df**) referred to as the *slaving principle* has been given an exact mathematical form by Haken for a large class of systems (for a discussion of these and many additional examples, see, e.g., Haken,[21,22] and references therein. A related approach is developed in a recent book by Nicolis and Prigogin.[57] The mathematically oriented text by Murray[54] contains a number of worked out examples. A good review of the slaving principle is provided by Wunderlin[81]).

The spontaneous formation of patterns in nonequilibrium systems may be understood as special solutions of the system's dynamics, Eq. (1), that allow for a much lower dimensional description. Patterns emerging at *nonequilibrium phase transitions* (the term preferred by physicists) or *bifurcations* (the mathematical term) are defined in terms of *attractors* of the collective variable dynamics. We shall define and discuss these terms more fully below. Suffice to note at this point that attractors of the collective variable dynamics exist because nonequilibrium systems are *dissipative*: many independent trajectories with different initial conditions converge in time to a certain limit set or attractor solution. Stable fixed point, periodic limit cycle, and chaotic solutions are thus all possible in the *same* system (e.g., Eq. (1)), depending on parameter values. We have a glimpse, then, of one of nature's themes for handling different kinds of complexity. Vast *compositional complexity* is compressed at critical points (as demonstrated by the slaving principle of nonequilibrium phase-transition theory). The resulting low-dimensional pattern dynamics are nonlinear thus giving rise to enormous *behavioral complexity*. This theme (cf. Kelso[36]) thus embraces both the disorder-order and order-order principles advocated by Schrödinger[68] in *What is Life?* and adds the evolutionary order-disorder principle.

Of course, from a scientific point of view, life (and its understanding!) is not so easy. The reason is that in most biological systems the state vector q, its dynamics, the nature of the parameters, and the noise sources are largely unknown. The path from the microscopic dynamics of Eq. (1) to the understanding of macroscopic pattern formation in terms of collective order parameters is, therefore, not accessible to theoretical analysis. Moreover, it is not clear that biological systems possess the hierarchy of time scales necessary for the slaving mechanism of order formation. Consider, for instance, the typical time for synaptic integration ($\sim$10 ms) and compare it to typical reaction times ($\sim$100 ms) that may involve macroscopic movement of the entire organism. These times are not clearly separated although they refer to processes on very different spatial scales. Biological order formation poses challenges to theory also with respect to flexibility, in that biological systems are able to synthesize a tremendous number of different patterns, often continuously (consider, for example, visual perception). As different such patterns are formed, the components of the biological system may stabilize qualitatively different relationships, e.g., the same neurons in a central pattern generator, may stabilize synchronous activity in one pattern and alternating activity in another (e.g., Mpitsos and Cohan[53]).

On the other hand, it has proven possible to learn from, and build upon the foundation provided by, theories of nonequilibrium pattern formation by using an alternative approach to biological order, namely one in which the nature and dynamics of the low-dimensional order parameters are first empirically determined.

Phase transitions (or bifurcations) are a key part of this strategy, constituting a special entry point for developing theoretical understanding. The reason is that qualitative change allows a clear distinction of one pattern from another, thereby allowing the identification of collective variables for different patterns and the pattern dynamics (stability, loss of stability, etc). *Near critical points the essential processes governing a pattern's stability, flexibility, and even its selection can be uncovered.* Theoretically motivated measures (fluctuations, relaxation times, dwell times near the critical point, etc; see below) are available to elucidate these processes and to allow tests of theoretical predictions. In addition, the *control* parameters that promote instabilities can be discovered. Different levels of description can be related through a study of component dynamics and their coupling. In a companion paper, we apply this strategy to our chief experimental model system which involves the study of coordination in humans as a way to discover laws and principles of biological self-organization. In what follows, we provide a more detailed explication of the approach, particularly its reliance on concepts of low-dimensional systems.[3]

2.1 GENERAL DEFINITIONS

Let $(x_1, x_2, \ldots, x_n)$ be the collective variables characterizing a complex biological system. Then the *phase space* is an n-dimensional Euclidean space spanned by the vector $\mathbf{x} = (x_1, x_2, \ldots, x_n)$ with $x_i (i = 1, 2, \ldots, n)$, assuming all permissible values. A *dynamical system* is a system of equations stipulating the temporal evolution of $\mathbf{x}$. If $\mathbf{x}$ is a continuous function of time, then the dynamics of $\mathbf{x}$ is typically defined by a set of first-order ordinary differential equations (ODEs),

$$\dot{\mathbf{x}} = \mathbf{F}(\mathbf{x}) \tag{2}$$

where $\dot{\mathbf{x}}$ denotes the derivative with respect to time and $\mathbf{F}(\mathbf{x})$ gives the vector field. ODEs are frequently encountered in biology, especially when $\mathbf{F}$ specifies a nonlinear oscillator. There is another important class of dynamical systems which appear as difference equations or maps

$$\mathbf{x}_{t+1} = \mathbf{G}(\mathbf{x}_t). \tag{3}$$

Here "time" t is discrete and assumes integer values. A trajectory (orbit) of the map is $\mathbf{x}_0, \mathbf{x}_1, \mathbf{x}_2, \ldots$. That is, given $\mathbf{x}_0$, the map gives $\mathbf{x}_1$; given $\mathbf{x}_1$, Eq. (3) gives $\mathbf{x}_2$; and so on. This type of system has found applications in ecology, most notably in insect population problems. Later we will argue that certain forms of Eq. (3) also provide a natural framework for understanding multifrequency processes in biology and behavior.[37,7]

[3] It is interesting to note that complex, biological systems containing very many df often appear to live (at least part of the time) in lower-dimensional spaces. Boundary conditions (e.g., functional context, environmental constraints, energetic requirements) appear to play the role of pinning the system onto a low-dimensional manifold without any slaving in the classical sense. Other examples are synchronization and entrainment in systems of nonlinear oscillators which can often be characterized in terms of phase coupling (see section 2.7).

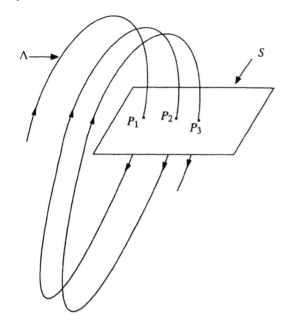

FIGURE 1   Poincaré surface section for an ODE.

Maps and ODEs are intrinsically related to each other. In particular, maps can arise in ODEs in the form of a Poincaré surface of section as illustrated in Figure 1. The plane S is the designated surface of section, and Λ denotes a trajectory of the ODE. Every time Λ pierces S going downward (points p_1, p_2 and p_3 in the figure), the corresponding coordinates in the plane are recorded. Clearly the coordinates of p_1 uniquely determine those of p_2, and the coordinates of p_2 uniquely determine those of p_3, and so on. Thus there exists a map G such that $p_{n+1} = G(p_n)$. In the special case of periodically forced ODEs, the Poincaré map can be interpreted as resulting from strobing the system at times $t_n = nT$ where T is the forcing period. Note that although maps and ODEs differ in respective analysis techniques, together they provide complementary means for studying nonlinear dynamical systems.

A *dissipative system* is one whose phase-space volume decreases (dissipates) in time. As a consequence of this, dissipative systems are usually characterized by the presence of attractors. An *attractor* is a subset of the phase space to which initial conditions asymptote as time $t \to \infty$. For example, for a damped pendulum, oscillations induced by an initial displacement from equilibrium will wind down and eventually come to a halt. The attractor in this case is a *fixed point* at rest (Figure 2). (Pictures like the one shown in Figure 2 are often referred to as *phase portraits*.)

A self-excitatory system usually exhibits a *limit cycle* as its attractor. Figure 3 shows one such attractor on which the dynamics are periodic. When a system is on a limit cycle, it oscillates with a certain frequency and amplitude that are a function of

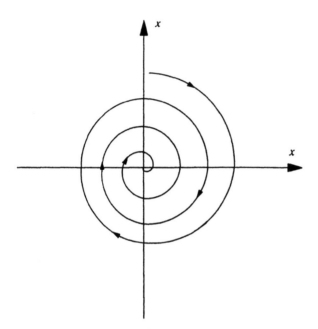

FIGURE 2 The origin is the fixed point attractor for a damped pendulum.

system parameters only, not of the initial conditions. The stability of this attractor is revealed by the fact that trajectories outside the limit cycle spiral inward, while trajectories inside spiral outward toward the limit cycle. *Quasi-periodicity* arises in higher-dimensional systems. Topologically, a quasi-periodic attractor is defined by an m-dimensional torus (T^m). Figure 4 shows an example for $m = 2$. The quasi-periodic dynamics on a torus exhibits two or more incommensurate frequencies.

Higher-dimensional ODEs ($n > 2$) or maps of any dimensionality may also exhibit *deterministic chaos*, a type of irregular dynamical behavior resembling that of random noise. The presence of chaos in physical systems is ubiquitous and has been demonstrated extensively (see Hao[25] and references therein). Recent evidence suggests that chaos may also play an important part in certain biological functions.[2,14] In the present primer we will briefly touch on this subject in section 2.8 but otherwise leave more detailed discussions to the burgeoning literature.

Related to the idea of attractors is the concept of *basin of attraction*. A given attractor's basin of attraction refers to the region in phase space in which almost all initial conditions converge to the attractor. Several attractors with different basins of attraction may also coexist, a feature called multistability. Multistability, the coexistence of several collective states for the same value of the control parameter, is an essential property of biological dynamics.

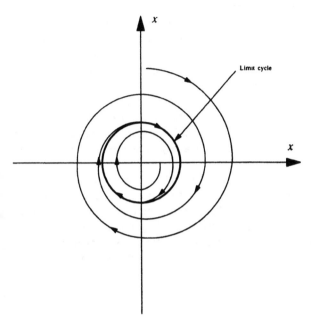

FIGURE 3   A limit cycle attractor.

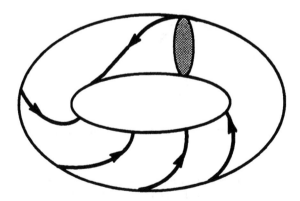

FIGURE 4   A torus attractor.

transients are generally infinite. The reason is because a typical trajectory approaches the attractor only in the limit $t \to \infty$. In practice, transients are bound to be finite due to finite experimental resolution. For example, let us assume our experimental resolution is ϵ. Then, for the damped pendulum shown in Figure 2, the transient may be regarded as that segment of the trajectory which occurs before the trajectory reaches the circle of radius ϵ around the origin.

Equations (2) and (3) often depend on *control parameters.* For a damped pendulum, the strength of the gravitational field and the friction coefficient may be considered as such control parameters. For biological systems, control parameters are usually not readily identified; hence, detailed analysis of the circumstances is required.

When a parameter p changes smoothly, the attractor, in general, also changes smoothly. When the parameter passes through a critical point $p = p_c$, however, an abrupt change in the attractor takes place. This phenomenon, as mentioned above, is called a *bifurcation* or *nonequilibrium phase transition* in physical theories of pattern formation. As we have stressed, bifurcations are particularly important for the investigation of complex biological systems, both methodologically and conceptually (see section 2.6 below).

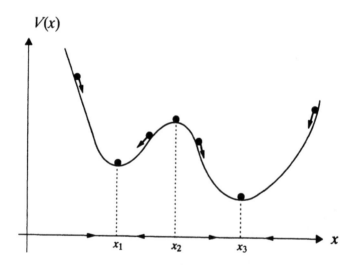

FIGURE 5   A bistable potential with point attractors at x_1 and x_3 and a repeller at x_2. On the X-axis (phase space), the flow directions are indicated.

2.2 A SIMPLE EXAMPLE

We begin with the simplest possible case, an overdamped one-dimensional oscillator. Simple though it is, it nevertheless is a useful vehicle for explaining most of the relevant concepts. The equation of motion is

$$\dot{x} = f(x) + \text{noise} \tag{4}$$

$$= -dV(x)/dx + \text{noise}. \tag{5}$$

The phase space in this case is the entire x-axis. In mechanical terms, x is the position of a point particle in the potential landscape, $f(x)$ is the force, and $V(x)$ is the potential producing the force. As will become clear, the presence of noise in Eq. (4) is of both conceptual and practical importance. For the moment, however, we concentrate on the deterministic part of Eq. (4).

A typical potential $V(x)$ is shown in Figure 5. The extrema of $V(x)$, x_1, x_2, and x_3 in Figure 5, are points of vanishing force, giving rise to *steady-state* solutions. For initial conditions near a minimum of the potential, the resulting trajectories approach the minimum in a fashion resembling that of a point particle moving in a "very sticky" well. Near a maximum, on the other hand, trajectories are repelled away from that point. Consequently, the maximum is called a *repeller*. For the potential in Figure 5, x_1 and x_3 are fixed-point attractors and x_2 is a fixed-point repeller.[4] The basin of attraction for x_1 is $-\infty < x < x_2$ and for x_3 is $x_2 < x < \infty$. Repellers always lie on the boundary of basins of attraction. On the x-axis of Figure 5, we have drawn the attractors, the repeller, and the directions of flow (the set of all possible trajectories) in the various regions. The whole picture constitutes a phase portrait, albeit in this case a very trivial one. Implementing a dynamical theory means mapping the reproducibly observed states of a system (i.e., those that occur independent of initial conditions) onto attractors of a corresponding dynamical model. Thus, *stability* is a central concept. How does one determine stability of dynamic patterns?

2.3 MEASURES OF STABILITY: RELAXATION TIME, FLUCTUATIONS

Stability can be measured in several ways. First, if a small perturbation applied to a system drives it away from its stationary state, x_f, the time for the system to return to its stationary state is independent of the size of the perturbation (as long as the

---

[4]Sometimes it is said that point attractors are boring and non-biological, but this, in our opinion, depends on what the point attractor refers to. Here the context refers to fixed points of the collective variable dynamics which may capture the essential aspects of the phenomena under study. In other contexts, whole patterns may be defined in terms of point attractors, as in artificial neural networks for pattern recognition and associative learning (for recent review, see Haken[24]).

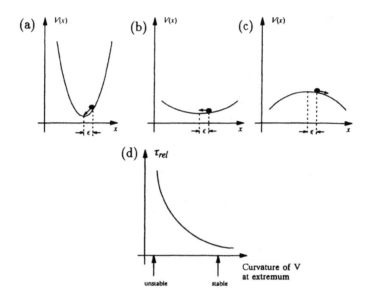

FIGURE 6   (a) In a steep potential, the system relaxes quickly from a small perturbation of size ϵ. (b) In a shallower potential (smaller curvature at the minimum), the relaxation after the same perturbation as in (a) takes longer due to the smaller restoring force exerted. (c) When the potential has a maximum, the system does not return to an unstable state after perturbation. (d) When the shape of the potential is changed by varying a parameter, thus changing the stability of a stationary state, τ_{rel} reflects this change.

latter is sufficiently small). This "local relaxation time" (local with respect to the attractor) is therefore an observable system property that measures the stability of the attractor state. Mathematically:

$$x(t) = x_f + \epsilon e^{-t/\tau_{\mathrm{rel}}} \tag{6}$$

where ϵ is the size of the perturbation.[5] The smaller τ_{rel} is, the more stable is the attractor. The case $\tau_{\mathrm{rel}} \to \infty$ corresponds to a loss of stability. Figure 6 illustrates these relationships.

When explored systematically, τ_{rel} reveals the critical parameter values at which one pattern loses stability and another pattern spontaneously emerges. A second measure of stability is related to the noise sources indicated in Eq. (4). Any real

---

[5] Time scales are generally defined with respect to time dependences governed asymptotically by exponential functions as indicated in Eq. (6). Exceptions are the time scale of observation and the time scale of parameter change defined below.

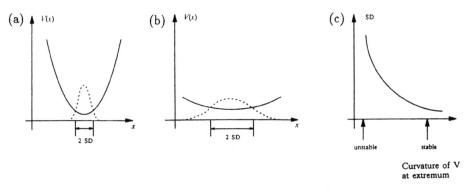

FIGURE 7   (a) The width of the probability distribution (dashed line), as measured by the standard deviation (SD), is a measure of stability. It is smaller for a more pronounced minimum of the potential (a) than for a shallower potential (b). (c) If one varies the shape of the potential experimentally, the SD exhibits the corresponding change in stability.

system described by low-dimensional dynamics will be composed of, and be coupled to, many subsystems. These act to a certain degree as *stochastic forces* on the collective variables (cf. Gardiner[10] and Haken[21], section 6.2).

The presence of stochastic forces and hence of *fluctuations* of the macroscopic variables is not merely a technical issue, but is of both fundamental and practical importance (cf., Haken,[21] section 7.3). In the present context, the stochastic forces act as continuously applied perturbations and therefore produce deviations from the attractor state. The size of these fluctuations as measured, for example, by the variance or standard deviation (SD) of x around the attractor state, is a measure of the stability of this state. The more stable the attractor, the smaller the mean deviation from the attractor state for a given strength of stochastic force. Such a situation is illustrated in Figure 7.

Without elaborating the details, it is worth mentioning that the relaxation time may also be determined from fluctuation measures (e.g., by measuring the line width of the spectral density function; see Kelso, Schöner, Scholz, and Haken[35]). All of the above statements can, of course, be made exact and quantitative (see Haken[22] for many examples; and Schöner, Haken, and Kelso[61] for a specific example related to biological coordination).[6]

[6]Deterministic dynamics in the chaotic régime may show typical times that do not converge exponentially, but rather as stretched exponential or powers of time with fractional exponents.[48,70] Regardless of the form convergence takes, τ_{rel} is still a key quantifiable property of any complex system.

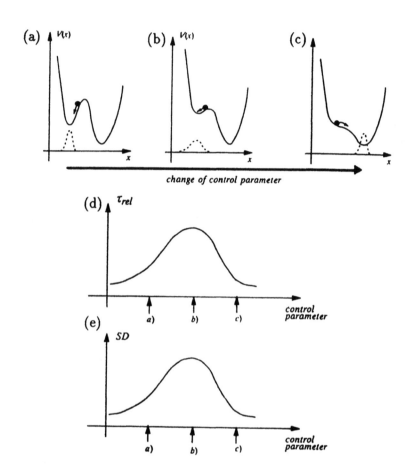

FIGURE 8   (a) A bistable potential (—) with a local probability distribution (- - -).
Local relaxation is fast. (b) One minimum has flattened out as a control parameter is
changed to a critical value. Local relaxation is slow, and the local probability distribution
is very wide. (c) As the control parameter is scaled beyond the transition, the system
switches to another available stable stationary state. Local relaxation and the probability
distribution now refer to this new stable state. (d) Relaxation time is plotted as a
function of the control parameter (the approximate locations of the situations (a), (b),
and (c) are indicated). The maximum, as the system goes through the transition,
indicates critical slowing down. (e) The standard deviation as a function of control
parameter, the maximum of which reveals critical fluctuations.

2.4 LOSS OF STABILITY: CRITICAL SLOWING DOWN, CRITICAL FLUCTUATIONS

How can a dynamical system change its behavior, defined in terms of its attractors, stabilities, etc.? As indicated in Eq. (1), the dynamical equations usually depend on parameters. As these parameters change, the dynamics change. Often a small change in some parameter results in a small change of the dynamics. However, at certain critical points, the dynamics may also change qualitatively, e.g., the stability of an attractor is lost.

Such changes of stability are, of course, reflected in stability measures. A simple example, schematically illustrated in Figure 8 shows that as the potential deforms due to scaling of its parameters, the minimum in question becomes shallower and shallower. Then, at the critical point, it becomes completely flat (having collided with a neighboring maximum). Beyond the critical point, the original minimum vanishes. Accordingly, relaxation time increases as the minimum flattens out, and then, as the system switches, decreases sharply to the small value corresponding to the other stable, stationary state. The strong increase in relaxation time is called *critical slowing down*.[7] The flattening of the potential is also seen as an enhancement of fluctuations (so-called *critical fluctuations*). Thus, the shallower potential has a less restraining influence on the fluctuations, thereby allowing the system to be, on the average, further away from the minimum.

These features of critical slowing down and critical fluctuations are characteristic of *instabilities*. Their experimental detection represents strong evidence not only that observed patterns correspond to attractor states, but also that the switching between attractors is due to loss of stability. Once again, these predictions have been worked out in quantitative detail in the case of biological coordination (for reviews, see Schöner and Kelso[63] and Jeka and Kelso[28]).

2.5 TIME SCALES RELATIONS

Obviously the presence of fluctuations is important for some aspects of critical behavior. In this section, we explain how they are quite fundamental to the consistency of a dynamic modeling approach, especially to transitions and the ability to change patterned states. We mentioned earlier that any realistic low-dimensional model of a complex system must include noise. The presence of noise, however, renders the interpretation of observed states as attractor states of a dynamical system nontrivial. This situation is most obvious in the bistable case illustrated in Figure 9. In the presence of noise, the system, inititially prepared in the state x_1, will—after sufficiently long time and with non-zero probability—switch over to state x_3. Indeed, the stationary probability distribution that describes the sytem after a sufficiently

---

[7]The term is borrowed from the field of critical phenomena or second-order equilibrium phase transitions.[46] Note, in contrast to equilibrium systems, here the growth of relaxation time is bounded by stochastic switching so that one deals with a strong increase but not a singularity of relaxation time.

long transient time will be a bimodal distribution with some probability mass at x_1 and some at x_3. Is, then, the state x_1 not really a stable state? To answer this question a discussion of three types of *time scales* is necessary. The first one is the previously discussed local relaxation time, τ_{rel}; a second is the typical time scale on which the system is *observed*, τ_{obs} (i.e., the time interval over which data are statistically analyzed); and a third time scale is the so-called *equilibration time* (or global relaxation time), τ_{equ}. The last is defined as the time it takes the system to achieve the stationary probability distribution from a typical initial distribution. In a bistable situation such as shown in Figure 9, τ_{equ} is determined mostly by the typical time it takes to cross the potential hill. If these time scales fulfill the following relation

$$\tau_{rel} \ll \tau_{obs} \ll \tau_{equ} \tag{7}$$

(local stationarity), then the interpretation of observed states as attractors is consistent. That is, the system has relaxed to an attractor on the observed time scale, but is not yet distributed over all coexisting attractors according to the stationary probability distribution. When stationary states in an experimental system are referred to, what is meant is that the time scales relation Eq. (7) is obeyed.

It is important to realize that much of the work in dynamical modeling of biological systems uses deterministic models only and thus implicitly makes the

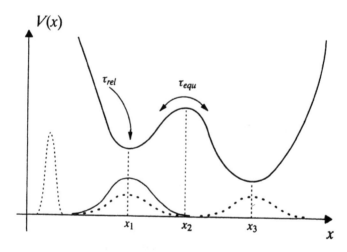

FIGURE 9 Illustration of time scales in a bistable potential. An initial distribution (- - -) relaxes to an intermediate local distribution (—) around a stationary state x_1 on the time scale τ_{rel}. This local distribution relaxes to the stationary distribution (o o o) on a time scale τ_{equ}, that is determined largely by the typical time it takes to climb over the potential hill.

assumption that Eq. (7) holds (see, for instance, the contributions to the "1982 Conference on Nonlinearities in Brain Function" edited by Garfinkel and Walter,[11] for typical examples). To neglect fluctuations and assume Eq. (7) throughout is dangerous, however, because (a) the relation Eq. (7) breaks down at critical points; (b) fluctuations are an important feature of bifurcation phenomena; and (c) fluctuations are essential in bringing about transitions. Before examining the first and third points in some detail, let us add as an aside that the analysis of fluctuations allows one (in cases where Eq. (7) is or is not valid) to test explicitly the consistency of dynamic modeling as well as certain non-trivial theoretical predictions. The analysis of fluctuations is thus crucial to employing dynamical language in a scientific way, that is, one in which constructs are operationally defined and open to experimental test.

We have seen how, when a transition is approached, local relaxation time increases while global relaxation time decreases (e.g., Figure 8). At the critical point, however, both are of the same order as the observed time and one can see the transition. Thus, at the transition point, the time scales relation (7) is violated and an additional time scale assumes importance, namely the *time scale of parameter change*, τ_p. This reflects the fact that in all biological systems, the control parameter that brings about the instability is itself changed in time. The relation of the time scale of parameter change to the other system times plays a decisive role in predicting the nature of the phase transition. If, for example,

$$\tau_{rel} \ll \tau_p \ll \tau_{equ}, \tag{8}$$

then the system changes state only as the old state actually becomes unstable, as was assumed in Figure 8. The features of critical slowing down and critical fluctuations discussed in connection with Figure 8 are predicted, as well as jumps and hysteresis (among other features).

If, on the other hand,

$$\tau_{rel} \ll \tau_{equ} \ll \tau_p, \tag{9}$$

then the system, with overwhelming probability, always seeks out the lowest potential minimum. It therefore switches state before the old state actually becomes unstable. [Note: In atastrophe theory, these two different transition behaviors are sometimes referred to as conventions, although they can, of course, be derived from the observationally accessible relations (8) and (9) (see, e.g., Gilmore[12]). Failure to treat fluctuations renders catastrophe theory incomplete in this respect].

Finally, let us clarify the role of fluctuations *per se* in effecting transitions. In so-called symmetry-breaking transitions (in which a minimum turns into a maximum, with two symmetrical new minima emerging), 'stochastic forces are obviously necessary if the sytem is to change state at all. Without such forces, the system may well stay in its now unstable state (the new maximum of V) "unaware," as it were, of its surroundings (cf. Figure 10). Aside from such a fundamental role, fluctuations also determine the speed at which a transition takes place. For example, it is possible to predict theoretically how long the switching behavior will actually take (the *mean switching time*, cf. Schöner, Haken, and Kelso[61]). A related concept is that of *mean first passage time*, that is, the length of time on average, before the system first changes state when the parameters are fixed. Again, detailed comparisons of theory (e.g., Schöner, Haken, and Kelso[61]) and experimental observations have been carried out.[32,35,33,60,64]

2.6 BIFURCATIONS

Bifurcations come in various forms and guises. For example, as parameters vary, *pitchfork* and *saddle-node* bifurcations may take place. We illustrate both bifurcations next .

A. PITCHFORK BIFURCATION A simple nontrivial potential exhibiting a pitchfork bifurcation is

$$V(x) = ax^2 + bx^4 \tag{10}$$

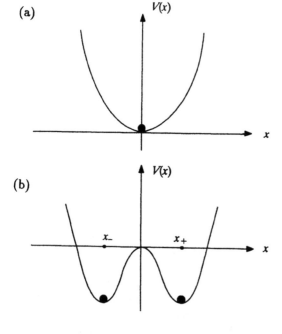

FIGURE 10 The potential $V(x) = ax^2 + bx^4$. (a) $b > 0$ and $a > 0$; (b) $b > 0$ and $a < 0$.

where a and b are control parameters and $V(x)$ possesses a *reflection symmetry* (i.e., the system is invariant under the transformation $x \to -x$). Let $b > 0$; then for $a > 0$, the origin $x = 0$ is a minimum and the only fixed-point attractor (Figure 10(a)). For $a < 0$, however, $x = 0$ becomes a local maximum and two new minima, $x_+ = \sqrt{-a/b}$ and $x_- = -x_+ = -\sqrt{-a/b}$, come into existence (Figure 10(b)). This qualitative change is called a pitchfork bifurcation in which $a = a_c = 0$ is the critical point. For $b < 0$, a pitchfork bifurcation with a flipped potential proceeds from $a < 0$ to $a > 0$. System dynamics for both $b > 0$ and $b < 0$ are collectively represented in Figure 11 where the entire parameter plane (a, b) is partitioned into four different regions according to the shape of the potential. The diagrams shown in Figure 11 are called *phase diagrams* which establish a one-to-one correspondence between the system dynamics and regions in the parameter plane. As a result, we can predict the behavior and the change of behavior in the system as parameters

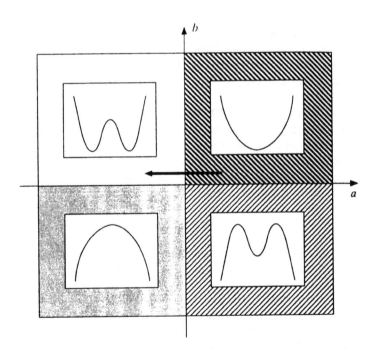

FIGURE 11 Phase diagram for the potential $V(x) = ax^2 + bx^4$ with parameters a and b. The different regions in the parameter plane (a,b) correspond to regimes with different forms of the potential, indicated schematically. The arrow corresponds to a path in parameter space (see text).

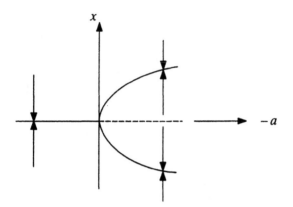

FIGURE 12 Bifurcation diagram along the arrow in Figure 11.

vary in the plane. Along the arrow shown in Figure 11, we obtain the *bifurcation diagram* (Figure 12) in which solid lines indicate stable fixed points and dashed lines unstable fixed points. Notice that the appearance of the bifurcation diagram in Figure 12 resembles that of a pitchfork.

B. SADDLE-NODE BIFURCATION AND HYSTERESIS The terms *saddle* and *node* have their origin in higher-dimensional systems. In the present situation, a saddle simply refers to a repeller and a node refers to an attractor. Unlike pitchfork bifurcations, saddle-node bifurcations occur in more generic situations where the presence of local reflection symmetry is not required. This asymmetry is easily accommodated by adding a linear term to the potential Eq. (10), i.e.,

$$V(x) = kx + ax^2 + bx^4 \tag{11}$$

where k may be considered as the parameter specifying the degree of asymmetry. If we fix $a = -1/2$ and $b = 1/4$ and consider the system behavior as a function of increasing k, then the following set of events take place. (1) When $k > 0$ is very small, the potential landscape becomes slightly tilted but otherwise remains unchanged in terms of the composition of attractors and repellers (Figure 13(a)). (2) The steady-state solutions drift in the phase space as k increases from 0. (3) At $k = k_c = \sqrt{4/27}$, an attractor and a repeller coalesce (Figure 13(b)). (4) For $k > k_c$, the coalescing solutions cease to exist and only one fixed-point attractor remains in the system (Figure 13(c)). Steps (2), (3), and (4) constitute the phenomenology of a saddle-node bifurcation and $k = k_c = \sqrt{4/27}$ is the critical point. Figure 14 shows the schematic bifurcation diagram. If we reverse the direction of k, then a reverse sequence of events takes place.

Now let us examine the *hysteresis effect*. Specifically, we consider the dynamics system for both increasing and decreasing k. Figure 15 schematically illustrates the collective states of the system, where the directions of parameter variation are indicated by the arrows. A striking feature revealed by the figure is that there exists an overlapping region, $-k_c < k < k_c$, in which, depending on the directions of k, the system can rest in either one of two possible states. At $k = k_c$ and $k = -k_c$, the system switches to a different state signifying the end of the coexistence of two distinct attractors.

From an empirical point of view, the coexistence of multiple attractors may be regarded as evidence of nonlinearity, and it is manifested by the hysteresis effect. In the quest for a better understanding of biological dynamics, the role of hysteresis can hardly be exaggerated. Finally, we note that there is only one further bifurcation that can be observed in one-dimensional ODEs, namely the transcritical bifurcation in which two fixed-point solutions exchange stability (cf. Guckenheimer and Holmes,[19] p. 145). To accomodate more sophisticated biological dynamics, we must look beyond one-dimensional phase space.

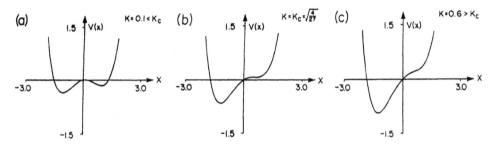

FIGURE 13   Potential landscape defined by Eq. (11) for three values of K.

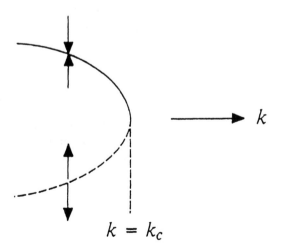

FIGURE 14   Schematic bifurcation diagram for a saddle-node bifurcation.

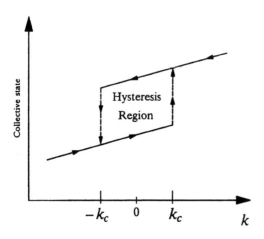

FIGURE 15   Schematic illustration of hysteresis.

C. HOPF BIFURCATIONS IN SELF-EXCITATORY SYSTEMS A self-excitatory system is one in which rhythmic motions can be generated spontaneously as a parameter passes a certain critical value. Mathematical mechanisms underlying such transitions are provided by Hopf bifurcations (see Marsden and McCracken[50] for a comprehensive review of Hopf bifurcations and their applications).

I.   **Supercritical Hopf Bifurcation.** Consider the following two-dimensional ODE in polar coordinates

$$\dot{r} = br(a - r^2) \tag{12a}$$
$$\dot{\theta} = 2\pi f \tag{12b}$$

where a and $b > 0$ are parameters and f is the frequency of revolution in the θ direction. The relation between the Cartesian coordinates x and y and polar coordinates r and θ can be expressed as (Figure 16)

$$x = r\cos\theta \tag{13a}$$
$$y = r\sin\theta. \tag{13b}$$

For $a \leq 0$, the origin $r = 0$ is a stable fixed point attracting all initial conditions in the plane. Trajectories starting from any nonzero initial conditions spiral toward the origin as shown in Figure 17(a). (The origin $r = 0$ is sometimes called a *stable focus*). Near the origin the relaxation dynamics obeys

$$r(t) = r(0)e^{-t/\tau_{rel}} \tag{14}$$

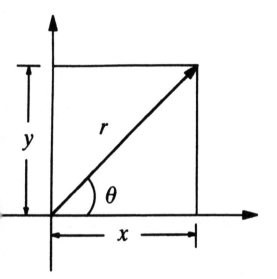

FIGURE 16 The relationship between the Cartesian coordinates x, y and polar coordinates r, θ.

(a) $a < 0$

(b) $a > 0$

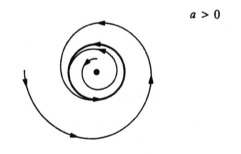

FIGURE 17 (a) Fixed-point attractor at the origin. (b) Limit-cycle attractor at $r = \sqrt{a}$ (cf. Eq. (12).

where $\tau_{rel} = -1/ab$ is the relaxation time ($a \neq 0$). For $a > 0$, the origin loses stability and a limit-cycle attractor appears at $r = \sqrt{a}$ (Figure 17(b)). This phenomenon is called a *supercritical Hopf bifurcation* or *soft excitation*. The term soft excitation refers to the fact that the limit-cycle attractor emerges gradually from the origin. Initial conditions inside and outside the circle, $r = \sqrt{a}$, spiral toward the circle. Trajectories on the circle execute periodic (rhythmic) oscillations of period $T = 1/f$. The local relaxation dynamics toward the limit cycle is determined by

$$\delta r(t) = \delta r(0)e^{-t/\tau_{rel}} \tag{15}$$

where $\delta r(t) = r(t) - \sqrt{a}$ and $\delta r(0) = r(0) - \sqrt{a}$ are both small quantities and $\tau_{rel} = 1/(2ab)$ is the relaxation time.

II. **Subcritical Hopf Bifurcation.** A different scenario for generating rhythmic oscillations occurs in the following system

$$\dot{r} = br(a + 2r^2 - r^4) \tag{16a}$$
$$\dot{\theta} = 2\pi f. \tag{16b}$$

Figure 18 shows the phase portraits for three different situations. For $a < -1$, $r = 0$ is a fixed-point attractor attracting all the initial conditions in the plane (Figure 18(a)). For $a > -1$, a stable limit cycle appears at $r = (1 + (1 + a)^{1/2})^{1/Kern}$, and for $-1 < a < 0$, an unstable limit cycle also exists at $r = (1 - (1 + a)^{1/2})^{1/2}$ (Figure 18(b)). At $a = 0$, the origin becomes unstable by absorbing the unstable limit cycle, and for $a \geq 0$, the system exhibits finite amplitude oscillations (Figure 18(c)). The bifurcation of the origin is called a *subcritical Hopf bifurcation*. From a global perspective we have a *hard excitation* transition to oscillation at $a = 0$. The term hard excitation refers to the fact that oscillations appear at $a = 0$ with finite amplitude. As a is decreased from above zero, the finite amplitude oscillations persist until $a = -1$. Below $a = -1$, the oscillation suddenly disappears. Thus, for $-1 < a < 0$, there exist two stable attractors, implying that the hysteresis effect will be observed if the parameter is systematically increased and then decreased.

Hopf bifurcations are important mechanisms for living systems to self-generate rhythmic oscillations from resting steady states. Interestingly, such self-generated rhythms can be turned on and off by varying a single parameter.

2.7. SYNCHRONIZATION AND RELATIVE PHASE DYNAMICS

Synchronization and entrainment are commonly observed phenomena.[9,44,45,75,79] In the present primer we are particularly interested in their occurrence in nonlinear *coupled* oscillators. In neurobiology, groups of interacting neurons may be

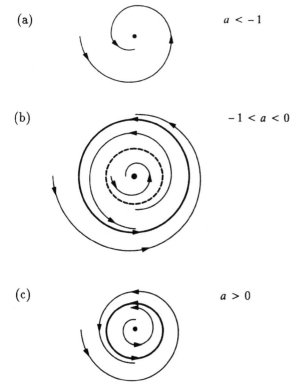

(a) $a < -1$

(b) $-1 < a < 0$

(c) $a > 0$

FIGURE 18 Phase portraits of Eq. (16) for three different values of a.

considered as coupled oscillators. One possible mechanism involves mutually inhibitory coupling with postsynaptic rebound, which may lead to sustained periodic
oscillations.[14] Limit-cycle oscillations arise also in populations of neurons that have
excitatory and inhibitory coupling.[78] Animal activities such as walking and running
require individual limbs to execute rhythmic motions. If we use coupled nonlinear
oscillators to model the coordinative limb dynamics, then the parameters reflecting
the coupling strength and the frequency (such as the pace of oscillation) stipulate
the formation of different patterns of coordination. Furthermore, smooth change in
the parameters may induce transitions (bifurcations) among the coordination patterns. For example, a horse may spontaneously switch from walking to trotting as
the speed of free running passes a certain threshold (for theory, see Schöner, Jiang,
and Kelso[62]; for experiments see Kelso and Jeka[38]).

An important aspect to note in the above examples is that all the participat-
ing oscillators evolve with a single frequency. This situation is usually referred to
as 1:1 frequency *synchronization* or *mode locking* (these two terms together with
entrainment are used interchangeably in the present primer). Dynamical patterns
are now defined by the relative phases among the oscillators. A typically observed

phenomenon is that the relative phases asymptote on certain fixed relationships (*phase locking*) which appear as attractors in the space of relative phases. We illustrate this with a simple idealized example. Let two oscillators be synchronized at a common frequency ω with sinusoidal trajectories

$$x_1 = A\sin(\omega t + \phi_1) \tag{17a}$$
$$x_2 = B\sin(\omega t + \phi_2). \tag{17b}$$

Under suitable conditions the relative phase $\phi = \phi_2 - \phi_1$ may be regarded as a collective variable obeying the following equation of motion:

$$\dot{\phi} = f(\phi) = -dV(\phi)/d\phi. \tag{18}$$

Different attractors of Eq. (18) then give rise to different phase-locked dynamical patterns. Switching among these patterns can be analyzed using the methods developed in section 2.6. Thus, with the aid of a collective variable such as relative phase, the higher-dimensional problem of coupled nonlinear oscillators reduces to the study of a one-dimensional ODE.

The study of coordination at both neural and behavioral levels demonstrates the value of relative phase dynamics. In the case of human bimanual coordination, the relative phase between the underlying oscillatory components has been rigorously shown to be a collective variable in the 1:1 frequency synchronized situation.[23] For more complex forms of coordination, symmetry arguments and group theory are employed to study the classification of various coordination patterns and transitions among these patterns (Schöner, Jiang, and Kelso[62]; for an amusing but insightful treatment, see Stewart[71]).

Much richer dynamics occur in multifrequency synchronization paradigms. Consider the case of two initially uncoupled self-excitatory oscillators. The difference between their respective natural frequencies, $\Delta\omega = \omega_1 - \omega_2$, can be viewed as a control parameter. If the coupling is weak and $\Delta\omega$ is very small, the oscillators are expected to be 1:1 entrained with each other. As $\Delta\omega$ increases, transition to desynchronization takes place at a critical point. In the desynchronization region, the system typically exhibits quasi-periodic dynamics. That is, the amplitude and the phase of the oscillation vary slowly and the Fourier spectra for the time series display two or more incommensurate fundamental frequencies. If ω_1 is close to $2\omega_2$, multifrequency synchronization at 2:1 may occur, in which the time interval for one oscillation in oscillator 2 is the same as that for two oscillations in oscillator 1 (Figure 19). If the coordinates of oscillator 1 are $(x, \dot{x})$ and those of oscillator 2 are $(y, \dot{y})$, then the plots x versus y are called *Lissajous figures*. For 1:1 frequency synchronization, the Lissajous figure is typically a circle (Figure 20(a)), whereas for 2:1 frequency synchronization, the corresponding Lissajous figure can exhibit a figure **8** (Figure 20(b)).

An important mathematical example displaying multifrequency mode locking is the periodically forced van der Pol oscillator

$$\ddot{x} - \epsilon(1 - x^2)\dot{x} + x = B\cos(\nu t). \tag{19}$$

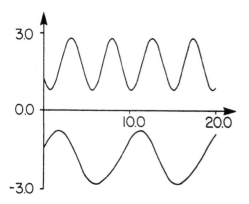

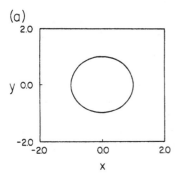

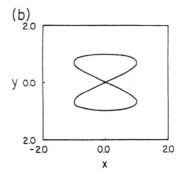

FIGURE 19　Oscillator 1 and Oscillator 2 exhibit a 2:1 frequency synchronization.

FIGURE 20　Lissajous figures for (a) 1:1 and (b) 2:1 frequency synchronization.

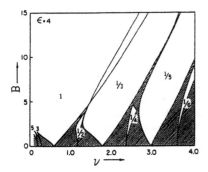

FIGURE 21　Phase diagram for the van der Pol oscillator (Eq. (19)).[27]

The term, $B\cos(\nu t)$, in this case may be regarded as the output of another oscillator whose dynamics is not affected by x. If $B = 0$, the two oscillators are totally decoupled, and the van der Pol oscillator possesses a unique limit-cycle attractor whose frequency is denoted by ω_0. For different values of ν and B, the attractors of Eq. (19) and the sinusoidal driving may be mode-locked at different frequency ratios as shown in Figure 21. Between the mode-locking regions, one may find parameter values for which the dynamics of Eq. (19) is quasi-periodic.

Of course, coupled nonlinear oscillators also give rise to spatial and spatiotemporal structures. A good example is the Belousev-Zhabotinski reaction which consists of a recipe of multiple reactants that, instead of relaxing to equilibrium, exhibit rhythmic alterations in the form of wave patterns. The latter correspond to gradients of phase (for excellent discussion and analysis, see Murray,[54] and for beautiful illustrations of phase gradients and other rhythmically generated structures, see Winfree[80]).

The B-Z reaction shares many common features with excitable media, including the nerve membrane. Traveling waves, e.g., of chemical concentration, appear to play a key role in developmental processes and reaction-diffusion systems in general. A famous example is the so-called Turing instability. Turing[74] showed that spatially heterogeneous patterns of chemical or morphogen concentration can arise in an otherwise homogeneous system, due to reaction and diffusion. A local perturbation to one of the (slowly diffusing) catalytic agents produces standing waves. Once again, low-level noise from the surroundings or internally generated molecular fluctuations are sufficient to "trigger" the development of pattern.[8] Recent experiments on the chlorite-iodide-malonic acid-starch reaction in a gel reactor [5] provide convincing evidence of the stable patterns predicted by Turing.

2.8 CIRCLE MAPS AND INTERMITTENCY

It is important to note that in the multifrequency regimes, Eq. (18) is no longer an adequate description for the relative phase dynamics. Alternative methods use discrete maps.

Consider the phase space $(x, \dot{x})$ of a nonlinear oscillator and the associated polar coordinates (r, θ). If the system is periodically driven, then discrete maps for the variable (r, θ) may be derived by strobing the system at time $t_n = nT$, where $T = 2\pi/\nu$ is the driving period (section 2.1). Specifically, let the point on a

---

[8]Turing[74] considered the case of a system just beginning to leave a homogeneous condition, hence the first appearance of pattern, as "the exception rather than the rule. Most of an organism, most of the time, is developing [he said] from one pattern into another, rather than from homogeneity into a pattern" (p.72). The dynamics we outline here emphasizes the latter case, as do our experimental studies of order-order transitions. Interestingly Schrödinger[68] advocated an order-order transition principle as the essential distinction between living and nonliving systems. In the context of open, nonequilibrium systems, this is no longer necessary.

trajectory at time t_n be (r_n, θ_n) and the point on the same trajectory at time t_{n+1} be (r_{n+1}, θ_{n+1}), then the following relations hold

$$r_{n+1} = R(r_n, \theta_n) \tag{20a}$$
$$\theta_{n+1} = \Theta(r_n, \theta_n). \tag{20b}$$

This two-dimensional map of (r, θ) may be further simplified if the dissipation in the system is very strong. In that case, the dependence of Θ on r may be neglected and Eq. (20(b)) reduces to a one-dimensional map on a circle (θ is an angular variable)

$$\theta_{n+1} = f(\theta_n) = \theta_n + \Omega + g(\theta_n). \tag{21}$$

The issue concerning the validity of reducing Eqs. (20) and (21) has been addressed in the literature. The results show that the mode-locking properties of coupled oscillators in certain regions of the parameter space are essentially encoded in the circle maps (see Glazier and Libchaber,[15] and references therein). The advantage of studying circle maps is that they are low dimensional and, hence, permit more detailed analysis.

A much studied example of circle maps takes the following concrete form

$$\theta_{n+1} = f(\theta_n) = \theta_n + \Omega - \frac{K}{2\pi} \sin(2\pi\theta_n) \tag{22}$$

where K represents the strength of nonlinearity, and Ω is the ratio between the natural frequencies of the uncoupled oscillators, i.e., $\Omega = \omega_0/\nu$. The θ-circle in Eq. (22) is assumed to be of unit length (instead of 2π). We henceforth refer to Eq. (22) as the circle map.

(A different form of circle maps is studied in the companion paper where the map is defined as

$$\theta_{n+1} = \theta_n + \Omega - \frac{K}{2\pi}(1 + A \cos 2\pi\theta_n) \sin 2\pi\theta_n \ . \tag{23}$$

Equation (23) is called the "phase attractive" circle map which has been used to model the properties of multifrequency coordination of behavior.[37,7]

To study the circle map (22), we need to introduce a few concepts. The *winding number* W for a trajectory $\{\theta_n\}$ of Eq. (22) is defined as

$$W \equiv \lim_{n \to \infty} \frac{\theta_n - \theta_0}{n} \tag{24}$$

which is, in fact, the measured frequency ratio of the two coupled oscillators. Specifically, W as a rational number implies that the underlying oscillators are mode locked or synchronized. If W is irrational, then the oscillators are desynchronized. More precisely, they exhibit quasi-periodicity or chaos. For $K \leq 1$, the circle map is characterized by a unique winding number W.[8] For $K > 1$, however, W is not always well defined. In particular, trajectories starting from different initial conditions may yield different winding numbers. In this elementary primer we will focus mainly on the region $K < 1$ (see also companion paper). The region $K > 1$ is explored in Glass et al.,[13] and Zeng and Glass.[82]

The *Lyapunov exponent* Λ is defined as

$$\Lambda = \lim_{k \to \infty} \frac{1}{k} \sum_{n=1}^{k} \ln |f'(\theta_n)|, \tag{25}$$

which measures the rate of exponential divergence or convergence of two trajectories with slightly different initial conditions (Figure 22), thus providing a criterion for delineating different types of attractors in the system. In particular, $\Lambda < 0$ implies that the map has a periodic attractor and, consequently, the underlying coupled oscillators are in a mode-locked state. $\Lambda = 0$ means that the dynamics is quasi-periodic and the attractor occupies the entire unit circle. $\Lambda > 0$ depicts a situation in which two nearby trajectories exponentially diverge from each other regardless of how closely situated are the respective initial conditions (Figure 22). In this case, the system is said to exhibit *sensitive dependence on initial conditions*, a fact often viewed as the hallmark of chaotic dynamics.

Now let us examine the parameter plane of the circle map (Figure 23). If $K = 0$ in Eq. (22), then $W = \Omega = \omega_0/\nu$, and the situation is simple. The map exhibits either periodic (synchronized) or quasi-periodic motions, depending on whether Ω is rational or irrational. Finite K introduces nonlinearity in the system and causes a given periodic orbit to persist for a range of Ω values. The result is the Arnold

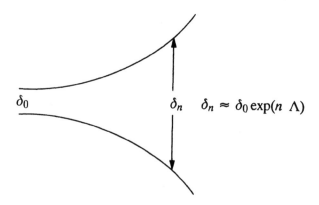

FIGURE 22 Schematic illustration of a Lyapunov exponent $\Lambda \cdot \Lambda > 0$ for the case shown in the picture.

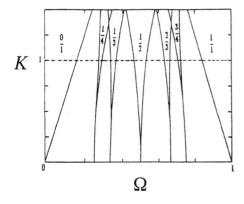

K

Ω

FIGURE 23   The $\Omega - K$ parameter plane for the circle map Eq. (22) (schematic).[25]

tongue structure of the $\Omega - K$ plane shown in Figure 23, which is the phase diagram for the circle map. Inside each Arnold tongue there is a unique attracting periodic trajectory and its winding number satisfies the mode-locking condition: $W = m/n$ with m and n being integers. The orbit of the periodic attractor is composed of n distinct θ values: $\theta_1, \theta_2, \ldots, \theta_n$, such that $\theta_2 = f(\theta_1)$, $\theta_3 = f(\theta_2), \ldots, \theta_n = f(\theta_{n-1})$, and $\theta_n - \theta_1 = m$.[9] Between the tongues the winding number W is irrational and the dynamics is quasi-periodic ($\Lambda = 0$). As K increases, the Arnold tongues widen and eventually touch each other on the critical curve $K = 1$. Above this critical curve, the tongues overlap and the system can display chaos.

It is interesting to note that the phase diagram of the circle map (Figure 23) and that of the van der Pol oscillator (Figure 21) exhibit remarkable similarity. In particular, the order of occurrence of various mode-locking tongues in Figure 21 surveyed from right to left is the same as that in Figure 23 surveyed from left to right. The reversal of order in the two diagrams is caused by the reciprocal relationship between Ω and ν, i.e., $\Omega \sim 1/\nu$. Mathematically, this order of occurrence of the mode-locking tongues is given by the so-called *Farey sequences*. An n-Farey sequence is the increasing succession of rational numbers p/q with $q \leq n$. An example for $n = 4$ is

$$0/1, 1/4, 1/3, 1/2, 2/3, 3/4, 1/1$$

(see Figure 23). As shown in the works of Herman,[4] the universality of frequency locking derives from the structural stability of the rational frequency ratios. *This accounts for similar behavior across very different systems* (e.g., clock mechanisms, moon-earth phase locking, walking and breathing in humans, externally stimulated nerve membranes, frequency locking in mammalian visual cortex, etc.).

[9] Two points with integer difference in their coordinates appear as the same point on the circle of unit length. Conventionally, periodic orbits in the circle map are discussed in the sense of *modulo 1* which means that we only consider the fractional part of θ. Under this convention, we may assert that $\theta_n = \theta_1$.

(a)

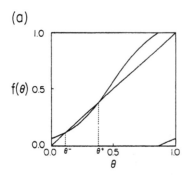

(b)

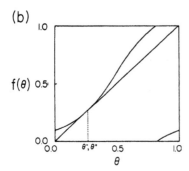

(c)

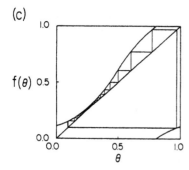

FIGURE 24   Function $f(\theta)$ for three values of Ω. (a) $\Omega = \Omega_c = 0.03$, (b) $\Omega = \Omega_c$, (c) $\Omega = \Omega_c + 0.01$ (see text).

Periodic attractors are created and destroyed in the circle map through saddle-node bifurcations (cf. section 2.6) which occur at the boundaries of the Arnold tongues. We show an example in Figure 24 in which we vary the values of Ω near the boundary of the Arnold tongue with winding number $W = 0/1$ while keeping K as a constant. The boundary in this case is defined by

$$K = 2\pi\Omega. \tag{26}$$

For $K = 0.6$, the saddle-node bifurcation occurs at $\Omega_c = 0.6/2\pi \approx 0.0955$. Figure 24(a) shows the function $f(\theta)$ intersecting the diagonal line at two points: θ^- and θ^+, where θ^- is a fixed-point attractor and θ^+ is a fixed-point repeller ($\Omega = \Omega_c - 0.03$ in Figure 24(a)). Initial conditions other than $\theta = \theta^+$ converge to θ^- as $n \to \infty$. As Ω increases, θ^- and θ^+ approach each other and coalesce when $\Omega = \Omega_c$ (Figure 24(b)). For $\Omega = \Omega_c + 0.01$ beyond the boundary of the Arnold tongue, θ^- and θ^+ cease to exist (Figure 24(c)), and the system exhibits either periodic orbits of higher period or quasi-periodic orbits, depending on the exact location of Ω. If Ω is decreased, then the reverse sequence of events is observed.

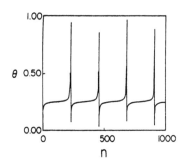

FIGURE 25 Time series for the intermittent trajectory ($\Lambda = 0$).

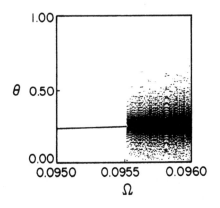

FIGURE 26 Bifurcation diagram for the circle map. The range of Ω straddles the boundary of the Arnold tongue 90/1). ($K = 0.6$).

(a)

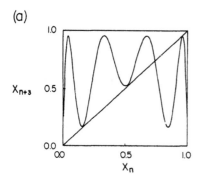

(b)

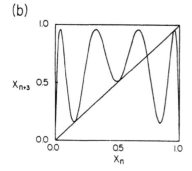

(c)

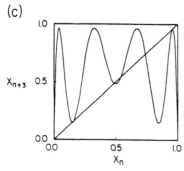

FIGURE 27   Saddle-node bifurcations in the logistic map. (a) $a < a_c$, (b) $a = a_c$, (c) $a > a_c$.

The narrow channel between the function $f(\theta)$ and the diagonal line in Figure 24(c) induces interesting dynamical behavior called *type I intermittency*.[59] The phenomenology is the following. Inside the channel, iterates of the map move very slowly (Figure 24(c)), giving rise to the impression that the fixed-point attractor was already in place (from the point of view of decreasing Ω). After exiting the channel, the trajectory takes large strides for a number of times before re-entering the channel. Figure 25 shows the time series θ versus n for a typical initial condition ($\Omega = \Omega_c + 0.0001$ in this case). During the time course the trajectory in

Figure 25 spends most of the time near the narrow channel, which is clearly shown in the bifurcation diagram (Figure 26) where the Ω values are plotted in the horizontal direction, and in the vertical direction we plot 500 iterates of the map for each value of Ω. The dark area in the figure corresponds to the θ interval most frequently visited by the trajectory and this interval is precisely where the channel is situated.

The intermittent dynamics seen in Figure 25 is quasi-periodic with $\Lambda = 0$. An often cited example of chaotic intermittency happens in the following one-dimensional logistic map

$$x_{n+1} = h(x_n) = ax_n(1 - x_n) \qquad (27)$$

where x may be considered as a variable proportional to the population of a certain type of insect and a is the control parameter representing the environment. For $0 < a < 4$, Eq. (27) maps the unit interval $[0,1]$ into itself.

At $a = a_c = 1 + \sqrt{8}$, the logistic map undergoes a saddle-node bifurcation in which a periodic attractor and a periodic repeller, both of period 3, are created. Figure 27 shows the function $h^3(x) = h(h(h(x)))$ for three values of a: $a = 3.82 < a_c$, $a = a_c \approx 3.8282$, and $a = 3.84 > a_c$. Each of the three narrow channels in Figure 27(a) resembles that in Figure 24(c). But the attractor corresponding to Figure 27(a) has a positive Lyapunov exponent, i.e., $\Lambda > 0$. As a result, the iterates of the map between two successive passages through the narrow channels are erratic and unpredictable. Figure 28 shows the time series x versus n of a typical trajectory for $a = 3.82819 < a_c$. Let us define the re-injection time t to be the number of iterates between exiting one of the channels and re-entering another (cf. Figure 27(a)). Then t can be treated as a stochastic variable whose probability distribution is schematically shown in Figure 29. A quantitative measure of intermittency is the average re-injection time $\langle \rangle$ which is found to scale with the parameter a as

$$\langle t \rangle \sim \frac{1}{|a - a_c|^\alpha} \qquad (28)$$

where $\alpha = 1/2$.

Intermittency is observed in a variety of different natural systems ranging from turbulent fluids to chemical reactions (see Bergé, Pomeau, and Vidal[4] for examples). We have found indications of intermittent dynamics in sensorimotor[39] and multifrequency[42,43] coordination. The broader implications of intermittency, where the system is poised near a fixed point but not actually in it, are explored in the companion paper.[37,40,42,43,47]

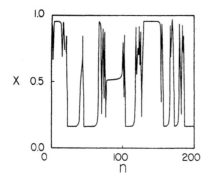

FIGURE 28 The time series for an intermittent trajectory ($\Lambda > 0$).

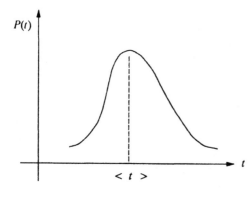

FIGURE 29 Schematic illustration of the probability distribution on the re-injection time t.

3. SYNOPSIS

In scientific modeling, a variety of different observations are often accounted for by a specific model. Here our considerations focus on very general principles that pertain to dynamic pattern formation and change on several different scales of observation. The main idea behind the present approach is to view pattern formation, on a given level of observation, in terms of coordination dynamics. Now we briefly summarize the essential, stripped-down aspects of these (abstract, level-independent) dynamics. Specific extensions are developed somewhat in the companion paper and elsewhere.[42,43]

1. Spatiotemporal patterns are characterized by low-dimensional collective variables or order parameters which define stable and reproducible relationships among the system's elements. What constitutes an element is based on a functional decomposition at a chosen level of description (section 2). Nonlinear interactions among the elements produce patterns.

2. As a first step, observable patterns are mapped onto attractors of the order parameter dynamics, using the strategy of phase transitions (section 2.1). Thus, the dynamics are dissipative.

3. The order parameter dynamics may give rise to pattern complexity, i.e., additional features not seen or understood before may appear. For example, partially ordered spatiotemporal patterns may correspond to the intermittent régime of the coordination dynamics (section 2.8; Kelso and DeGuzman,[41] and see companion paper).

4. Biological boundary conditions act as parameters on the coordination dynamics. A parameter that moves the system through different patterns is called a *control parameter*, which may be quite unspecific to the resulting patterns (section 2.1). [On the other hand, specific parametric effects are also expected in biological systems, e.g., specific environmental or learning requirements. These are treated in mathematical detail elsewhere, however; e.g., Schöner and Kelso.[65,66]

5. The spontaneous (self-organized) emergence of pattern arises through instability. Fluctuations are always probing the stability of patterned states in open systems and are conceptually crucial (section 2.3). Similarly, switching from one pattern to another is often connected to instability.

6. Competitive processes lead to pattern selection. For example, the relative values of parameters in Eq. (23) determine the patterns observed. Interestingly, one of the parameters (A) represents intrinsic constraints (e.g., attraction to certain "preferred" relative phases) and the other (K) represents external influences.

On a final, somewhat speculative note, we are reminded of recent work in evolution which suggests that it is the dismantling of intrinsic (genotypic) constraints (described by words such as "internal balance," "genetic homeostasis") that effects speciation.[51] For example, rapid changes in mollusc sequence under the stress of an evaporating Lake Turkana produced a major "developmental instability," a breakup of genetic homeostasis. Accompanying these points of rapid taxa change was a significant elevation of phenotypic variance.[77] If Lake Turkana's drought is viewed (in dynamical language) as an environmental control parameter, and the "cohesion of the genotype" corresponds to the intrinsic, collective variable dynamics characterizing morphologically stable states, then speciation may be viewed as a self-organized pattern formation process.

Under appropriate time scale assumptions, the theory of nonequilibrium pattern formation predicts critical fluctuations near instabilities. It is tempting to interpret the enormous phenotypic variance preceding the emergence of new species as a striking confirmation of this prediction. In any case, it is hard to ignore the analogy between speciation and the present (elementary) concepts of dynamic pattern formation. This does not say, of course, that the conventional neo-Darwinian dogma in which new species (forms) arise by accumulating slight, successive variations in the genotype is wrong. But it does suggest that autonomous self-organization processes (the basic principles and language of which we present here) may play an important role.[27]

ACKNOWLEDGEMENTS

Research supported by NIMH (Neurosciences Research Branch) Grant MH 42900, BRS Grant RR07258, and the U.S. Office of Naval Research Contract N00014-88-J-119. Our thanks to Bill McLean for help with the figures and graphics, and Pamela Case for assistance with manuscript preparation.

REFERENCES

1. Babloyantz, A. *Molecule, Dynamics, and Life*. New York: Wiley, 1986.
2. Basar, E. *Chaos in Brain Function*. Berlin: Springer-Verlag, 1990.
3. Baskin, A., R. E. Reinke, and J. E. Mittenthal. "Exploring the Role of Finiteness in the Emergence of Structure." This volume.
4. Bergé, P., Y. Pomeau, and C. Vidal. *Order Within Chaos*. Paris: Hermann, 1984.
5. Castets, V., E. Dulos, J. Boissonade, and P. De Kepper. "Experimental Evidence of a Sustained Standing Turing-Type Nonequilibrium Chemical Pattern." *Phys. Rev. Lett.* **64** (1990): 2953–2956.
6. Collet, P., and J. P. Eckmann. *Instabilities and Fronts in Extended Systems*. Princeton, NJ: Princeton University Press, 1990.
7. DeGuzman, G. C., and J. A. S. Kelso. "Multifrequency Behavioral Patterns and the Phase Attractive Circle Map." *BioCyber* **64/6** (1991): 485–495.
8. Devaney, R. L. *An Introduction to Chaotic Dynamical Systems*. Redwood City, CA: Addison-Wesley, 1989.
9. Ermentrout, G. B., and J. Rinzel. "Beyond a Pacemaker's Entrainment Limit: Phase Walk-Through." *Amer. J. Phys.* **246** (1984): R102–R106.
10. Gardiner, C. W. *Handbook of Stochastic Methods for Physics, Chemistry and the Natural Sciences*. New York: Springer-Verlag, 1983.
11. Garfinkel, A., and D. O. Walter. "Proceedings of Symposium on Nonlinearity in Brain Function." *Amer. J. Phys.* **245** (1983): R450.
12. Gilmore, R. *Catastrophe Theory for Scientists and Engineers*. New York: Wiley, 1981.
13. Glass, L., M. R. Guevara, A. Shrier, and R. Perez. "Bifurcation and Chaos in a Periodically Stimulated Cardiac Oscillator." *Physica* **7D** (1983): 89–101.
14. Glass, L., and M. C. MacKey. *From Clocks to Chaos—The Rhythms of Life*. Princeton, NJ: Princeton University Press, 1988.
15. Glazier, J. A., and A. Libchaber. "Quasi-Periodicity and Dynamical Systems." *IEEE Transactions on Circuits and Systems* **35** (1988): 790–809.
16. Gleick, J. *Chaos—Making a New Science*. New York: Viking Penguin, 1987.
17. Goodwin, B. C. "Structuralism in Biology," preprint.
18. Goodwin, B. C., and S. A. Kauffman. "Deletions and Mirror Symmetries in *Drosophila* Segmentation Mutants Reveal Generic Properties of Epigenetic Mappings." This volume.
19. Guckenheimer, J., and P. Holmes. *Nonlinear Oscillations, Dynamical Systems, and Bifurcations of Vector Fields*. New York: Springer-Verlag, 1983.
20. Haken, H. "Cooperative Phenomena in Systems Far From Thermal Equilibrium and in Non-Physical Systems." *Rev. Mod. Phys.* **47** (1975): 67–121.
21. Haken, H. *Synergetics, an Introduction: Non-Equilibrium Phase Transitions and Self-Organization in Physics, Chemistry and Biology*. 3rd edition. Berlin: Springer, 1983.

22. Haken, H. *Advanced Sunergetics: Instability Hierarchies of Self-Organizing Systems and Device*. Berlin: Springer, 1983.
23. Haken, H., J. A. S. Kelso, and H. Bunz. "A Theoretical Model of Phase Transitions in Human Hand Movements." *Biol. Cyber.* **51** (1985): 347–356.
24. Haken, H. *Synergetics, Computers and Cognition*. Berlin: Springer, 1991.
25. Hao, B. L. *Chaos II*. Singapore: World Scientific, 1990.
26. Harrison, L. G. "Kinetic Theory of Living Pattern and Form and Its Possible Relationship to Evolution." In *Entropy, Information, and Evolution*, edited by B. H. Weber, D. J. Depew, and J. D. Smith, 53–74. Cambridge: MIT Press, 1988.
27. Hayashi, C. *Nonlinear Oscillations in Physical Systems*. New York: McGraw-Hill, 1964.
28. Jeka, J. J., and J. A. S. Kelso,. "The Dynamic Pattern Approach to Coordinated Behavior: A Tutorial Review." In *Perspectives on the Coordination of Movement*, edited by S. A. Wallace, 3–45. Amsterdam: North-Holland, 1989.
29. Kauffman, A. S. "Developmental Logic and its Evolution." *Bio. Essays.* **6** (1990): 82–86.
30. Kelso, J. A. S. "On the Oscillatory Basis of Movement." *Bull. of the Psychonomic Soc.* **18** (1981): 63.
31. Kelso, J. A. S. "Phase Transitions and Critical Behavior in Human Bimanual Coordination." *Amer. J. Phys.: Reg. Integrative anc Comparative Physiology* **15** (1984): R1000–R1004.
32. Kelso, J. A. S., and J. P. Scholz. "Cooperative Phenomena in Biological Motion." In *Complex Systems: Operational Approached in Neurobiology, Physical Systems and Computers*, edited by H. Haken, 124–149. Berlin: Springer, 1985.
33. Kelso, J. A. S., J. P. Scholz, and G. S. Schøner. "Dynamics Governs Switching Among Patterns of Coordination in Biological Movement." *Phys. Lett.* *A134* **1** (1988): 8–12.
34. Kelso, J. A. S., and G. S. Schöner. "Toward a Physical (Synergetic) Theory of Biological Coordination." *Springer Proceedings in Physics* **19** (1987): 224–237.
35. Kelso, J. A. S., G. S. Schöner, J. P. Scholz, and H. Haken. "Phase-Locked Modes, Phase Transitions and Component Oscillators in Biological Motion." *Phys. Scr.* **35** (1987): 79–87.
36. Kelso, J. A. S. "Introductory Remarks: Dynamic Patterns." In *Dynamic Patterns in Complex Systems*, edited by J. A. S. Kelso, A. J. Mandell, and M. F. Shlesinger, 1–5. Singapore: World Scientific, 1988.
37. Kelso, J. A. S., and DeGuzmann. "Order in Time: How Cooperation Between the Hands Informs the Design of the Brain." In *Neural and Synergetic Computers*, edited by H. Haken, 180–196. Berlin: Springer, 1988.
38. Kelso, J. A. S., and J. J. Jeka. "Symmetry Breaking Dynamics of Human Multilimb Coordination." *J. Exper. Psychol.: Human Perception & Performance*, in press.

39. Kelso, J. A. S., J. D. DelColle, and G. S. Schöner. "Action-Perception as a Pattern Formation Process." In *Attention and Performance XIII*, edited by M. Jeannerod, 139–169. Hillsdale, NJ: Erlbaum, 1990.

40. Kelso, J. A. S. "Anticipatory Dynamical Systems, Intrinsic Pattern Dynamics and Skill Learning." *Human Movement Science* **10** (1991): 93–111.

41. Kelso, J. A. S., and G. C. DeGuzman. "An Intermittency Mechanism for Coherent and Flexible Brain and Behavioral Function." In *Tutorials in Motor Neuroscience*, edited by J. Requin and G. E. Stelmach. Dordrecht: Kluwer, 1991.

42. Kelso, J. A. S., G. C. DeGuzman, and T. Holroyd. "The Self-Organized Phase Attractive Dynamics of Coordination." In *Self-Organization, Emerging Properties and Learning*, edited by A. Babloyantz. New York: Plenum, 1991.

43. Kelso, J. A. S., G. C. DeGuzman, and T. Holroyd. "Synergetic Dynamics of Biological Coordination with Special Reference to Phase Attraction and Intermittency." In *Synergeitcs of Rhythms*, edited by H. Haken and H. P. Höepchen. Berlin: Springer, 1991.

44. Kopell, N. "Toward a Theory of Modelling Central Pattern Generators." In *Neural Control of Rhythmic Movements in Vertebrates*, edited by A. V. Cohen, S. Rossignol, and S. Grillner, 369–413. New York: John Wiley, 1988

45. Kuramoto, Y. *Chemical Oscillations, Waves, and Turbulence*. Berlin: Springer, 1984.

46. Ma, S-K *Theory of Critical Phenomena*. Reading, MA: Benjamin/Cummings, 1976.

47. Mandell, A. J. "From Intermittency to Transitivity in Neuropsychobiological Flows." *Amer. J. Phys.* **245** (1983): R484–R494.

48. Mandell, A. J., and J. A. S. Kelso. "Dissipative and Statistical Mechanics of Amine Neuron Activity." In *Essays on Classical and Quantum Dynamics*, edited by J. A. Ellison and H. Uberall, 203–235. New York: Gordon-Beach, 1991.

49. Marder, E. "Modulation of Neural Networks Underlying Behavior." *Sem. Neuro.* **1(1)** (1989): 3–4.

50. Marsden, J. E., and M. M. McCracken. *The Hopf Bifurcation and Its Applications*. New York: Springer-Verlag, 1976.

51. Mayr, E. "Questions Concerning Speciation." *Nature* **296** (1982): 609.

52. Mittenthal, J. E., A. B. Baskin, and R. E. Reinke. "Patterns of Structure and Their Evolution in the Organization of Organisms: Modules, Matching, and Compaction." This volume.

53. Mpitsos, G. J., and C. S. Cohan. "Convergence in a Distributed Nervous System: Parallel Processing and Self-Organization." *J. Neurobiol.* **7** (1986): 517–545.

54. Murray, J. D. *Mathematical Biology*. Berlin: Springer-Verlag, 1989.

55. Newman, S. A. "Generic Physical Mechanisms of Morphogenesis and Pattern Formation as Determinants in the Evolution of Multicellular Organization." This volume.

56. Nicolis, G. G., and I. Prigogine. *Self-Organization in Nonequilibrium Systems.* New York: John Wiley, 1977.

57. Nicolis, G., and I. Prigogine. *Exploring Complexity: An Introduction.* San Francisco: Freeman, 1989.

58. Poincaré, H. *Science and Hypothesis.* New York: Dover, 1905/1952

59. Pomeau, Y., and P. Manneville. "Intermittent Transitions to Turbulence in Dissipative Dynamical Systems." *Comm. in Math. Phys.* **74** (1980): 189.

60. Scholz, J. P., J. A. S. Kelso, and G. S. Schöner. "Non-Equilibrium Phase Transitions in Coordinated Biological Motion: Critical Slowing Down and Switching Time." *Phys. Lett.* **A123** (1987): 390–394.

61. Schöner, G. S., H. Haken, and J. A. S. Kelso. "A Stochastic Theory of Phase Transitions in Human Hand Movement." *Bio. Cyber.* **53** (1986): 442–452.

62. S. Schöner, G., W. Y. Jiang, and J. A. S. Kelso. "A Synergetic Theory of Quadrupedal Gaits and Gait Transitions." *J. Theor. Biol.* **142(3)** (1990): 359–393.

63. Schöner, G. S., and J. A. S. Kelso. "Dynamic Pattern Generation in Behavioral and Neural Systems." *Science* **239** (1988): 1513–1520.

64. Schöner, G. S., and J. A. S. Kelso. "Dynamic Patterns in Biological Coordination: Theoretical Strategy and New Results." In *Dynamic Patterns in Complex Systems*, edited by J. A. S. Kelso, A. J. Mandell, and M. F. Shlesinger, 77–102. Singapore: World Scientific, 1988.

65. Schöner, G. S., and J. A. S. Kelso. "A Synergetic Theory of Environmentally Specified and Learned Patterns of Movement Coordination. I. Relative Phase Dynamics." *Bio. Cyber.* **58** (1988): 71–80.

66. Schöner, G. S., and J. A. S. Kelso. "A Synergetic Theory of Environmentally Specified and Learned Patterns of Movement Coordination. II. Component Oscillator Dynamics." *Bio. Cyber.* **58** (1988): 81–89

67. S. Schöner, G., P. G. Zanone, and J. A. S. Kelso. "Learning as a Change of Coordination Dynamics: Theory and Experiment." *J. Mot. Behav.*, in press.

68. Schrödinger, E. *What is Life?* Cambridge: Cambridge University Press, 1945.

69. Selverston, A. I. "Switching Among Functional States by Means of Neuromodulators in the Lobster Stomatogastric Ganglion." *Experientia* **44** (1988): 376–383

70. Shlesinger, M. F. "Fractal Time in Condensed Matter." *Ann. Rev. Phys. Chem.* **38** (1988): 269–290.

71. Stewart, I. "Mathematical Recreations: Why Tarzan and Jane Can Walk in Step with the Animals that Roam the Jungle." *Sci. Amer.* (1991): 158–161.

72. Thompson, D. W. *On Growth and Form.* Cambridge: Cambridge University Press, 1917.

73. Thompson, D'Arcy. *On Growth and Form*, abridged by J. T. Bonner. Abridged edition. Cambridge: Cambridge University Press, 1961.

74. Turing, A. M. "The Chemical Basis of Morphogenesis." *Phil. Trans. Roy. Soc. London* **B237** (1952): 37–72.

75. van der Pol, B. "An Oscillation-Hysteresis in a Triode-Generator." *Phil. Mag.* **43** (1922): 177.

76. von Holst, E. "Relative Coordination as a Phenomenon and as a Method of Analysis of Central Nervous Function." In *The Collected Papers of Erich von Holst*, edited by R. Martin, 33–135. Coral Gables, FL: University of Miami, Reprinted, 1973.
77. Williamson, P. G. "Paleontological Documentation of Specification in Cenozoic Molluscs from Turkana Basin." *Nature* **294** (1981): 214.
78. Wilson, H. R., and J. D. Cowan. "A Mathematical Theory of the Functional Dynamics of Cortical and Thalamic Nervous Tissue." *Kybernetik* **13** (1973): 55–80.
79. Winfree, A. T. *The Geometry of Biological Time.* New York: Springer, 1980
80. Winfree, A. T. *The Timing of Biological Clocks.* New York: W. H. Freeman, 1987.
81. Wunderlin, A. "On The Salving Principle." *Springer Proceedings in Physics* **19** (1987): 140–147.
82. Zeng, W., and L. Glass. "Symbolic Dynamics and Skeletons of Circle Maps." *Physica D* **40** (1989): 218–234.

Index

A

absolute coordination, 16
accessibility, 359
adaptation, 35
adaptabion, 36, 38, 47
adaptation, 95, 178, 309
adaptive evolution, 175
adaptive radiation, 178
aggregation, 324
Alchemy, 312
allosteric macromolecular dynamics, 237
allostery, 225
alternative splicing, 104, 328
Amphiumidae, Sirenidae, 188
Aneides, 183
Arnol'd tongues, 24
attenuation, 188
attractor, 369
autoregulation, 254
autotomy plane, 189
availability, 359

B

Baldwin effect, 237
basal constriction, 190
basin of attraction, 404
basket of criteria, 372
Baskin, Arthur, 328
Batrachoseps, 180
bauplan, 177, 179
Belousev-Zhabotinski reaction, 424
bicaudal, 271-272, 283
bifurcation, 183, 406, 414
    diagram, 416
biological coordination, 398
biological development, 241
biological organization, 51
blastema, 189
Bolitoglossa, 178
Bonner, J.T., 327
Boolean networks, 326, 329
bottle cells, 76
brain, 195, 225
    enzymes, 227
bronchial airway, 44

C

caenogenesis, 179
cAMP, 324
canalization, 254
cardiac activation, 54
cardiac structures, 47
cat, 144, 161-162
catastrophe theory, 413
caudosacral vertebrae, 190
cell, 195
    adhesion molecules, 212
    number, 180
    packing, 195
    size, 180
centrum, 184
cerebellum, 43
Changeux, J. P., 127
chaos, 305
    edge of, 305
Chiropterotriton, 179, 183
circle maps, 425
cladistic, 176
cladogram, 180
clustering, 329
cohort, 354
collective variables, 13
color wheel, 277, 284
    model, 288
compaction, 321, 328
comparative studies, 73-75, 82
competence interface, 367, 371
competition, 147
complexity, 177, 340, 401
    compositional, 400
compromises, 195
conciseness, 341
condensation, 183
conduction, 55
congruence, 176
conjunction, 176
connection machine, 216
connectionism, 109
connectionist model, 109
constraints, 176, 323-324
continuity, 292
control genes, 212
control parameters, 406
convergence, 77, 175-176, 359

coordination, 11-13, 21
  dynamics, 13-14, 16
correlation functions, 152
coupling, 325
crayfish, 205
Creutzfeldt, O., 130
critical fluctuations, 411
critical slowing down, 411
Cryptobranchidae, 194

D

Danchin, A., 127
Darwin, C., 312
Dawkins, R., 312
default, 180
delamination, 251
deletion, 104, 269, 276
  processes of, 104
Dendrotriton, 179
deprivation experiments, 153, 162
design, 176
  limitations, 175, 180, 188
Desmognathinae, 178
development, 73-75, 97, 109, 111, 123, 141, 188, 260
  constraints, 103
  dynamics, 110
  embryonic, 95
  speed of, 97
developmental regulation, 180
dicephalic, 272
differential adhesion, 250
differentiation, 257
diffusion, 41, 245
digit, 180
  development, 180
dimension reduction, 143
dinitrophenol, 87-88
direct development, 179
discrete maps, 424
DNA replication, 99
downward causation, 178
Drosophila, 96-97, 99, 102-104, 109-110, 112-113, 117, 269
Drosophila, melanogaster, 95, 213
dynamic, 11
  module, 321-322, 329
  pattern approach, 13
  systems, 19, 21, 110
  systems theory, 397

E

E. coli, 99
ecomorphology, 177, 188
edge of chaos, 326, 329, 362
elongation, 188
embryogenesis, 98
embryology, 73
embryonic development, 95
engrailed, 274
Ensatina, 190
entrainment, 420
environment, 249
enzyme kinetics, 225
epiboly, 251
epigenesis, 291
epigenetic, 290
  space, 290
  mapping, 269, 290
equilibrium time, 412
Eurycea, 180
events, 111
evolution, 43, 75, 82, 110, 206, 241, 260, 325, 328
evolutionary biology, 227
evolutionary precursors, 206
evolvability, 307
exon hypothesis, 104
exon theory of genes, 327
extension, 77
externalist, 175
eyes, 195

F

Farey sequences, 427
feature space, 143
feature vector, 145
ferret, 144
fertilization, 97
finiteness, 337
  limits, 368
fitness landscapes, 308
fitness-proportional reproduction, 214
flexibility, 19, 328-329, 400
fluctuations, 413
foci, 143
foot webbing, 190
fossoriality, 188
Fourier transform, 151

fractal, 36, 38
    dimension, 39
fractures, 158
frog, 74
functional adaptation, 176
functionalist, 175
fusion, 179

G

gap, 272
    genes, 281-282
gastrulation, 73-79, 81-82
Gaul, U., 282
gene, 257
    duplication, 257
    networks, 333
generation, 176
generic, 260
    development, 260
    physical mechanisms, 246
    generic-genetic coupling, 252
    assimilation, 244
    development, 260
    networks, 333
    program concept, 243
geniculocortical afferents, 121, 124
genome, 241
    size, 177
genotype, 210, 290
Gergen, J. P., 276
germ band, 97
gigantism, 194
global patterning, 282
Goodwin, Brian, 274, 329
gravity, 245, 248
ground plans, 177

H

hairy phenotypes, 286
Haken's synergetics, 12
hard excitation, 420
head, 194
heat shock response, 104
Hebb synapse model, 131
Hebb, D. O., 127
Hemidactyliini, 190
Hemidactylium, 180
Hess, R., 130
heterochrony, 87-89, 177, 258, 329

heterostasy, 258
heterotopy, 329
hierarchical organization, 178
hierarchy of processes, 322
high-dimensional network model, 156
His-Purkinje system, 47
historical contingency, 176
holomorphology, 180
homology, 176
homoplasy, 175-176, 188, 196, 258
Hopf bifurcations, 418
Hubel, D. H., 121, 123
human genome initiative, 243
hunchback, 272, 274, 281
Hunding, A., 285
hydroids, 87-88, 91
    hydractiniid, 88
Hynobiidae, 180
hyobranchial skeleton, 192
hysteresis, 16, 24, 417
    effect, 417

I

ingression, 251
integrated systems, 177
interaction, 398
interdigital webbing, 190
intermittency, 13
internalist, 175
interphase, 113
intervertebral cartilage, 184
intervertebral joints, 184
intrinsic dynamics, 14
invagination, 251
involution, 73, 76, 251

J

Jäckle, H., 282
joints, 186

K

K selection, 102
Kauffman, Stuart A., 274, 329
Keller, J. B., 128
Kelso, J. A. S., 14
key innovation, 190
knirps, 114, 281

Kolmogorov complexity, 346
Krüppel, 272, 274, 281

L

Lacalli, T. C., 293
lambda calculus, 313
laminar region, 27
larvae, 193
LeVay, S., 130
limb reduction, 188
Lineatriton, 188
Lissajous figure, 422
locomotion, 178, 190
lung, 45
lungless salamanders, 177

M

macaque, 143-144, 161-162
maladaptive organization, 190
map formation, 146, 160
Marshall, W. H., 120
matching, 321, 323-324
mechanochemical process, 285
mesopodia, 183
metabolism, 91
metadynamics, 309
metamorphosis, 180
microfingering, 251
microhabitat specialization, 178
microtubule, 101
migration, 324
Miller, K. D., 128, 130
mind, 315
miniaturization, 178, 194
mirror, 271
    duplication, 271, 273, 277
    symmetries, 269
    mirror-symmetric mapping, 279, 286
    mirror-symmetric patternin, 283
mitosis, 100, 113
Mittenthal, Jay, 328
mobile genetic elements, 105
mode locking, 29, 421
modeling development, 132
modular organization, 328
modularity, 333-334, 359
modules, 321, 325-326
molecular data, 197

monophyletic taxon, 177
morphogenesis, 73, 75-76, 177, 241
morphogenetic mechanisms, 73-75, 82
morphogenetic options, 188
morphological simularities, 175
multidimensional analytical approach, 186
multistability, 404
muscimol, 133
mutant, 105
mutant bicaudal phenotypes, 271
mutations, 105, 269, 272

N

natural selection, 176
Negishi, K., 130
neo-Darwinian, 176, 238, 244
neural canal, 184
neural network, 144
Newman, S. A., 329, 103
niches, 327
nonequilibrium phase transition, 401
    attractor, 401
    bifurcation, 401
nonlinear recursion, 51
noodling, 370
notochord, 184
Nototriton, 179

O

ocular dominance, 119, 126, 155
    columns, 124, 128, 132
    plasticity, 132
Oedipina, 188
ontogenetic repatterning, 193
ontogenetic trajectories, 177
ontogeny, 177
optic tectum, 195
order, 341
    parameters, 13, 400
ordinary differential equations, 402
organic selection, 244
organic unity, 321
organism point, 366
organization, 35, 341
organogenesis, 257
orientation, 119, 143

orientation (cont'd.)
  columns, 141
  gradient, 158
  preference, 144
  specificity, 144
oskar, 283

P

P-elements, 105
paedomorphosis, 177
pair-rule, 272
  genes, 272-273, 279
  gene products, 277
parallelism, 103, 175-176
Parvimolge, 181
patch, 273
pattern, 400
  complexity, 400, 433
  distribution, 162
  formation, 11-12, 91, 241, 243, 286, 292, 397
  specification, 277
peramorphosis, 177
performance, 328, 359
  criteria, 95
Phaeognathus, 184
phase attraction circle map, 21-22, 28
phase diagrams, 415
phase transitions, 13
phaseless point, 277
phenotype, 213, 290
photoreceptors, 195
phylogenetic analysis, 175
pitchfork bifurcation, 415
pleiotropy, 210
plesiomorphous, 179
Plethodontidae, 176, 178
Plethodontini, 190
Poincaré surface of section, 403
point estimate, 14
pre-pattern formation, 270
preadaptation, 205-206, 217
  hypothesis, 209
precise definitions, 343
premaxillary bone, 179
principle of matching, 323
problem size, 339
process deletion, 104

Pseudoeurycea, 183

Q

quasi-periodicity, 404

R

r selection, 102
radial intercalation, 77
reaction-diffusion processes, 285
recapitulatory evolution, 177
recombination, 328
redundant processes, 103-104
regeneration, 189
Reinke, R. E., 328
Reiter, H. O., 132-133
relative coordination, 12-13, 16, 19, 25, 27
relative phase dynamics, 422
reliability, 333-334
repeller, 407
replication, 99, 328
repressors, 286
resources, 325
  limits, 326
response properties, 119
retinal ganglion cells, 195
retinotopic maps, 141, 143
reversal, 175
rhodamine, 91
ribosomal RNA, 197
rules of development, 111
runt, 279
  alleles, 279
  gene, 276
  mutants, 286

S

saddle-node bifurcations, 416
salamanders, 175, 178, 180
scale invariancy, 229
scaling, 36, 232
  kinetics, 227
scenariogram, 180
segment polarity genes, 273
segmentation, 97, 183, 188, 323
  genes, 109, 113-114, 273
selection, 214, 325, 327, 329
self-excitatory system, 418
self-organization, 11, 176, 307, 327, 400
self-organizing feature map algorithm, 144

sensitive dependence, 426
separation, 103
SFM algorithm, 150, 158
Sherk, Helen, 121
silencing cells, 121
similarity, 176
Simon, H., 322, 328
size, 104
slime mold, 323
smell, 374
soft excitation, 420
somatotopic map, 142
sonication, 232
spatial periodities, 274
specialization, 96
spliceosome, 328
splicing, 328
stability, 400, 407, 411
stable focus, 418
Stent, G. S., 127
Stern, C., 291
stochastic forces, 409
striping, 246
structural fractals, 41
structural genes, 212
structuralism, 175-176, 194
structure, 337-338, 340, 371
    space, 350-351
Stryker, M. P., 128, 130, 132-133
sturgeon, 73-77, 79, 81
symmetry breaking, 16
synapomorphy, 179
synchronization, 421

T

tail autotomy, 189
tailflip escape maneuver, 208
Talbot, S. A., 120
tarsus, 183
tesselation, 148
Thompson, D'Arcy, 244, 291

Thorius, 184
tissue specificity space, 277, 290
toes, 180
tongue, 192
    evolution of, 192
topographic maps, 141
topology, 119
transformation, 176
transient, 406
translation, 328
tree shrew, 144
tree-like structure, 44
tributyltin, 91
truncation, 177
turbulence, 41
Turing instability, 424
Turing machine, 343
Turing, A., 285, 12
type I intermittency, 429

U

universal Turing machine, 344
universality, 399
upward causation, 178

V

vertebrae, 188
vertebral joints, 184
vision, 195, 324, 374
visual cortex, 119, 143
visual deprivation, 125, 127
visual system, 119
voltage-sensitive dye, 143

W

Waitzman, D. M., 132
Wiesel, T. N., 121, 123
Wieschaus, E. F., 276
Winfree, A. T., 277
wound-healing specialization, 189